T0257896

IET CONTROL ENGINEERING SERIES 71

# Advances in Cognitive Systems

# Other volumes in this series:

# Advances in Cognitive Systems

Edited by S. Nefti and J.O. Gray

The Institution of Engineering and Technology

Published by The Institution of Engineering and Technology, London, United Kingdom

© 2010 The Institution of Engineering and Technology

First published 2010

The Institution of Engineering and Technology
Michael Faraday House
Six Hills Way, Stevenage
Herts, SG1 2AY, United Kingdom

www.theiet.org

**British Library Cataloguing in Publication Data**
A catalogue record for this product is available from the British Library


**ISBN 978-1-84919-075-6 (hardback)**
**ISBN 978-1-84919-076-3 (PDF)**



Typeset in India by MPS Ltd, A Macmillan Company
Printed in the UK by CPI Antony Rowe, Chippenham

# Contents

**3 Compliant actuation: enhancing the interaction ability of cognitive robotics systems**     **37**

*Nikolas G. Tsagarakis, Matteo Laffranchi, Bram Vanderborght and Darwin G. Caldwell*

**4 Toward a dynamical enactive cognitive architecture**     **57**

*Boris A. Durán*

**9    More from the body: embodied anticipation for swift
      readaptation in neurocomputational cognitive architectures
      for robotic agents                                              249**
*Alberto Montebelli, Robert Lowe and Tom Ziemke*

**10   The strategic level and the tactical level of behaviour         271**
*Fabio Ruini, Giancarlo Petrosino, Filippo Saglimbeni
and Domenico Parisi*

# Preface

The study of the nature and origins of our cognitive processes is a human challenge similar to the challenge of understanding the fundamental nature of the physical world. This study is motivated not just by a natural curiosity to probe the structure and workings of the cognitive phenomenon but also by a drive to explore the application of such knowledge to enhance and make more effective the vast array of machines and processes with which we interact during our daily life.

Earlier studies beginning towards the middle of the past century were generally associated with the quest to create so-called artificial intelligence which could with advantage be embedded in machines and systems. These approaches were generally computer orientated and focussed mainly, but not exclusively, on the logical manipulation of symbolic representations of physical and other phenomena. Later approaches were more anthropomorphic and sought to incorporate human knowledge within software structures and develop so-called expert systems which manipulated and deployed such knowledge through inference mechanisms. More recent paradigms are based on the concept of cognition emerging from physical interaction with the environment and being influenced by the physiology of the human neurone motor architecture. This new concept has greatly broadened the field of study and encouraged a much wider interdisciplinary approach to the subject.

By its nature the subject is diverse, but it can, with some simplification, be divided into two principal paradigms, namely cognitivism and emergence. For cognitivists, cognition is based on a now traditional approach where the focus is on the brain and central processing and is generally concerned with rational planning and reasoning and where the cognition is purely internal and symbolic. For emergence, which embraces dynamical, connectionist and enactive systems, the focus is mainly on embodiment where physical instantiation plays an important role in the cognitive process.

This book has been inspired by the portfolio of recent scientific outputs from a range of European and national research initiatives in cognitive science. Its aim is to present an overview of recent developments in cognition research, and to compile the various emerging research strands within a single text as a reference point for progress in the subject. It also provides guidance for new researchers on the breadth of the field and the interconnections between the various research strands identified here.

Although this book does not offer to deliver a comprehensive review of modelling and building cognitive systems, it does bring together a wide range of material from leading workers in the field as well as the outputs from research groups around the world, covering the two principal cognition paradigms of cognitivism and emergence. Furthermore, it suggests some interesting lines of thought about the challenge and opportunities for future work in the field for promoting the various research agendas highlighted within.

This book will be of interest not only to those working in the fields of computer science and AI but also to psychologists, neural scientists, researchers in information science, and engineers involved in the development of advanced robotics, mechatronic systems and HCI systems.

The book comprises 17 chapters. The first chapter introduces the two principal cognition paradigms of cognitivism and emergence. In this chapter, the authors highlight one specific emergent approach to embodied cognition – that of enaction – and discuss the challenges it poses for the advancement of both robotics and cognition.

The second chapter explores the consequences, concerned with connecting brain, body and environment, or more generally with the relation between physical and information processes. The authors present a number of case studies to illustrate how 'morphological computation' can be exploited for designing intelligent, adaptive robotic systems and the understanding of natural systems.

In Chapter 3, the authors propose biologically compliant actuation concepts for the new generation of the 'iCub' robot, a child humanoid with 53 DOF. This actuation morphology has been designed to allow the robot to interact safely within a less constrained and a more human living environment.

Chapter 4 introduces the concept of Dynamical Field Theory (DFT) as a plausible way of combining the power of non-linear dynamical systems theory and the autonomous developmental aspects of enactive systems. The proposed approach is demonstrated by replicating in an anthropomorphic humanoid robotic platform (the iCub) the responses found in 7- to 12-month-old infants when exhibiting a classic simple cognitive behaviour in developmental psychology.

Chapter 5 proposes a computational model capable of encoding object affordances during exploratory learning trials and the authors show the application of their proposed model in a real world task in which a humanoid robot interacts with objects and uses the acquired knowledge and learns from demonstrations.

Chapter 6 describes three lines of research related to the issue of helping robots imitate people which are based on observed human behaviour, technical metrics and implemented technical solutions. In this merging exercise, the authors propose a possible solution pathway that could lead to the design of a Human–Robot Interaction (HRI) system able to be taught by humans to robot via demonstration.

Chapter 7 presents models of game players which account for the behaviour of human participants in experiments, where several aspects of the classic Prisoner's Dilemma game playing have been considered both theoretically and empirically and have been shown to be important for explaining the human decision-making process and the role of cooperation in human interaction.

Chapter 8 introduces the Pogamut 3 toolkit, a freeware software platform for building the behaviour of virtual humans embodied in a 3D virtual world. This toolkit allows the rapid building of simple reactive agents, and it facilitates development of technically advanced agents by exploiting advanced artificial intelligence techniques.

In Chapter 9, the authors outline a novel neurocomputational cognitive architecture for robotic agents which supports the behaviour of a simple simulated robotic agent and demonstrates the modulatory effect of non-neural bodily dynamics on the production of adaptive behaviour. They posit that non-neural internal dynamics can be powerfully involved in the anticipatory process.

Chapter 10 introduces the distinction between a strategic (or motivational) level of behaviour and a tactical (or cognitive) level. The authors describe three sets of simulations with artificial organisms that evolve in an environment that addresses some simple aspects linked to the strategic level of behaviour. Other phenomena such as the operation of the organism's brain in processing information from its own internal body sensors have also been investigated.

In Chapter 11, the authors propose an approach for building a physiologically and physically motivated model of the speech production system. In their model, the authors focus exclusively on the essential dynamical properties of an embodied motor action system incorporating only high-level features of the human vocal tract that constrain speech production.

Chapter 12 presents a novel methodology for analysing the dynamics of emotional responses to music in terms of computational representations of perceptual processes. Aspects of psychoacoustic features and the self-perception of physiological activation (peripheral feedback) are discussed.

Chapter 13 presents a cognitive conceptual model of the individual financial investor. By taking a descriptive point of view, the authors focus on how investors make their investment decisions in a real world setting, as opposed to rational/optimal behaviour proposed by normative financial theories. This conceptual model incorporates results of the research on investment decisions from fields of behavioural finance and cognitive psychology.

In Chapter 14, the authors propose a significant modification to agent (Particle) reasoning processes employed so far in conventional swarm intelligence techniques. They show that by endowing each agent with some descriptive behavior from the field of psychology named Prospect Theory (PT), they can improve considerably the efficiency of global searching procedures. The efficiency of the technique is illustrated by numerical results obtained from applications to two classical problems quoted in the literature.

Chapter 15 provides a contribution to the Cognitive Dimensions Framework, highlighting the distinction between cognitive effort and the constraints arising at the notational level and they suggest that interaction with symbols and notations may be seen as a type of embodiment, and believe that development of a cognitive model of interaction with notations may be of use in better understanding of the embodiment concept.

In Chapter 16, the authors address the adjudication problem which emerges as a key element in distributed cognition systems where the decision-making process is inferred from a set of multiple agents and they investigate the different criteria that help the designer in choosing an appropriate adjudication function. They out-line some interesting properties that characterise an optimal adjudication function, if a specific utility function and failure hypothesis are given for the system under test.

The last chapter reviews in some detail all the previous chapters and attempts to put the collected work within a philosophical framework that stresses the inter-connections of the various disparate strands identified here. Suggestions are made for future work, including how to address the key challenges presented in the first chapter. It does this by focusing on the issue of diversity itself, an issue which is amply demonstrated by the profile of this collection of papers.

S. Nefti and J.O. Gray, Editors

# Contributors

Ahmed I. Al-Dulaimy
: Advanced Robotics Research Centre, The University of Salford, Manchester, UK

Aris Alissandrakis
: Department of Computational Intelligence & Systems Science, Interdisciplinary Graduate School of Science & Engineering, Tokyo Institute of Technology, Japan

Andrew Basden
: Informatics Research Institute (IRIS), University of Salford, UK

Alexandre Bernardino
: Instituto Superior Técnico, Instituto de Sistemas e Robótica, Portugal

Michal Bída
: Department of Software and Computer Science Education, Faculty of Mathematics and Physics, Charles University in Prague, Czech Republic

Cyril Brom
: Department of Software and Computer Science Education, Faculty of Mathematics and Physics, Charles University in Prague, Czech Republic

May N. Bunny
: Advanced Robotics Research Centre, University of Salford, Manchester, UK

Ondřej Burkert
: Department of Software and Computer Science Education, Faculty of Mathematics and Physics, Charles University in Prague, Czech Republic

Darwin G. Caldwell
: Italian Institute of Technology (IIT), Genova, Italy

A. Cangelosi
: Instituto Superior Técnico, Instituto de Sistemas e Robótica, Portugal

E. Coutinho
: Instituto Superior Técnico, Instituto de Sistemas e Robótica, Portugal

Fred Cummins
: School of Computer Science and Informatics, University College Dublin, Ireland

Kerstin Dautenhahn
: Adaptive Systems Research Group, University of Hertfordshire, UK

Boris A. Durán
: Italian Institute of Technology (IIT), Genova, Italy

Jakub Gemrot — Department of Software and Computer Science Education, Faculty of Mathematics and Physics, Charles University in Prague, Czech Republic

Gabriel Gómez — Humanoid Robotics Group, MIT Computer Science and Artificial Intelligence Laboratory, Cambridge, MA, USA

John Gray — Control Systems Centre, University of Manchester, UK

Maurice Grinberg — Central and East European Center for Cognitive Science, New Bulgarian University, Sofia, Bulgaria

Evgenia Hristova — Central and East European Center for Cognitive Science, New Bulgarian University, Sofia, Bulgaria

Rudolf Kadlec — Department of Software and Computer Science Education, Faculty of Mathematics and Physics, Charles University in Prague, Czech Republic

Uzay Kaymak — Department of Econometrics, Erasmus School of Economics, Erasmus University Rotterdam, The Netherlands

Maria Kutar — Informatics Research Institute (IRIS), University of Salford, UK

Matteo Laffranchi — Italian Institute of Technology (IIT), Genova, Italy

Emilian Lalev — Central and East European Center for Cognitive Science, New Bulgarian University, Sofia, Bulgaria

Manuel Lopes — Instituto Superior Técnico, Instituto de Sistemas e Robótica, Portugal

Milan Lovric — Department of Finance, Erasmus School of Economics, Erasmus University Rotterdam, The Netherlands

Robert Lowe — School of Humanities and Informatics, University of Skövde, Sweden

Francisco Melo — Instituto Superior Técnico, Instituto de Sistemas e Robótica, Portugal

Giorgio Metta — Italian Institute of Technology (IIT), Genova, Italy

Alberto Montebelli — School of Humanities and Informatics, University of Skövde, Sweden

Luis Montesano — Instituto Superior Técnico, Instituto de Sistemas e Robótica, Portugal

Samia Nefti — Advanced Robotics Research Centre, University of Salford, Manchester, UK

| Chrystopher CL. Nehaniv | Adaptive Systems Research Group, University of Hertfordshire, UK |
| Nuno Otero | Department of Information Systems, University of Minho, Portugal |
| Mourad Oussalah | Department of Electronic, Electrical and Computer Engineering (EECE), University of Birmingham, UK |
| Domenico Parisi | Laboratory of Autonomous Robotics and Artificial Life, Institute of Cognitive Sciences and Technologies, National Research Council (CNR), Rome, Italy |
| Giancarlo Petrosino | Laboratory of Autonomous Robotics and Artificial Life, Institute of Cognitive Sciences and Technologies, National Research Council (CNR), Rome, Italy |
| Rolf Pfeifer | Artificial Intelligence Laboratory, Department of Informatics, University of Zurich, Switzerland |
| Radek Píbil | Department of Software and Computer Science Education, Faculty of Mathematics and Physics, Charles University in Prague, Czech Republic |
| Tomáš Plch | Department of Software and Computer Science Education, Faculty of Mathematics and Physics, Charles University in Prague, Czech Republic |
| Fabio Ruini | Adaptive Behaviour and Cognition Research Group, Centre for Robotics and Neural Systems, School of Computing and Mathematics, University of Plymouth, UK |
| Filippo Saglimbeni | Cognition & Interaction Lab, School of Humanities and Informatics, University of Skövde, Sweden |
| Giulio Sandini | Italian Institute of Technology (IIT), Genova, Italy |
| Jose Santos-Victor | Instituto Superior Técnico, Instituto de Sistemas e Robótica, Portugal |
| Joe Saunders | Adaptive Systems Research Group, University of Hertfordshire, UK |
| Juraj Simko | School of Computer Science and Informatics, University College Dublin, Ireland |
| Jaap Spronk | Rotterdam School of Management, Erasmus University Rotterdam, The Netherlands |
| L. Strigini | Centre for Software Reliability, City University, London, UK |
| Nikolas G. Tsagarakis | Italian Institute of Technology (IIT), Genova, Italy |

Bram Vanderborght        Italian Institute of Technology (IIT), Genova, Italy

David Vernon             LIRA-Lab, DIST, University of Genoa, Italy

Michal Zemčák            Department of Software and Computer Science
                         Education, Faculty of Mathematics and Physics,
                         Charles University in Prague, Czech Republic

Tom Ziemke               School of Humanities and Informatics, University of
                         Skövde, Sweden

*Chapter 1*

# Embodiment in cognitive systems: on the mutual dependence of cognition and robotics

*David Vernon, Giorgio Metta and Giulio Sandini*

## Abstract

Cognitive systems anticipate, assimilate, and adapt. They develop and learn, predict, explain, and imagine. In this chapter, we look briefly at the two principal paradigms of cognition, cognitivism and emergence, to determine what embodied form each entails, if any. We highlight one specific emergent approach to embodied cognition – enaction – and discuss the challenges it poses for the advancement of both robotics and cognition.

## 1.1 Introduction

In recent years, robotic systems have increasingly been deployed in the empirical study of cognition. In this introductory chapter, we will explore the basis for the relationship between these two areas, with the specific goal of teasing out exactly what role robotics plays in cognition research and whether or not it is a necessary role. To do this, we briefly identify the different approaches that have been taken to modelling and realising cognitive systems and the relevance of embodiment to each approach. We then examine what it actually means to be embodied and identify different forms of embodiment. By considering the different pairings of modes of cognition and modes of embodiment, we can reveal the different roles of robotics in cognition research. We focus on one particular pairing which forms the cornerstone of enaction, a far-reaching paradigm of adaptive intelligence founded on generative (i.e., self-constructed) autonomous development and learning [1, 2]. Our goal here in highlighting enactive systems is not to present a comprehensive review of the field but to identify its chief tenets and anchor the important relationship between cognition and robotics in a significant research paradigm. We conclude with an overview of the challenges facing both robotics and cognitive systems in pushing forward their related research agendas.

## 1.2   Cognition

Cognitive systems anticipate, assimilate, and adapt; they develop and they learn [3]. Cognitive systems predict future events when selecting actions, they subsequently learn from what actually happens when they do act, and thereby they modify subsequent predictions, in the process changing how things are perceived and what actions are appropriate. Typically, they operate autonomously. To do this, cognitive systems generate and use knowledge which can be deployed not just for prediction, looking forward in time, but also for explanation, looking back in time, and even for imagination, exploring counterfactual scenarios.

The hallmark of a cognitive system is that it can function effectively in circumstances that were not planned for explicitly when the system was designed. That is, it has a degree of plasticity and is resilient in the face of the unexpected [4]. This adaptive, anticipatory, autonomous viewpoint reflects the position of Freeman and Núñez who, in their book *Reclaiming Cognition* [5], assert the primacy of action, intention, and emotion in cognition. In the past, however, cognition was viewed in a very different light as a symbol-processing module of mind concerned with rational planning and reasoning. Today, however, this is changing and even proponents of these early approaches now see a much tighter relationship between perception, action, and cognition (e.g., see [6, 7]).

So, if cognitive systems anticipate, assimilate, and adapt; if they develop and learn, predict, explain, and imagine, the question is how do they do this? Unsurprisingly, many approaches have been taken to address this problem. Among these, however, we can discern two broad classes: the *cognitivist* approach based on symbolic information-processing representational systems, and the *emergent systems* approach, embracing connectionist systems, dynamical systems, and enactive systems, all based to a lesser or greater extent on principles of self-organisation [8, 9].

Cognitivist approaches correspond to the classical and still common view that cognition is a type of computation [10], a process that is symbolic, rational, encapsulated, structured, and algorithmic [11]. In contrast, advocates of connectionist, dynamical, and enactive systems approaches argue in favour of a position that treats cognition as emergent, self-organising, and dynamical [12].

Although the use of symbols is often presented as definitive difference between the cognitivist and emergent positions, they differ more deeply and more fundamentally, well beyond a simple distinction based on symbol manipulation. For example, each adopts a different stance on computational operation, representational framework, semantic grounding, temporal constraints, inter-agent epistemology, embodiment, perception, action, anticipation, adaptation, motivation, and autonomy [3].

Cognitivism asserts that cognition involves computations defined over internal representations of knowledge, in a process whereby information about the world is abstracted by perception, and represented using some appropriate symbolic data-structure, reasoned about, and then used to plan and act in the world. The approach has also been labelled by many as the *information processing* (or symbol manipulation) approach to cognition [8, 11–17].

For cognitivists, cognition is representational in a strong and particular sense: it entails the manipulation of explicit symbolic representations of the state and behaviour of the external world and the storage of the knowledge gained from experience to reason even more effectively in the future [18]. Perception is concerned with the abstraction of faithful spatio temporal representations of the external world from sensory data.

In most cognitivist approaches concerned with the building artificial cognitive systems, the symbolic representations are initially at least the product of a human designer: the designer's description of his view of the world. This is significant because it means that these representations can be directly accessed and interpreted by others. It has been argued that this is also the key limiting factor of cognitivist systems as these programmer-dependent representations effectively blind the system [19], constraining it to an idealised description that is a consequence of the cognitive requirements of human activity. This approach works well as long as the assumptions that are implicit in the designer model are valid. As soon as they are not, the system fails. Sometimes, cognitivist systems deploy machine learning and probabilistic modelling in an attempt to deal with the inherently uncertain and incomplete nature of the sensory data that is being used to drive this representational framework. However, this doesn't alter the fact that the representational structure is still predicated on the descriptions of the designers.

Emergent approaches take a very different view of cognition. They view cognition as the process whereby an autonomous system becomes viable and effective in its environment. It does so through a process of self-organisation through which the system is continually reconstituting itself in real-time to maintain its operational identity. For emergent approaches, perception is concerned with the acquisition of sensory data to enable effective action [20] and is dependent on the richness of the action interface [21]. It is not a process whereby the structure of an absolute external environment is abstracted and represented in a more or less isomorphic manner.

In contrast to the cognitivist approach, many emergent approaches assert that the primary model for cognitive learning is anticipative skill construction rather than knowledge acquisition, and that processes which both guide action and improve the capacity to guide action while doing so are taken to be the root capacity for all intelligent systems [22]. Cognitivism entails a self-contained abstract model that is disembodied in principle and the physical instantiation of the systems plays no part in the model of cognition [4, 23]. In contrast, emergent approaches are intrinsically embodied and the physical instantiation plays a pivotal role in cognition.

## 1.3   Embodiment

What exactly it is to be embodied? One form of embodiment, and clearly the type envisaged by proponents of the emergent systems approach to cognition, is a physically active body capable of moving in space, manipulating its environment, altering the state of the environment, and experiencing the physical forces associated with that manipulation [24]. But there are other forms of embodiment. Ziemke introduced a framework to characterise five different types of embodiment [25, 26]:

1. *Structural coupling* between agent and environment in the sense a system can be perturbed by its environment and can in turn perturb its environment;
2. *Historical embodiment* as a result of a history of structural coupling;
3. *Physical embodiment* in a structure that is capable of forcible action (this excludes software agents);
4. *Organismoid embodiment*, i.e., organism-like bodily form (e.g., humanoid or rat-like robots); and
5. *Organismic embodiment* of autopoietic living systems.

This list is ordered by increasing specificity and physicality. Structural coupling entails that the system can influence and be influenced by the physical world. A computer controlling a power plant or a computerised cruise control in a passenger car would satisfy this level of embodiment. A computer game would not necessarily do so. Historical embodiment adds the incorporation of a history of structural coupling to this level of physical interaction so that past interactions shape the embodiment. Physical embodiment is most closely allied to conventional robot systems, with organismoid embodiment adding the constraint that the robot morphology is modelled on specific natural species or some feature of natural species. Organismic embodiment corresponds to living beings.

  Despite the current emphasis on embodiment in cognitive system research, Ziemke argues that many current approaches in cognitive robotics and epigenetic robotics still adhere to the functionalist hardware/software distinction in the sense that the computational model does not in principle involve a physical instantiation and, furthermore, that all physical instantiations are equivalent as long as they support the required computations. Ziemke notes that this is a significant problem because the fundamental idea underpinning embodiment is that the morphology of the system is actually a key component of the systems dynamics. The morphology of the cognitive system not only matters, it is a constitutive part of the cognitive process.

## 1.4   The role of robotics in cognition

Cognitivist systems do not need to be embodied. They are functionalist in principle [5]: cognition comprises computational operations defined over symbolic representations and these computational operations are not tied to any given instantiation. Although any computational system requires some physical realisation to effect its computations, the underlying computational model is independent of the physical platform on which it is implemented. For this reason, it has also been noted that cognitivism exhibits a form of mind–body dualism [11, 24].

Cognitivism is also positivist in outlook: all cognitive systems – designer and designed – share a common universally accessible and universally representable world that is apprehended by perception. Consequently, symbolic knowledge about this world, framed in the concepts of the designer, can be programmed in directly and doesn't necessarily have to be developed by the system itself through exploration of the environment. Some cognitivist systems exploit learning to augment or even supplant the *a priori* designed-in knowledge and thereby achieve a greater degree of adaptiveness and robustness. Embodiment at any of the five levels identified in the previous section may offer an additional degree of freedom to facilitate this learning, but it is by no means necessary.

The perspective from emergent systems is diametrically opposed to the cognitivist position. Emergent systems, by definition, must be embodied and embedded in their environment in a situated historical developmental context [11]. Furthermore, the physical instantiation plays a direct constitutive role in the cognitive process [4, 27, 28].

To see why embodiment is a necessary condition of emergent cognition, consider again what cognition means in the emergent paradigm. It is the process whereby an autonomous system becomes viable and effective in its environment. In this, there are two complementary things going on: one is the self-organisation of the system as a distinct entity, and the second is the coupling of that entity with its environment. 'Perception, action, and cognition form a single process' [24] of *self-organisation in the specific context of environmental perturbations of the system*. This gives rise to the ontogenic development of the system over its lifetime. This development is identically the cognitive process of establishing the space of mutually consistent couplings. These environmental perturbations don't control the system since they are not components of the system (and, by definition, don't play a part in the self-organisation) but they do play a part in the ontogenic development of the system. Through this ontogenic development, the cognitive system develops its own epistemology, i.e., its own system-specific knowledge of its world, knowledge that has meaning exactly because it captures the consistency and invariance that emerges from the dynamic self-organisation in the face of environmental coupling. Thus, we can see that, from this perspective, cognition is inseparable from 'bodily action' [24]: *without physical embodied exploration, a cognitive system has no basis for development.*

## 1.5 Enaction

Enactive systems [8, 19, 20, 29–32] take the emergent paradigm one step further. The five central concepts of enactive cognitive science are embodiment, experience, emergence, autonomy, and sense-making [2, 33]. In contradistinction to cognitivism, which involves a view of cognition that requires the representation of a given objective predetermined world [8, 34], enaction asserts that cognition is a process whereby the issues that are important for the continued existence of a cognitive entity are brought out or enacted: co-determined by the entity as it interacts with the environment in which it is embedded. The term co-determination [20] is laden with meaning. It implies that the cognitive agent is deeply embedded in the environment and specified by it. At the same time, it implies that the process of cognition determines what is real or meaningful for the agent. In other words, the system's actions define the space of perception. This space of perceptual possibilities is predicated not on an objective environment, but on the space of possible actions that the system can engage in whilst still maintaining the consistency of the coupling with the environment. Co-determination means that the agent constructs its reality (its world) as a result of its operation in that world. In this context, *cognitive behaviour is inherently specific to the embodiment of the system and dependent on the system's history of interactions, i.e., its experiences.* Thus, nothing is 'pre-given'. Instead there is an enactive interpretation: a real-time context-based choosing of relevance (i.e., sense-making).

For cognitivism, the role of cognition is to abstract objective structure and meaning through perception and reasoning. For enactive systems, the purpose of cognition is to uncover unspecified regularity and order that can then be construed as meaningful because they facilitate the continuing operation and development of the cognitive system. In adopting this stance, the enactive position challenges the conventional assumption that the world *as the system experiences it* is independent of the cognitive system ('the knower'). Instead, knower and known 'stand in relation to each other as mutual specification: they arise together' [8].

For an enactive system, knowledge is the effective use of sensorimotor contingencies grounded in the structural coupling in which the nervous system exists. Knowledge is particular to the system's history of interaction. If that knowledge is shared among a society of cognitive agents, it is not because of any intrinsic abstract universality, but because of the consensual history of experiences shared between cognitive agents with similar phylogeny and compatible ontogeny.

The knowledge possessed by an enactive system is built on sensorimotor associations, achieved initially by exploration, and affordancees.[1] However, this is only the beginning. The enactive system uses the knowledge gained to form new knowledge which is then subjected to empirical validation to see whether or not it is warranted (we, as enactive beings, imagine many things but not everything we imagine is plausible or corresponds well with reality, i.e., our phenomenological experience of our environment).

---

[1] For true enaction, everything is affordance since everything that is experienced is contingent upon the systems own spatiotemporal experience and embodiment.

One of the key issues in cognition, in general, and enaction, in particular, is the importance of internal simulation in accelerating the scaffolding of this early developmentally acquired sensorimotor knowledge to provide a means to:

- predict future events;
- explain observed events (constructing a causal chain leading to that event);
- imagine new events.

Crucially, there is a need to focus on (re-)grounding predicted, explained, or imagined events in experience so that the system – the robot – can *do* something new and interact with the environment in a new way.

The dependence of a cognitive system's perceptions and knowledge on its history of coupling (or interaction) with the environment and on the very form or morphology of the system itself has an important consequence: there is no guarantee that the resultant cognition will be consistent with human cognition. This may not be a problem, as long as the system behaves as we would wish it to. On the other hand, if we want to ensure compatibility with human cognition, then we have to admit a stronger humanoid form of embodiment and adopt a domain of discourse that is the same as the one in which we live: one that involves physical movement, forcible manipulation, and exploration [35].

## 1.6 Challenges

The adoption of an embodied approach to the development of cognitive systems poses many challenges. We highlight just a few in the following text.

The first challenge is the identification of the phylogenetic configuration and ontogenetic processes. Phylogeny – the evolution of the system configuration from generation to generation – determines the sensorimotor capabilities that a system is configured with at the outset and that facilitate the system's innate behaviours. Ontogenetic development – the adaptation and learning of the system during its lifetime – gives rise to the cognitive capabilities that we seek. To enable development, we must somehow identify a minimal phylogenic state of the system. In practice, this means that we must identify and effect perceptuo-motor capabilities for the minimal behaviours that ontogenetic development will subsequently build on to achieve cognitive behaviour.

The requirements of real-time synchronous system–environment coupling and historical, situated, and embodied development pose another challenge. Specifically, the maximum rate of ontogenic development is constrained by the speed of coupling (i.e., the interaction) and not by the speed at which internal processing can occur [19]. Natural cognitive systems have a learning cycle measured in weeks, months, and years and, while it might be possible to condense these into minutes and hours for an artificial system because of increases in the rate of internal adaptation and change, it cannot be reduced below the timescale of the interaction. You cannot short-circuit ontogenetic development because it is the agent's own experience that defines its cognitive understanding of the world in which it is embedded. This has serious implications for the degree of cognitive development we can practically expect of these systems.

Development implies the progressive acquisition of predictive anticipatory capabilities by a system over its lifetime through experiential learning. Development depends crucially on motivations which underpin the goals of actions. The two most important motives that drive actions and development are social and explorative. There are at least two exploratory motives, one focussing on the discovery of novelty and regularities in the world, and one focussing on the potential of one's own actions. A challenge that faces all developmental embodied robotic cognitive systems is that of modelling these motivations and their interplay, and identifying how they influence action.

Enactive systems are founded on the principle that the system discovers or constructs for itself a world model that supports its continued autonomy and makes sense of that world in the context of the system's own morphology-dependent coupling or interaction with that world. The identification of such generative self-organising processes is pivotal to the future progress of the field. While much current research concentrates on generative processes that focus on sensorimotor perception–action invariances, such as learning affordances, it is not clear at present how to extend this work to generate the more abstract knowledge that will facilitate the prediction, explanation, and imagination that is so characteristic of a true cognitive system.

Finally, development in it fullest sense represents a great challenge for robotics. It is not just the state of the system that is subject to development but also the very morphology, physical properties, and structure of the system – the kinematics and dynamics – that develop and contribute to the emergence of embodied cognitive capabilities. To realise this form of development, we will need new adaptive materials and a new way of thinking to integrate them into our models of cognition.

## 1.7  Conclusions

Cognitivist systems are dualist, functionalist, and positivist. They are dualist in the sense that there is a fundamental distinction between the mind (the computational processes) and the body (the computational infrastructure and, where required, the devices that effect any physical interaction). They are functionalist in the sense that the actual instantiation and computational infrastructure is inconsequential: any instantiation that supports the symbolic processing is sufficient. They are positivist in the sense that they assert a unique and absolute empirically accessible external reality that can be modelled and embedded in the system by a human designer. Consequently, embodiment of any type plays no necessary role.

In the enactive paradigm, the situation is the reverse. The perceptual capacities are a consequence of an historic embodied development and, consequently, are dependent on the richness of the motoric interface of the cognitive agent with its world. That is, the action space defines the perceptual space and thus is fundamentally based in the frame-of-reference of the agent. Consequently, the enactive position is that cognition can only be created in a developmental agent-centred manner, through interaction, learning, and co-development with the environment. It follows that, through this ontogenic development, the cognitive system develops its own epistemology, i.e., its own system-specific knowledge of its world, knowledge that has meaning exactly because it captures the consistency and invariance that emerges from the dynamic self-organisation in the face of environmental coupling.

Despite the current emphasis on embodiment in cognitive systems research, many current approaches in cognitive and epigenetic robotics still adhere to the functionalist dualist hardware/software distinction. It is not yet clear that researchers have embraced the deep philosophical and scientific commitments of adopting an enactive approach to embodied robotic cognitive systems: the non-functionalist, non-dualist, and non-positivist stance of enaction. It is non-functionalist since the robot body plays a constitutive part in the cognitive process and is not just a physical input–output device.

It is non-dualist since there is no distinction between body and mind in the dynamical system that constitutes a cognitive system. It is non-positivist since knowledge in an enactive system is phenomenological and not directly accessible; the best we can hope for is a common epistemology deriving from a shared history of experiences.

There are many challenges to be overcome in pushing back the boundaries of cognitive systems research, particularly in the area of enaction. Foremost among these is the difficult task of identifying the necessary phylogeny and ontogeny of an artificial cognitive system: the requisite cognitive architecture that facilitates both the system's autonomy (i.e., its self-organisation and structural coupling with the environment) and its capacity for development and self-modification. To allow true ontogenetic development, this cognitive architecture must be embodied in a way that allows the system the freedom to explore and interact and to do so in an adaptive physical form that enables the system to expand its space of possible autonomy-preserving interactions. This in turn creates a need for new physical platforms that offer a rich repertoire of perception–action couplings and a morphology that can be altered as a consequence of the system's own dynamics. In meeting these challenges, we move well beyond attempts to build cognitivist systems that exploit embedded knowledge and which try to see the world the way we designers see it. We even move beyond learning and self-organising systems that uncover for themselves statistical regularity in their perceptions. Instead, we set our sights on building enactive phenomenologically grounded systems that construct their own understanding of their world through adaptive embodied exploration and social interaction.

## Acknowledgements

This work was supported by the European Commission, Project IST-004370 RobotCub, under Strategic Objective 2.3.2.4: Cognitive Systems.

## References

1. T. Froese and T. Ziemke. 'Enactive artificial intelligence: Investigating the systemic organization of life and mind'. *Artificial Intelligence*, 173:466–500, 2009.
2. E. Di Paolo, M. Rohde, and H. De Jaegher. 'Horizons for the enactive mind: Values, social interaction, and play'. In J. Stewart, O. Gapenne, and E. Di Paolo, eds., *Enaction: Towards a New Paradigm for Cognitive Science*. MIT Press, Cambridge, MA, 2008.
3. D. Vernon, G. Metta, and G. Sandini. 'A survey of artificial cognitive systems: Implications for the autonomous development of mental capabilities in computational agents'. *IEEE Transaction on Evolutionary Computation*, 11(2):151–180, 2007.
4. D. Vernon. 'The space of cognitive vision'. In H. I. Christensen and H. H. Nagel, eds., *Cognitive Vision Systems: Sampling the Spectrum of Approaches*, LNCS, pp. 7–26. Springer-Verlag, Heidelberg, 2006.
5. W. J. Freeman and R. Núñez. 'Restoring to cognition the forgotten primacy of action, intention and emotion'. *Journal of Consciousness Studies*, 6(11–12):ix–xix, 1999.
6. J. R. Anderson, D. Bothell, M. D. Byrne, S. Douglass, C. Lebiere, and Y. Qin. 'An integrated theory of the mind'. *Psychological Review*, 111(4):1036–1060, 2004.
7. P. Langley. 'An adaptive architecture for physical agents'. *IEEE/WIC/ACM International Conference on Intelligent Agent Technology*, pp. 18–25. IEEE Computer Society Press, Compiegne, France, 2005.
8. F. J. Varela. 'Whence perceptual meaning? A cartography of current ideas'. In F. J. Varela and J.-P. Dupuy, eds., *Understanding Origins – Contemporary Views on the Origin of Life, Mind and Society*, Boston Studies in the Philosophy of Science, pp. 235–263. Kluwer Academic Publishers, Dordrecht, 1992.
9. A. Clark. *Mindware – An Introduction to the Philosophy of Cognitive Science*. Oxford University Press, New York, 2001.
10. Z. W. Pylyshyn. *Computation and Cognition* (2nd edition). Bradford Books, MIT Press, Cambridge, MA, 1984.
11. E. Thelen and L. B. Smith. *A Dynamic Systems Approach to the Development of Cognition and Action*. MIT Press/Bradford Books Series in Cognitive Psychology. MIT Press, Cambridge, MA, 1994.
12. J. A. S. Kelso. *Dynamic Patterns – The Self-Organization of Brain and Behaviour* (3rd edition). MIT Press, Cambridge, MA, 1995.

13. D. Marr. 'Artificial intelligence – A personal view'. *Artificial Intelligence*, 9:37–48, 1977.
14. A. Newell and H. A. Simon. 'Computer science as empirical inquiry: Symbols and search'. *Communications of the Association for Computing Machinery*, vol. 19, pp. 113–126, March 1976. Tenth Turing award lecture, ACM, 1975.
15. J. Haugland. 'Semantic engines: An introduction to mind design'. In J. Haugland, ed., *Mind Design: Philosophy, Psychology, Artificial Intelligence*, pp. 1–34. Bradford Books, MIT Press, Cambridge, MA, 1982.
16. S. Pinker. 'Visual cognition: An introduction'. *Cognition*, 18:1–63, 1984.
17. J. F. Kihlstrom. 'The cognitive unconscious'. *Science*, vol. 237, pp. 1445–1452, September 1987.
18. E. Hollnagel and D. D. Woods. 'Cognitive systems engineering: New wind in new bottles'. *International Journal of Human-Computer Studies*, 51:339–356, 1999.
19. T. Winograd and F. Flores. *Understanding Computers and Cognition – A New Foundation for Design*. Addison-Wesley Publishing Company, Inc., Reading, MA, 1986.
20. H. Maturana and F. Varela. *The Tree of Knowledge – The Biological Roots of Human Understanding*. New Science Library, Boston and London, 1987.
21. G. H. Granlund. 'The complexity of vision'. *Signal Processing*, 74:101–126, 1999.
22. W. D. Christensen and C. A. Hooker. 'An interactivist-constructivist approach to intelligence: Self-directed anticipative learning'. *Philosophical Psychology*, 13(1):5–45, 2000.
23. D. Vernon. 'Cognitive vision: The case for embodied perception'. *Image and Vision Computing*, 26(1):127–141, 2008.
24. E. Thelen. 'Time-scale dynamics and the development of embodied cognition'. In R. F. Port and T. van Gelder, eds., *Mind as Motion – Explorations in the Dynamics of Cognition*, pp. 69–100. Bradford Books, MIT Press, Cambridge, MA, 1995.
25. T. Ziemke. 'Are robots embodied?' In C. Balkenius, J. Zlatev, K. Dautenhahn, H. Kozima, and C. Breazeal, eds., *Proceedings of the First International Workshop on Epigenetic Robotics – Modeling Cognitive Development in Robotic Systems*, Lund University Cognitive Studies, vol. 85, pp. 75–83. Lund, Sweden, 2001.
26. T. Ziemke. 'What's that thing called embodiment?' In R. Alterman and D. Kirsh, eds., *Proceedings of the 25th Annual Conference of the Cognitive Science Society*, Lund University Cognitive Studies, pp. 1134–1139. Lawrence Erlbaum, Mahwah, NJ, 2003.
27. J. L. Krichmar and G. M. Edelman. 'Principles underlying the construction of brain-based devices'. In T. Kovacs and J. A. R. Marshall, eds., *Proceedings of AISB '06 – Adaptation in Artificial and Biological Systems, Symposium on Grand Challenge 5: Architecture of Brain and Mind,* vol. 2, pp. 37–42. University of Bristol, Bristol, 2006.

28.  H. Gardner. *Multiple Intelligences: The Theory in Practice*. Basic Books, New York, 1993.
29.  H. Maturana. 'Biology of cognition'. Research Report BCL 9.0, University of Illinois, Urbana, IL, 1970.
30.  H. Maturana. 'The organization of the living: A theory of the living organization'. *International Journal of Man-Machine Studies*, 7(3):313–332, 1975.
31.  H. R. Maturana and F. J. Varela. *Autopoiesis and Cognition – The Realization of the Living*. Boston Studies on the Philosophy of Science. D. Reidel Publishing Company, Dordrecht, Holland, 1980.
32.  F. Varela. *Principles of Biological Autonomy*. Elsevier North Holland, New York, 1979.
33.  E. Thompson. *Mind in Life: Biology, Phenomenology, and the Sciences of Mind*. Harvard University Press, Boston, MA, 2007.
34.  T. van Gelder and R. F. Port. 'It's about time: An overview of the dynamical approach to cognition'. In R. F. Port and T. van Gelder, eds., *Mind as Motion – Explorations in the Dynamics of Cognition*, pp. 1–43. Bradford Books, MIT Press, Cambridge, MA, 1995.
35.  R. A. Brooks. *Flesh and Machines: How Robots Will Change Us*. Pantheon Books, New York, 2002.

*Chapter 2*

# Intelligence, the interaction of brain, body and environment: design principles for adaptive systems

*Rolf Pfeifer and Gabriel Gómez*

## Abstract

There have been two ways of approaching intelligence: the traditional approach, where the focus has been on the study of the control or the neural system itself and a more recent approach that is centered around the notion of 'embodiment,' the idea that intelligence always requires a body, a complete organism that interacts with the real world. This chapter[1] explores the deeper and more important consequences, concerned with connecting brain, body, and environment, or more generally with the relation between physical and information (neural, control) processes. Often, morphology and materials can take over some of the functions normally attributed to the brain (or the control), a phenomenon called 'morphological computation.' A number of case studies are presented to illustrate how 'morphological computation' can be exploited for designing intelligent, adaptive robotic systems, and to understand natural systems. We conclude with a theoretical scheme that can be used to embed the diverse case studies and that captures the essence of embodiment and morphological computation.

## 2.1 Introduction

The overall 'philosophy' of our research program is centered around the notion of 'embodiment,' the idea that intelligence always requires a body, a complete organism that interacts with the real world. The implications of this concept can hardly be overestimated and are summarized in Pfeifer and Scheier (1999); Pfeifer and Bongard (2007).

---

[1] Parts of the ideas presented in this chapter have appeared in previous publications that will be referenced throughout the text.

a                                                          b

*Figure 2.1    Two ways of approaching intelligence. (a) The classical approach.
The focus is on the brain and central processing. (b) The embodied
approach. The focus is on the interaction with the environment.
Cognition is emergent from the system–environment interaction*

Figure 2.1 depicts a clear distinction between a traditional or classical
approach to intelligence where the focus has been the study of the control or
the neural system itself, and the more recent, embodied one that focuses on the
relation between physical and information (neural and control) processes.

In the traditional approach there has been a focus on the study of the control
or the neural system, the brain, itself. In robotics one usually starts with a particular given hardware design (the morphology) and then the robot is programmed, to perform certain tasks. In other words, there is a clear separation
between hardware and software. In computational neuroscience, the focus is
essentially on the simulation of certain brain regions. For example, in the 'Blue
Brain' project (Markram, 2006), the focus is, for the better part, on the simulation of cortical columns – the organism into which the brain is embedded does
not play a major role in these considerations (see Figure 2.1a).

In the modern 'embodied' approach, the focus is on the interaction with
the environment. Cognition is emergent from the system–environment interaction. Recently, this approach has attracted increasing interest from all disciplines dealing with intelligent behavior, including psychology, philosophy,
artificial intelligence, linguistics, and to some extent neuroscience. In this
chapter, we explore the far-reaching and often surprising implications of
embodiment and argue that if we want to understand the function of the brain
(or the control in the case of robots), we must understand how the brain is
embedded into the physical system, and how this system interacts with the real
world (see Figure 2.1b).

It can be shown that through this embodied interaction with the environment, in particular through sensory-motor coordination, information structure is induced in the sensory data, thus facilitating perception and learning, a phenomenon called 'information self-structuring' (Lungarella *et al.*, 2005; Pfeifer *et al.*, 2007, 2008). It is interesting to note that this phenomenon is the result of a physical interaction with the real world – it cannot be understood by looking at internal brain (or control) mechanisms only. The advantage of using robots is that embodiment can be investigated quantitatively. While a robot interacts with its environment, all its data (e.g., sensory stimulation, motor signals, and internal state) can be recorded as time series for further analysis, which is much harder to do with biological systems.

In the following case studies, we illustrate how embodiment, in particular morphological computation, can be exploited for designing intelligent systems and for understanding natural systems. Finally, we will try to integrate the diverse case studies into a general overarching scheme that captures the essence of embodiment and morphological computation.

## 2.2   Case studies

We start with a case study on sensory morphology that is followed by how morphology and materials can be exploited for grasping using an artificial hand. We then turn to sensory-motor coordination and information self-structuring. This is followed by an example of how to control complex bodies by exploiting morphological changes. We then present a few cases that illustrate the subtle relations between morphology, materials, and dynamics in the generation of diverse behavior. We also show how, by capitalizing on the interaction with the environment, two seemingly unsolvable problems, can be resolved (e.g., reach any position in 3D space with only one degree-of-freedom (DOF) of actuation, an autonomous reaching and grasping without relying on stereo vision or tactile sensors).

### 2.2.1   *Exploitation of sensory morphology*

#### 2.2.1.1   Eyebot

The morphology of the sensory system has a number of important implications. In many cases, more efficient solutions can be found by having the proper morphology for a particular task (Pfeifer, 2000, 2003). For example, it has been shown that for many objectives (e.g., obstacle avoidance) motion detection is all that is required. Motion detection can often be simplified if the light-sensitive cells are not spaced evenly, but if there is a nonhomogeneous arrangement.

For instance, Franceschini and his co-workers found that in the compound eye of the house fly the spacing of the facets is denser toward the front of the animal (Franceschini *et al.*, 1992). This nonhomogeneous arrangement, in a

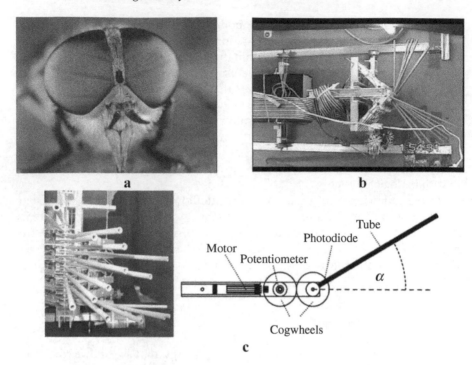

*Figure 2.2    The Eyebot. The specific nonhomogeneous arrangement of the facets compensates for motion parallax, thereby facilitating neural processing. (a) Insect eye. (b) The experiment seen from the top. The robot has to maintain a constant lateral distance to an obstacle (indicated by the vertical light tube) by modifying its morphology, i.e., the positioning of the facet tubes. This is under the control of an evolutionary strategy. (c) Front view: the Eyebot consists of a chassis, an on-board controller, and 16 independently controllable facet units, which are all mounted on a common vertical axis. A schematic drawing of the facet is shown on the right. Each facet unit consists of a motor, a potentiometer, two cogwheels and a thin tube containing a sensor (a photodiode) at the inner end. These tubes are the primitive equivalent of the facets in the insect compound eye*

sense, compensates for the phenomenon of motion parallax, i.e., the fact that at constant speed, objects on the side travel faster across the visual field than objects toward the front. The morphology of the sensory system performs the 'morphological computation,' so to speak.

Lichtensteiger performed experiments to evolve the morphology of an 'insect eye' on a real robot, 'the Eyebot' (Figure 2.2) and showed that keeping a constant lateral distance to an obstacle, can be solved by proper morphological arrangement of the ommatidia, i.e., frontally denser than laterally without changing anything inside the neural controller (Lichtensteiger, 2004). Because the sensory stimulation is only induced when the robot (or the insect) moves in a particular way, this is also called self-structuring of the sensory stimulation (see Section 2.2.2.2 – Physical dynamics and information self-structuring).

## 2.2.2   *Exploitation of morphological and material properties*

### 2.2.2.1   Cheap grasping

In this case study, we discuss how morphology, materials, and control interact to achieve grasping behavior. The 18 DOF tendon driven 'Yokoi hand' (Yokoi *et al.*, 2004; Figure 2.3) which can be used as a robotic and as a prosthetic hand, is partly built from elastic, flexible, and deformable materials. For grasping an object, a simple control scheme, a 'close' is applied. When the hand is closed, the fingers will, because of its anthropomorphic morphology, automatically come together and self-adapt to the shape of the object it is grasping (Figures 2.3b and c). The shape adaptation is taken over by the morphology of the hand, the elasticity of the tendons, and the deformability of the fingertips. Thus, there is no need for the robot to 'know' beforehand what the shape of the to-be-grasped object will be (which is normally the case in robotics, where the contact points are calculated before a grasping action can be attempted; Molina-Vilaplana *et al.*, 2007). In this setup, control of grasping is very simple, or in other words, very little 'brain power' is required.

Another way of achieving decentralized adaptation is to add pressure sensors to the hand and bending the fingers until a certain threshold is reached. By placing additional sensors on the hand (e.g., angle, torque, pressure) the

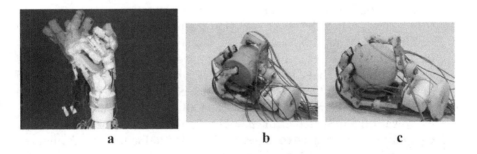

a                              b                              c

*Figure 2.3*   *'Cheap' grasping: exploiting system–environment interaction. (a) The Yokoi hand exploits deformable and flexible materials to achieve self-adaptation through the interaction between environment and materials. (b–c) Final grasp of different objects, the control is the same, but the behavior is very different*

abilities of the hand can be improved and feedback signals can be provided to the agent (the robot or the human), which can then be exploited by the neural system for learning and categorization purposes (Takamuku *et al.*, 2007; Gómez *et al.*, 2006; Gómez and Eggenberger Hotz, 2007). Clearly, these designs only work for a class of simple hand movements. For fine manipulation more sophisticated sensing, actuation, and control are required, for instance a general purpose robot with the dexterity needed to manipulate the stones used in the Chinese game GO was reported in Torres-Jara and Gómez (2008).

For prosthetics, there is an interesting implication. Electromyographic (EMG) signals can be used to interface the robot hand noninvasively to a patient; even though the hand has been amputated, he or she can still intentionally produce muscle innervations which can be picked up on the surface of the skin by EMG electrodes. If EMG signals, which are known to be very noisy, are used to steer the movement of the hand, control cannot be very precise and sophisticated. But by exploiting the self-regulatory properties of the hand, there is no need for such precise control, at least for some kinds of grasping: the relatively poor EMG signals are sufficient for the basic movements (Yu *et al.*, 2006; Hernandez *et al.*, 2008).

### 2.2.2.2    Physical dynamics and information self-structuring

Imagine that you are standing upright and your right arm is just loosely swinging from the shoulder; only the shoulder joint is slightly moved whereas the elbow, the ankle joint, and the finger joints are mostly passive. The brain does not directly control the trajectories of the elbow: its movements come about because the arm acts a bit like a pendulum that swings due to gravity and the dynamics, which is set by the muscle tone (which is in turn controlled by the brain). If we look at the trajectory of the hand, it is actually quite complex, a curved movement in 3D space. In spite of this (seeming) behavioral complexity, the neural control is comparatively simple (which is an instance of the notorious frame-of-reference problem, e.g., Pfeifer and Bongard, 2007).

Although it is nice to be able to say that the neural control (or the control in general) for a complex movement (e.g., loosely swinging arm, grasping) is simple, it would be even nicer if this movement were also useful. What could it be good for? Assuming that evolution typically comes up with highly optimized solutions, we can expect these movements to benefit the organism. The natural movements of the arm and hand are – as a result of their intrinsic dynamics – directed toward the front center of the body, in this way, proper morphology and exploitation of intrinsic dynamics, facilitate the induction of information.

For instance, if the hand is swinging from the right side toward the center and it encounters an object, the palm will be facing left, the object can, because of the morphology of the hand, be easily grasped. Through this act of grasping, sensory stimulation is generated in the palm and on the fingertips. Moreover, the object is brought into the range of the visual field thereby inducing visual stimulation. Because of gravity, proprioceptive stimulation is also induced so that cross-modal associations can be formed between the visual, the haptic, and the proprioceptive modality (a process that is essential for development). Thus, through sensory-motor coordinated behaviors such as foveation, reaching, and grasping, sensory stimulation is not only induced but it also tends to contain information structure, i.e., stable patterns of stimulation characterized by correlations within and between sensory channels, which can strongly simplify perception and learning (Lungarella and Sporns, 2006). Furthermore, Sporns suggests that this information structure could be measured and used as a cost function to drive artificial evolutionary processes, to minimize designer's intervention. Using an information–theoretic measure such as complexity as a cost function does not involve the specification of desired goal states (traditionally supplied by the designer), except that the sensori-motor-neural system should incorporate statistical structure that combines information that is both segregated and integrated (Sporns, 2009).

### 2.2.3   Managing complex bodies: adaptation through morphological changes

#### 2.2.3.1   The Fin Ray® effect

Inspired by observations of the internal structure of fish fins, Bannasch and co-workers have developed a robot based on the so-called 'Fin Ray'® effect. The basic structure derived from the anatomy of a fish's fin, consists of two ribs interconnected by elastic connective tissue that bends toward the direction of an external force, instead of moving away as one would normally expect.

These elastic structures provide an equal distribution of pressure in a dynamic system and have been used to build the 'Aqua ray' robot by Festo (http://www.festo.com/cms/en-us_us/5007_6541.htm). This robot, which resembles a manta ray, has a minimum setup with only three motors, one for each wing and one for the tail. Although each motor is just driven back and forth, the 'Fin Ray' effect provides the 'Aqua ray' robot with a perfect and cheap way to bend the wings and exploit the fluid dynamics (Shelley, 2007).

Applying the same principle to manipulation, Festo has introduced a new type of gripper (FinGripper; [http://www.festo.com/cms/en-us_us/4981.htm]) able to handle fragile objects (e.g., light bulbs).

#### 2.2.3.2   Exploiting self-adaptation: cockroaches climbing over obstacles

Cockroaches cannot only walk fast over relatively uneven terrain, but they can also, with great skill, negotiate obstacles that exceed their body height. They have complex bodies with wings and three thoracic segments, each one with a pair of legs. There is one thoracic ganglion at each segment (i.e., prothoracic, mesothoracic, and metathoracic, see Figure 2.4) controlling each pair of legs. Although both the brain and the thoracic ganglion have a large number of neurons, only around 250 neurons – a very small number – descend from the brain to the body (Staudacher, 1998).

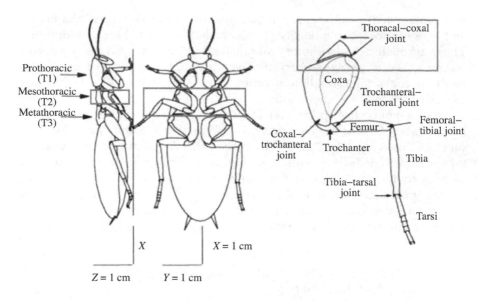

*Figure 2.4    Schematic representation of cockroach anatomy. As the cockroach is climbing over an obstacle, the configuration of the middle (mesothoracic) shoulder joint is reconfigured. The region where the change is taking place is roughly marked with a rectangle [picture courtesy Roy Ritzmann, Case Western Reserve University]*

The legs and feet are deformable to various extents, each leg has more than seven DOF, some joints in the legs are very complicated and there is also an intricate set of joints in the feet.

Now, how is it possible to control all these DOF with such a small number of descending neurons? In what follows, we discuss one potential solution that, although entirely hypothetical, does have a certain plausibility and it further illustrates morphological computation.

Suppose that a cockroach is walking on the ground using its local neural circuits which account for stable forward locomotion. If it now encounters an obstacle – whose presence and approximate height it can sense by means of its antennae – the brain, rather than changing the movement pattern by evoking a different neural circuit, 'reconfigures' the shoulder joint (i.e., thoracic–coxa, see Figure 2.4) by altering local control circuits in the thoracic ganglion. As a result, the middle legs are rotated downward so that extension now generates a rearing motion that allows it to easily place its front legs on top of the block with typical walking movements (Figure 2.5). The local neural circuits continue doing essentially the same thing – perhaps with increased torques on the joints – but because now the configuration of the shoulder joint is different, the effect of the neural circuits on the behavior, the way in which the joints are actuated, changes (see Watson and Ritzmann, 1998; Watson *et al.*, 2002).

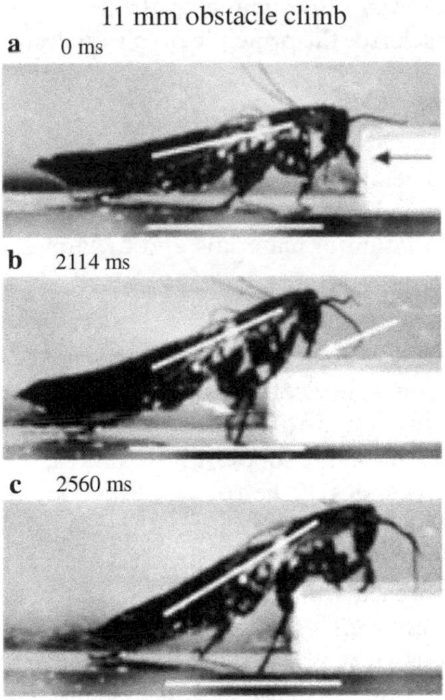

11 mm obstacle climb

*Figure 2.5* *A cockroach climbing over an obstacle. (a) Touching block with antenna. (b) Front leg put on block, middle legs pushing down, and posture changing to reared-up position. (c) Front legs pushing down on block, middle leg fully stretched [picture courtesy Roy Ritzmann, Case Western Reserve University]*

Interestingly, this change is not so much the result of a different neural circuit, but of an appropriately altered joint morphology. Instead of manipulating the movements in detail, the brain orchestrates the dynamics of locomotion by 'outsourcing' some of the functionality to local neural circuits and morphology through an ingenious kind of cooperation with body and decentralized neural controllers. To modify the morphology, only a few global parameters need to be altered to achieve the desired movement.

This kind of control through morphological computation has several advantages.

1. the control problem in climbing over obstacles can be solved efficiently with relatively few neurons.
2. this is a 'cheap solution,' because much of what the brain would have to do is delegated to the morphology, thereby freeing it from unnecessary control tasks.
3. it takes advantage of the inherent stability of the local feedback circuits rather than working in opposition to them.

4.  it illustrates a new way of designing controllers by exploiting mechanical change and feedback (see the principle of 'cheap design' Pfeifer *et al.*, 2007).

## 2.2.4   Exploitation of dynamics

In this section, inspired by the passive dynamic walkers, we introduce three locomotion robots developed at our lab, the walking and hopping robot 'Stumpy,' the quadruped robot 'Puppy,' and 'Crazy bird.' These robots demonstrate the exploitation of materials and dynamics.

### 2.2.4.1   Passive dynamic walkers

The passive dynamic walker that goes back to McGeer (1990a, 1990b), is a robot (or, if you like, a mechanical device) capable of walking down an incline without any actuation and without control. In other words, there are no motors and there is no microprocessor on the robot; it is 'brainless,' so to speak. To achieve this task the passive dynamics of the robot, of its body and its limbs, must be exploited. This kind of walking is very energy efficient – the energy is supplied by gravity only – and there is an intrinsic naturalness to it. However, its 'ecological niche' (i.e., the environment in which the robot is capable of operating) is extremely narrow: it only consists of inclines of certain angles.

Energy-efficiency is achieved because in this approach the robot is – loosely speaking – operated near one of its Eigenfrequencies. To make this work, a lot of attention was devoted to morphology and materials. For example, the robot is equipped with wide feet of a particular shape to guide lateral motion, soft heels to reduce instability at heel strike, counter-swinging arms to compensate for the yaw induced by leg swinging, and lateral-swinging arms to stabilize side-to-side lean (Collins *et al.*, 2005).

To walk over flat terrain, a passive dynamic walker has to be augmented with some kind of actuation, for instance adding electrical motors (e.g., 'Stumpy' designed by Raja Dravid at the University of Zurich), or adding pneumatic actuators at the hip (e.g., 'Denise' developed at the Technical University of Delft by Martijn Wisse (2005) or the biped robot with antagonistic pairs of pneumatic actuators designed at Osaka University by Koh Hosoda (Takuma and Hosoda, 2006).

### 2.2.4.2   Stumpy

The walking and hopping robot 'Stumpy' can be seen in Figure 2.6. Stumpy's lower body is made of an inverted 'T' mounted on wide springy feet. The upper body is an upright 'T' connected to the lower body by a rotary joint, the 'waist' joint, providing one DOF in the frontal plane. Weights are attached to the ends of the horizontal beam to increase its moment of inertia.

The horizontal beam is connected to the vertical one by a second rotary joint, providing one rotational DOF, in the plane normal to the vertical beam, the 'shoulder' joint. Stumpy's vertical axis is made of aluminum, while both its horizontal axes and feet are made of oak wood.

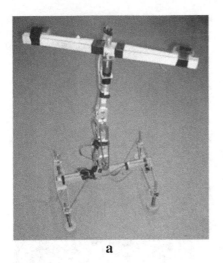
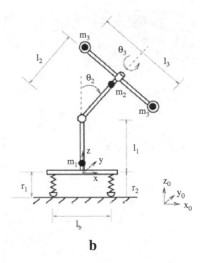

a
b

*Figure 2.6   The walking and hopping robot 'Stumpy'. (a) Photograph of the robot. (b) Schematic drawing*

Although Stumpy has no real legs or feet, it can locomote in many interesting ways: it can move forward in a straight or curved line, it has different gait patterns, it can move sideways, and it can turn on the spot. Interestingly, this can all be achieved by actuating only two joints with one DOF. In other words, control is extremely simple – the robot is virtually 'brainless.' The reason this works is because Stumpy has the right morphology (an upper body), which includes the proper weight distribution and the right materials (elastic feet).

If its morphology or material properties were different, it would exhibit less behavioral diversity. For instance, a robot without a heavy enough upper body will not generate enough momentum to get its feet off the ground. A robot with no elasticity in its feet will not move properly or will fall over because the forces are not adequately propagated through the robot to the ground for locomotion.

There is a delicate interplay of momentum exerted on the feet by moving the two joints in particular ways (for more details, see Paul *et al.*, 2002; Iida and Pfeifer, 2004a).

### 2.2.4.3   Puppy

We now present a case study where a very simple kind of artificial 'muscle' in the form of a normal spring is used to achieve rapid locomotion. One of the fundamental problems in rapid locomotion is that the feedback control loops can no longer be employed because the response times are too slow. In contrast, 'Puppy' (Figure 2.7a) can achieve not only fast but also robust locomotion without sensory feedback (i.e., it cannot distinguish between the stance and the flight phases, and it cannot measure acceleration, or inclination).

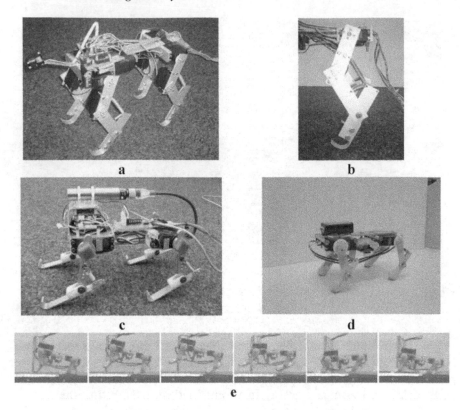

*Figure 2.7    Cheap rapid locomotion. (a) The quadruped robot 'Puppy.' (b) The spring system in the hind legs. (c) A version of puppy equipped with an ultrasound sensor. (d) A simplified version of puppy, the quadruped 'mini dog.' (e) Mini dog running on a treadmill*

The design of 'Puppy' was inspired by biomechanical studies. Each leg has two standard servomotors and one springy passive joint as a simple kind of artificial 'muscle' (Figure 2.7b).

The robot controller is very simple: we applied a synchronized oscillation-based control to the four motors – two in the 'hip' and two in the 'shoulder' – where each motor oscillates through sinusoidal position control (i.e., amplitude and frequency of the motor oscillation). No sensory feedback is used for this controller (Iida and Pfeifer, 2004b, 2006). Even though the legs are actuated by simple oscillations, in the interaction with the environment, through the interplay of the spring system, the flexible spine, and gravity, a natural running gait emerges (Figures 2.7d and e). Nevertheless, the robot maintains a stable periodic gait, which is achieved by properly exploiting its intrinsic dynamics (Iida, 2005).

The behavior is the result of the complex interplay of agent morphology, material properties (in particular the 'muscles,' i.e., the springs), control (amplitude, frequency, phase difference between the front and the hind legs), and environment (friction and slippage, shape of the ground, gravity). Exploiting morphological properties and the intrinsic dynamics of materials makes 'cheap' rapid locomotion possible because physical processes are fast and they are for free (for further references on cheap locomotion, see, e.g., Kubow and Full, 1999; Blickhan *et al.*, 2003; Buehler, 2002). Because stable behavior is achieved without control – simply due to the intrinsic dynamics of the physical system – we use the term 'self-stabilization.'

Now, if sensors – for example, pressure sensors on the feet, angle sensors in the joints, an ultrasound sensor on the head – are mounted on the robot, structured, i.e., correlated – sensory stimulation will be induced that can potentially be exploited for learning about the environment and its own body dynamics (for instance, the robot can learn to run in the middle of the tread-mill, see Figure 2.7c; Schmitz *et al.*, 2007a, 2007b). For more details, see Pfeifer and Bongard (2007, Chapter 5).

### 2.2.4.4 Crazy bird

To investigate how to gain controllability and increase the behavioral diversity of robots like 'Puppy,' a quadruped robot was built with the LEGO MINDSTORMS NXT kit (see Figure 2.8a).

It only has two motors and no sensors, the two motors are synchronized, turn at the same speed and in the same direction. The front legs are attached eccentrically to the wheels driven by the motors and are physically connected to the hind legs. Four experimental setups were tested with different phase delay between left and right legs (i.e., 0°, 90°, 180°, and 270°). Despite its simplicity, the robot could reach every point on a table by changing a single control parameter (i.e., the phase delay between its legs). The controller only

a          b

*Figure 2.8 Controllability and behavioral diversity. (a) The quadruped robot. (b) 'Crazy bird'*

has to induce a global bias rather than exactly controlling the joint angles or other parameters of the agent's movement (Rinderknecht *et al.*, 2007).

If we keep the controller of the quadruped robot as is, but modify the morphology by removing the hind legs and adding two loosely attached rubber feet, which can turn and move a bit, we obtain the 'Crazy bird' (see Figure 2.8b), that gets its name due to its rich behavioral diversity.

## 2.2.5    *Exploitation of system–environment interaction*

### 2.2.5.1    Leg coordination in insect walking

Leg movements in insects are controlled largely by independent local neural circuits that are connected to their neighbors, there is no central controller that coordinates the legs during walking.

The leg coordination comes about by the exploitation of the interaction with the environment (Cruse, 1990; Cruse *et al.*, 2002). If the insect stands on the ground and tries to move forward pushing backwards with one of its legs, as an unavoidable implication of being embodied, all the joint angles of the legs standing on the ground will instantaneously change. The insect's body is pushed forward, and consequently the other legs are also pulled forward and the joints will be bent or stretched.

This fact can be exploited to the animal's advantage. All that is needed is angle sensors in the joints – and they do exist – for measuring the change, and there is global communication between the legs! But the communication is through the interaction of the agent with the environment, not through neural processing.

Inspired by the fact that the local neural leg controllers need only exploit this global communication, a neural network architecture called WalkNet has been developed which is capable of controlling a six-legged robot (Dür *et al.*, 2003). This instance of morphological computation takes over part of the task that would have to be done by the brain – the communication between the legs and the calculation of the angles on all the joints – is performed by the inter-action between the insect and the world.

### 2.2.5.2    Wanda

The artificial fish, 'Wanda,' developed by Marc Ziegler and Fumiya Iida (Ziegler *et al.*, 2005; Pfeifer *et al.*, 2006; Figure 2.9) shows how the interaction with the environment can be exploited in interesting ways to achieve moving to any position in 3D space with only one DOF of actuation.

All 'Wanda' can do is wiggle its tail fin back and forth. The tail fin is built from elastic materials such that it will on average produce a high forward thrust if properly chosen. It can move forward, left, right, up, and down. Turning left and right is achieved by setting the zero-point of the wiggle movement either left or right at a certain angle. The buoyancy is such that if it moves forward slowly, it will sink, i.e., move down gradually. The speed is controlled by the wiggling frequency. If it moves fast and turns, its body

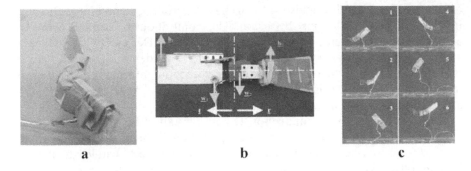

a                         b                         c

*Figure 2.9   The artificial fish. (a) 'Wanda' with one degree-of-freedom for
              wiggling the passive tail fin. (b) The forces acting on its body are
              illustrated by arrows. (c) A typical sequence of an upward
              movement*

because of the weight distribution will tilt slightly to one side, which produces
up thrust, so that it will move upwards. In other words, through its own
movements, it induces turbulences that it can then exploit to move upwards.
The fascinating point about this robot is that its high behavioral diversity is
achieved through 'morphological computation.' If material properties and the
interaction with the environment were not properly exploited, one would need
more complicated actuation, for example, additional fins or a flexible spine and
thus more complex control.

### 2.2.5.3   Cheap reaching and grasping

We have already looked at reaching in the context of physical dynamics and
information self-structuring, and grasping in the context of the Yokoi hand; let
us now dig into the issue of reaching and grasping a bit more deeply. The
difficulties of robot manipulation in human environments exist mainly because
robots are not good enough at recognizing generic objects and dexterously
manipulating them. As pointed out by Brooks (2004), the requirements for
autonomous manipulation will be the vision capabilities of a two-year-old and
the manipulation skills (manual dexterity) of a six-year-old child.

Traditionally in robotics, autonomous reaching and grasping of an object
require stereo vision (i.e., to calculate the distance) and tactile sensors (i.e., to
detect the contact with the object).

The embodied approach tells us that a robot is not merely a passive
observer of the external world. Perception is an active process, where an agent
actively structures its own sensory input by manipulating the world (information self-structuring). To disambiguate objects in the environment, Metta and
Fitzpatrick developed a reaching behavior based on poking actions using the
robot 'Cog' (Fitzpatrick and Metta, 2003). Even simple interactions such as
poking objects, can facilitate the categorization task, because in this way

objects can be easily distinguished from the background, something difficult if not impossible to realize just by vision. The point about perception being an active process has been made by many people (e.g., Dewey, 1896; Edelman, 1987; Aloimonos, 1993; to mention but a few). The idea put forward here is that through a sensory-motor coordinated interaction; information structure is induced (for a detailed discussion, see e.g., Pfeifer and Scheier, 1999; Pfeifer and Bongard, 2007; Lungarella and Sporns, 2006).

The upper torso humanoid robot 'Domo' developed by Aaron Edsinger at MIT (Edsinger, 2007) is able to see people and track them, segment its own self-motion from motion in the world, grasp unknown objects handed to it by a human collaborator, carry out elementary bimanual manipulation, use controlled force in putting objects down, respond to forces applied to it, interact with people safely, and respond to simple verbal commands. But it has neither stereo vision nor tactile sensors. Inspired by the experiments by Metta and Fitzpatrick (Fitzpatrick and Metta, 2003), Gabriel Gómez enhanced Domo with the capability of autonomous reaching and grasping objects based on poking actions.

When the robot detects a colored object on its table, the stiffness of the closest arm to the object is increased and a reaching trajectory is executed, the robot tries to poke the object with its open hand, during this motion, the hand is aligned with the direction of the gravity to make sure it does not collide with the table. By measuring the velocity of the colored object over time, the robot can detect if its own movement – the poking action – is responsible for the motion of the object, in which case a successful poke will be detected. Then a grasping action is attempted by increasing the stiffness of the hand and closing the fingers over the object. The success of this grasping action can be detected without relying on tactile sensors by inspecting the following three conditions: (a) if the net torque applied by the fingers is positive and above a threshold, then the hand is performing a grasp, (b) if the net angular velocity of the fingers is close to zero, it can be assumed that the fingers are in a stable state, and (c) if the grasp aperture, which is the distance between the thumb and the forefinger is more than 30 mm, then it is assumed that the fingers are wrapped around an object and not resting on the palm of the hand. If these three conditions are met, the hand is considered to be holding an object; otherwise, the attempted grasp would be considered as a failure and the robot's arm will return to its initial relaxed position.

By exploiting its interaction with the environment, Domo does not need to maintain a model of the world (i.e., it only reacts to color and motion), it does not have to rely on stereo vision or tactile sensors on its fingers to perform autonomous reaching and grasping of colored objects on its table (Figure 2.10).

## 2.3   Discussion

We have seen a large variety of case studies. The question that immediately arises is whether there are general overarching principles governing all of them.

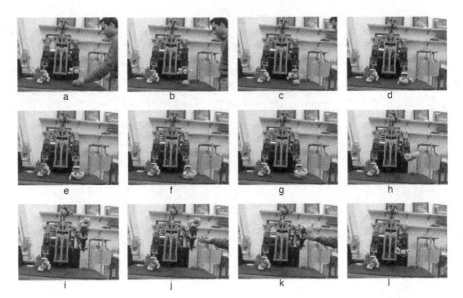

*Figure 2.10*   *Frontal view of a complete sequence of Domo grasping an object*
*autonomously. (a) The robot detects a colored object on its table;*
*(b–c) a poking action is started; (d) the object is moved as a result*
*of the poking action and its velocity exceeds a given threshold; (e) a*
*grasping action is attempted; (f) the robot detects that is holding*
*the object in its hand; (g–i) Domo brings the object closer to its*
*cameras and explores the object visually; (j–k) after a given*
*threshold is reached, Domo looks for a human collaborator and*
*delivers the object to him; (l) the robot returns to the relaxed*
*position*

A recently published scheme (Pfeifer *et al.*, 2007) shows a potential way of
integrating all of these ideas. We will use Figure 2.11 to summarize the most
important implications of embodiment and to embed our case studies into a
theoretical context. Driven by motor commands, the musculoskeletal system
(mechanical system) of the agent acts on the external environment (task
environment or ecological niche). The action leads to rapid mechanical feedback characterized by pressure on the bones, torques in the joints, and passive
deformation of skin tissue. In parallel, external stimuli (pressure, temperature,
and electromagnetic fields) and internal physical stimuli (forces and torques
developed in the muscles and joint-supporting ligaments, as well as accelera-
tions) impinge on the sensory receptors (sensory system). The patterns
thus induced depend on the physical characteristics and morphology of the
sensory systems and on the motor commands. Especially if the interaction is
sensory-motor coordinated, as in foveation, reaching, or grasping movements,
information structure is generated.

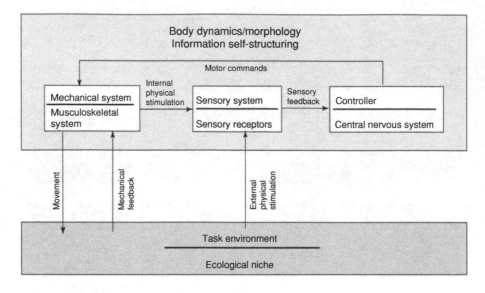

*Figure 2.11    Overview of the implications of embodiment. The interplay of
information and physical processes [After Pfeifer et al., 2007;
see text for details]*

The effect of the motor command strongly depends on the tunable morphological and material properties of the musculoskeletal system, where by
tunable we mean that properties such as shape and compliance can be changed
dynamically. During the forward swing of the leg in walking, the muscles
should be largely passive, whereas when hitting the ground, high stiffness is
required, so that the materials can take over some of the adaptive functionality
on impact, which leads to the damped oscillation of the knee joint. All parts of
this diagram are crucial for the agent to function properly, but only one part
concerns the controller or the central nervous system. The rest can be seen as
'morphological computation.'

Let us now go through the case studies, starting with the Eyebot. Given
certain behavioral patterns, for example, moving straight, the robot induces
sensory stimulation which, because of the specific morphology of the facet
distribution, is highly structured and easy to process for the nervous system
(information self-structuring). This process corresponds to the outer loop from
the controller via mechanical system to task environment, back to sensory
system and controller. The loosely swinging arm benefits from the mechanical
feedback generated through a particular movement (left side of Figure 2.9),
internal stimulation (forces generated in the muscles through gravity), and
external sensory stimulation whenever the hand encounters and object or
touches the agent's own body; the behavior cannot be understood if only the
motor commands are taken into account. Note that the control of the joint

angles in the arm is largely passive for this motion. In grasping, the movement is constrained by the mechanical system, which in turn provides mechanical feedback as the fingers wrap around an object. This leads on the one hand to internal physical stimulation (forces in the muscles and torques in the joints of the fingers), and on the other to external sensory stimulation from the touch sensors on the skin, which closes the loop to the controller. The Fin Ray effect is a very direct consequence of mechanical feedback, demonstrating that mechanical feedback can indeed take over nontrivial tasks (such as bending in the direction from where it is poked). In the cockroaches climbing over obstacles, the morphology of the mechanical system is changed as the insect encounters an obstacle, which changes the effect of the motor commands. The passive dynamic walker, Stumpy, the quadruped Puppy, and Crazy bird are all examples demonstrating the exploitation of intrinsic dynamics. Because in their basic versions, sensors are entirely missing, the pathway back to the controller on the right in Figure 2.11 simply does not exist: control is open loop and the stabilization is solely achieved through the mechanical feedback loops shown in the lower left of the figure. In these examples, the feedback is generated through ground reaction forces. Insects, when walking, also exploit mechanical feedback generated through ground reaction forces, but rather than exploiting it for gait stabilization, they capitalize on exploiting the internal sensory stimulation generated in the joint angles as one leg pushes back (thus inducing changes in the joint angles of all the other legs that are standing on the ground). This process corresponds to the lower left part of Figure 2.11 and the arrow pointing from the mechanical system to the sensory system. Finally, in the artificial fish Wanda, and presumably in biological fish as well, the mechanical feedback to the agent is provided through the induction of turbulences rather than ground reaction forces, a strategy that is no less effective and leads to natural and energy-efficient movement.

There are three main conclusions that can be drawn from these case studies.

1.  it is important to exploit the dynamics to achieve energy-efficient and natural kinds of movements. The term 'natural' not only applies to biological systems, but artificial systems also have their intrinsic natural dynamics.
2.  there is a kind of trade-off or balance: the better the exploitation of the dynamics, the simpler the control, the less neural processing will be required.
3.  if the agent is actually behaving in the real world – and it has to, otherwise all this does not work – it is generating sensory stimulation. Once again, we see the importance of the motor system for the generation of sensory signals, or more generally for perception. It should also be noted that motor actions are physical processes, not computational ones, but they are computationally relevant, or put differently, relevant for neural processing, which is why we use the term 'morphological computation.'

Having said all this, it should be mentioned that there is an additional trade-off. The more the specific environmental conditions are exploited – and the passive dynamic walker is an extreme case – the more the agent's success will be contingent upon them. Thus, if we really want to achieve brain-like intelligence, the brain (or the controller) must have the ability to quickly switch to different kinds of exploitation schemes either neurally or mechanically through morphological change.

In conclusion, it can be stated that because of the tight link of sensory-motor coordinated processes with the information processing of the brain – including categorization – there may not be a clear distinction between what people have called 'low-level sensory' motor processes, and 'high-level' cognitive processes: In embodied systems, cognition can be bootstrapped in natural ways from sensory-motor processes. It has been suggested that even very abstract mathematical concepts such as transitivity or a limit are firmly grounded in sensory-motor processes (Lakoff and Núñez, 2000).

## Acknowledgments

Rolf Pfeifer would like to thank the European Project: 'ROBOTCUB: ROBotic Open Architecture Technology for Cognition Understanding and Behavior,' No. IST-004370. For Gabriel Gómez funding has been provided by the Swiss National Science Foundation through a scholarship for advanced young researchers No. PBZH2-118812. The authors would like to thank Dr Nefti-Meziani for the opportunity to contribute to the book.

## References

Aloimonos, Y. (1993). *Active Perception.* Hillsdale, NJ, Lawrence Erlbaum Associates.

Blickhan, R., Wagner, H., and Seyfarth, A. (2003). 'Brain or muscles?'. *Recent Research Developments in Biomechanics*, vol. 1, pp. 215–245.

Brooks, R. (2004). 'The robots are here'. *Technology Review*, **107**(1):30.

Buehler, M. (2002). 'Dynamic locomotion with one, four and six-legged robots'. *Journal of the Robotics Society of Japan*, **20**(3):15–20.

Collins, S., Ruina, A., Tedrake, R., and Wisse, M. (2005). 'Efficient bipedal robots based on passive dynamic walkers'. *Science*, vol. 307, pp. 1082–1085.

Cruse, H. (1990). 'What mechanisms coordinate leg movement in walking arthropods?' *Trends in Neurosciences*, vol. 13, pp. 15–21.

Cruse, H., Dean, J., Durr, V., Kindermann, T., Schmitz, J., and Schumm, M. (2002). 'A decentralized, biologically based network for autonomous control of (hexapod) walking' (pp. 384–400). *Neurotechnology for Biomimetic Robots*. Cambridge, MIT Press.

Dewey, J. (1896). 'The reflex arc in psychology'. *Psychological Review*, 3(1896): 357–370. Reprinted in J. J. McDermott (ed.), *The Philosophy of John Dewey* (pp. 136–148). Chicago, IL, University of Chicago Press.

Dür, V., Krause, A. F., Schmitz, J., and Cruse, H. (2003). 'Neuroethological concepts and their transfer to walking machines'. *International Journal of Robotics Research*, **22**(3–4):151–167.

Edelman, G. (1987). *Neural Darwinism: The Theory of Neuronal Group Selection.* New York, Basic Books.

Edsinger, A. L. (2007). 'Robot manipulation in human environments'. Ph.D. dissertation, Massachusetts Institute of Technology, Department of Electrical Engineering and Computer Science, Cambridge, MA.

Franceschini, N., Pichon, J. M., and Blanes, C. (1992). 'From insect vision to robot vision'. *Philosophical Transactions of the Royal Society of London, Series B*, vol. 337, pp. 283–294.

Fitzpatrick, P. and Metta, G. (2003). 'Grounding vision through experimental manipulation'. *Philosophical Transactions of the Royal Society: Mathematical, Physical, and Engineering Sciences*, vol. 361, pp. 2165–2185.

Gómez, G. and Eggenberger Hotz, P. (2007). 'Evolutionary synthesis of grasping through self-exploratory movements of a robotic hand'. *IEEE Congress on Evolutionary Computation (CEC2007).* Paper Nr. 1098, pp. 3418–3425.

Gómez, G., Hernandez, A., and Eggenberger Hotz, P. (2006). 'An adaptive neural controller for a tendon driven robotic hand'. In T. Arai, R. Pfeifer, T. Balch, and H. Yokoi (eds.), *Proceedings of the 9th International Conference on Intelligent Autonomous Systems (IAS-9)*, pp. 298–307. Tokyo, Japan, IOS Press.

Hernandez, A., Dermitzakis, C., Damian, D., Lungarella, M., and Pfeifer, R. (2008). 'Sensory-motor coupling in rehabilitation robotics' (pp. 21–36). *Handbook of Service Robotics.* Vienna, Austria, I-Tech Education and Publishing.

Iida, F. (2005). 'Cheap design and behavioral diversity for autonomous adaptive robots'. *Ph.D. dissertation, Faculty of Mathematics and Science,* University of Zurich, Switzerland.

Iida, F. and Pfeifer, R. (2004a). 'Self-stabilization and behavioral diversity of embodied adaptive locomotion' (pp. 119–128). *Embodied Artificial Intelligence*, ser. *Lecture Notes in Computer Science*/Artificial Intelligence, vol. 3139. Springer, Berlin, Germany.

Iida, F. and Pfeifer, R. (2004b). ' "Cheap" rapid locomotion of a quadruped robot: Self-stabilization of bounding gait'. In F. Groen, N. Amato, A. Bonarini, E. Yoshida, and B. Krse (eds.), *Proceedings of the 8th International Conference on Intelligent Autonomous Systems (IAS-8)*, pp. 642–649. IOS Press, Amsterdam, The Netherlands.

Iida, F. and Pfeifer, R. (2006). 'Sensing through body dynamics'. *Robotics and Autonomous Systems*, **54**(8):631–640.

Kubow, T. M. and Full, R. J. (1999). 'The role of the mechanical system in control: A hypothesis of self-stabilization in hexapedal runners'. *Philosophical Transactions of the Royal Society, Series B*, vol. 354, pp. 849–861.

Lakoff, G. and Núñez, R. (2000). *Where Mathematics Come from: How the Embodied Mind Brings Mathematics into Being*. New York, Basic Books.

Lichtensteiger, L. (2004). 'On the interdependence of morphology and control for intelligent behavior'. *Ph.D. dissertation, University of Zurich*, Switzerland.

Lungarella, M. and Sporns, O. (2006). 'Mapping information flow in sensorimotor networks'. *PLoS Computational Biology*, vol. 2, p. e144.

Lungarella, M., Pegors, T., Bulwinkle, D., and Sporns, O. (2005). 'Methods for quantifying the information structure of sensory and motor data'. *Neuroinformatics*, 3(3):243–262.

Markram, H. (2006). 'The blue brain project'. *Nature Reviews, Neuro-Science*, vol. 7, pp. 153–159.

McGeer, T. (1990a). 'Passive dynamic walking'. *The International Journal of Robotics Research*, 9(2):62–82.

McGeer, T. (1990b). 'Passive walking with knees'. *IEEE Conference on Robotics and Automation*, vol. 2, pp. 1640–1645.

Molina-Vilaplana, J., Feliu-Batlle, J., and López-Coronado, J. (2007). 'A modular neural network architecture for step-wise learning of grasping tasks'. *Neural Networks*, 20(5):631–645.

Paul, C., Dravid, R., and Iida, F. (2002). 'Design and control of a pendulum driven hopping robot'. *IEEE/RSJ International Conference on Intelligent Robots and Systems, IROS-2002*, Lausanne, Switzerland.

Pfeifer, R. (2000). 'On the role of morphology and materials in adaptive behavior'. *Sixth International Conference on Simulation of Adaptive Behavior (SAB)*, pp. 23–32.

Pfeifer, R. (2003). 'Morpho-functional machines: basics and research issues'. *Morpho-Functional Machines: The New Species*. Tokyo, Springer.

Pfeifer, R. and Bongard, J. (2007). *How the Body Shapes the Way we Think*. Cambridge, MA, MIT Press.

Pfeifer, R. and Scheier, C. (1999). *Understanding Intelligence*. Cambridge, MA, MIT Press.

Pfeifer, R., Iida, F., and Gómez, G. (2006). 'Morphological computation for adaptive behavior and cognition'. *International Congress Series*, vol. 1291, pp. 22–29.

Pfeifer, R., Lungarella, M., and Iida, F. (2007). 'Self-organization, embodiment, and biologically inspired robotics'. *Science*, vol. 318, pp. 1088–1093.

Pfeifer, R., Lungarella, M., Sporns, O., and Kuniyoshi, Y. (2008). 'On the information theoretic implications of embodiment – principles and methods'. *Proceedings of the 50th Anniversary Summit of Artificial Intelligence*, pp. 76–86.

Rinderknecht, M., Ruesch, J., and Hadorn, M. (2007). 'The lagging legs exploiting body dynamics to steer a quadrupedal agent'. *International Conference on Morphological Computation*, Venice, Italy.

Schmitz, A., Gómez, G., Iida, F., and Pfeifer, R. (2007a). 'Adaptive control of dynamic legged locomotion'. *Concept Learning Workshop, IEEE International Conference on Robotics and Automation (ICRA 2007)*.

Schmitz, A., Gómez, G., Iida, F., and Pfeifer, R. (2007b). 'On the robustness of simple speed control for a quadruped robot'. *Proceedings of International Conference on Morphological Computation Workshop 2007*, pp. 88–90.

Shelley, T. (2007). *Novel actuators copy structures from fish* [online]. Findlay Media. Available from: http://www.eurekamagazine.co.uk/article/9658/Novel-actuators-copy-structures-from-fish.aspx [Accessed 12 Jan 2010].

Sporns, O. (2009). 'From complex networks to intelligent systems'. In B. Sendhoff, E. Körner, O. Sporns, H. Ritter and K. Doya (eds.), *Creating Brain-like Intelligence: From Basic Principles to Complex Intelligent Systems*. Springer, Berlin, Germany.

Staudacher, E. (1998). 'Distribution and morphology of descending brain neurons in the cricket *Gryllus bimaculatus*'. *Cell Tissues Research*, vol. 294, pp. 187–202.

Takamuku, S., Gómez, G., Hosoda, K., and Pfeifer, R. (2007). 'Haptic discrimination of material properties by a robotic hand'. In *Proceedings of the 6th IEEE International Conference on Development and Learning (ICDL 2007)*, Paper Nr. 76.

Takuma, T. and Hosoda, H. (2006). 'Controlling the walking period of a pneumatic muscle walker'. *The International Journal of Robotics Research*, 25(9), 861–866.

Torres-Jara, E. and Gómez, G. (2008). 'Fine sensitive manipulation'. *Proceedings of the Australian Conference on Robotics and Automation (ACRA 2008)*.

Watson, J. and Ritzmann, R. (1998). 'Leg kinematics and muscle activity during treadmill running in the cockroach, *Blaberus discoidalis*: I. Slow running'. *Journal of Comparative Physiology* A, vol. 182, pp. 11–22.

Watson, J., Ritzmann, R., and Pollack, A. (2002). 'Control of climbing behavior in the cockroach, *Blaberus discoidalis*: II. Motor activities associated with joint movement'. *Journal of Comparative Physiology* A, vol. 188, pp. 55–69.

Wisse, M. (2005). 'Three additions to passive dynamic walking: Actuation, an upper body, and 3d stability'. *International Journal of Humanoid Robotics*, 2(4):459–478.

Yokoi, H., Hernandez Arieta, A., Katoh, R., Yu, W., Watanabe, I., and Maruishi, M. (2004). 'Mutual adaptation in a prosthetics application'. In Iida *et al.* (2004), pp. 147–159.

Yu, W., Yokoi, H., and Kakazu, Y. (2006). Chapter 4: 'An interaction based learning method for assistive device systems'. In *Focus on Robotics Research*, eds: John X. Liu *et al.*, ISBN: 1-59454-594-4, pp. 123–159, Nova Science Publishers.

Ziegler, M., Iida, F., and Pfeifer, R. (2005). 'Cheap underwater locomotion: Morphological properties and behavioral diversity'. *IROS 2005 Workshop on Morphology, Control, and Passive Dynamics.*

1. [Online]. Available: http://www.festo.com/cms/en-us_us/5007_6541.htm
2. [Online]. Available: http://www.festo.com/cms/en-us_us/4981.htm

*Chapter 3*

# Compliant actuation: enhancing the interaction ability of cognitive robotics systems

*Nikolas G. Tsagarakis, Matteo Laffranchi,*
*Bram Vanderborght and Darwin G. Caldwell*

## Abstract

The growth of cognitive capabilities in biological systems and particularly in humans is assisted by the extensive and safe interaction of their bodies with their living environment. This ability to interact safely with the environment is mainly due to a set of inherent behaviours/properties that allow the biological systems to intelligently react to the interaction at the body/actuation level with minimum active control. One fundamental intrinsic characteristic of the biological systems is the compliance. Looking at the diverse range of actuation techniques, the natural world appears to have developed a much more ubiquitous actuation design, with the organic muscle that provides power to animals ranging from the largest whales to microbes with adaptation to cope with environmental extremes. The potential benefits, which can be gained in any mechanism with the incorporation of biologically inspired compliant actuation concepts, are well known. However, the majority of today's robots lack these characteristics. This chapter reviews some of the solution adopted to implement the compliant behaviour in robotic systems and introduces a case study of a new compact soft actuation unit intended to be used in multi-degree of freedom and small-scale robotic systems such as the cognitive child humanoid robot 'iCub' [1].

## 3.1 Introduction

Biological systems and particularly humans develop their cognitive capabilities with the assistance of extensive and safe interaction of their bodies with their living environment. This ability to interact safely with the environment is mainly due to their body morphology, which integrates a set of inherent behaviours/properties that allow the biological systems to intelligently react to the interaction at the body/actuation level with minimum active control. Although the potential benefits, which can be gained in any mechanism with the incorporation of biologically inspired compliant actuation concepts, are well known, the majority of today's robots lack these characteristics. Until the past decade, the main approach in robotic actuation was the use of heavy, stiff position/velocity and torque actuation units coupled with rigid non-back-drivable transmission mechanisms. However, the creation of mechatronic systems with natural compliant properties from the actuation point of view has been the inspiration to the robotic community for many years. This replication though has failed so far mainly due to the complexity and inherent properties of the organic actuation/transmission system (muscle–tendon–joint–bone), which have prevented its emulation from the traditional engineered actuation paradigms. As a result the majority of today's robots are optimized for precision and speed and are highly repeatable, acting within constrained and well-defined industrial environments, and are therefore suitable for conventional automation.

Recently, new demands were placed on the available actuation systems and on the ability of the robotic machines to interact safely within a less constrained and more human living environment. These demands are due to the new areas for technical exploitation that have increasingly made clear that this traditional stiff actuation approach has significant performance limitations related to safety and the ability to interact with the environment. One new area of technical exploitation that has gained a lot of attention is the development of the cognitive capabilities in robotic systems. The body/actuation morphology of biological systems and particularly humans allows the extensive interaction of their bodies in a safe manner during the cognitive development phase. This is not possible in the majority of the existing robots.

The interaction capability of the biological systems is mainly due to a set of intrinsic properties found in their body, which allow the biological systems to inherently react to the interaction at the body/actuation level. One fundamental intrinsic characteristic is the compliance at the actuation level. In contrast to the biological systems, the majority of the robots today lack the compliance characteristic, and as a result the capability of these systems still remains a major deficiency when interaction with the environment is considered. As a consequence the contribution of interactions in robots to the cognitive development of the robot is very limited to the cases of well-defined pre-programmed interaction scenarios. To address this, new mechatronic actuation solutions, which will form the motion units of the next generation of robotic systems, are needed.

This chapter is organized as follows: Section 3.2 introduces the state of the art of compliant actuation, while Section 3.3 presents a case study of a series elastic actuator (SEA). This actuator shows particular improvements over the existing SEA implementations based on ball screw/die spring, torsion springs or extension springs/cable assemblies. The compact design of the actuator is due to the novel mechanical implementation of the compliant module. Despite its small size, the actuator still retains a high level of performance and is a potential generic solution for small-scale multi-degree of freedom systems. The work on this actuation unit was motivated by the need for a soft, highly integrated actuation unit suitable for small-scale multi-degree of freedom robots. In particular the actuator was developed for the new generation of the 'iCub' robot, a child-sized humanoid (1 m height) with 53 degrees of freedom developed in [1]. By combining the passive compliance of this actuator with active compliance control, a wide range of compliance adjustment can also be obtained that can ensure the safe interaction of the unit with its environment.

Section 3.3 also focuses on the description of the actuator mechanical design, while Section 3.4 presents the stiffness model of the compliant module. A discussion of the overall dynamic model and control of the actuator follows in Section 3.5, while Section 3.6 contains experimental results for the performance of the actuator and Section 3.7 addresses the conclusions.

## 3.2   Brief review of compliant actuation

The ability of the stiff actuated robot units to safely interact with their surroundings can be increased by means of software techniques based on impedance/admittance regulation and joint torque control [2–4]. However, delays at all stages of the software system make these conventionally stiff actuated systems unsafe during fast impacts and generally incapable of managing high-speed contact transients.

To address this problem, a wide range of experimental novel compliant actuation systems have been developed during the past 15 years. The SEA family is an early development towards the realization of actuator units with inherent compliance [5–8]. Such actuators employ a fixed compliance element between a high impedance actuator and the load. The elastic element used in most realizations is based on the use of torsion/die linear springs or viscoelastic materials usually combined with ball screw reduction drives (linear SEA) or planetary gearbox and cable assemblies (rotary systems). Revolute actuators have been also produced using 'X shaped' cross section metal torsion springs or even rubber balls held between alternating teeth [9]. Although these designs provide successful solutions, they are relatively big (ball screw and die springs or cables and extension springs arrangements) or suffer from small passive deflection ranges (designs using torsion springs).

Actuation units with the ability to modulate compliance have also been developed [10–19]. These variable compliance actuation systems typically employ two actuator units in combination with passive stiffening elastic elements to control, independently, the compliance and the equilibrium position of the actuated joint. The variable compliant actuation systems can be implemented by both series and antagonistic configurations (Figure 3.1).

The series configuration is characterized by the mechanical series, motor–gear–compliant element, and it can be implemented both in linear and rotational designs. Antagonistically actuated variable stiffness joints employ two compliant elements to provide power to the joint. This design is biologically

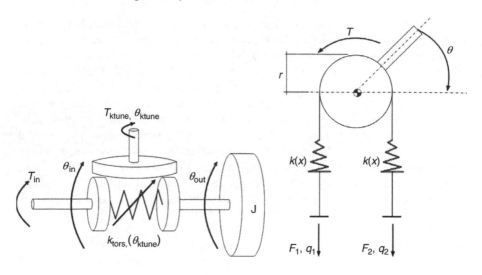

*Figure 3.1    Serial and antagonistic variable compliance actuation concept schemes*

inspired, since mammalian anatomy follows the same concept, i.e. a joint actuated by two muscles arranged in an antagonistic manner. The muscle–tendon cooperation gives the driven link (arm, leg, etc.) a controllable and variable compliance. In addition to using biological muscle, this type of antagonistic compliance control can be achieved using both conventional two-motor electric drive designs and other more biologically inspired forms such as pneumatic muscle actuators (pMA). In the latter case compliance is an inherent characteristic of the actuator, while for an electric design, compliant elements (generally springs) have to be embodied into the system.

Obvious advantages that variable stiffness implementations offer when compared with fixed passive compliance units are the ability to regulate stiffness and position independently and the wide range of stiffness and energy storage capabilities. On the other hand, the mechanical complexity, size, weight, cost and integration are still open issues in variable passive compliance realizations. As a result, their application to multi-degree of freedom or small-scale robotic machines still remains a challenging task. In this respect, the SEA family clearly has an advantage when compared with variable stiffness implementations. In addition, its main disadvantage of the preset passive mechanical compliance can be, to some extent, minimized by combining the unit with active stiffness control.

An alternative approach to safe human–robot interaction is the distributed macro-mini actuation approach (DM2) proposed by Zinn et al. [20]. This actuation method uses an actuation structure consisting of low- and high-frequency parallel actuators to create a unified high-performance and safe robotic system.

A further alternative to electric motor-based variable compliance actuation is the use of fluidic actuation. In particular, pneumatic actuation has been employed in various actuation forms ranging from pneumatic cylinders [21] to flexible muscles, including the McKibben muscle actuator [22], its variations [23, 24] and chamber structures [25]. Thanks to its inherent compliance that pneumatic actuation has intrinsically higher safety levels and has been employed in various setups from antagonist configurations [26, 27] to hybrid actuation schemas [28, 29]. Indeed, versions of the macro-mini actuation concept have sought to use braided pMA as the high-force, low-bandwidth actuator in combination with a smaller electric motor. In these configurations, independent control of torque and compliance becomes feasible by the co-activation of the antagonistic actuators. However, low bandwidth, friction, nonlinearities and indeed the lack of stiffness are some of the issues that have limited the widespread use of pneumatic actuation in traditional industrial inspired robotic designs.

## 3.3   Case study: the mechanics of a compact soft actuation unit

The mechanical realization of a soft actuation unit presented in this chapter is based on the serial elastic actuator concept, but particular attention has been paid to satisfying the dimensional and weight requirements of the 'iCub' robot [1] (Figure 3.2). The development of the 'iCub' robot is a research initiative dedicated to the realization of embodied cognitive systems and the creation of an advanced robotic platform for neuroscientific study. The two main goals are to create an open hardware/software humanoid robotic platform for research in embodied cognition and to advance our neural understanding of cognitive systems by exploiting this platform in the study of the development of cognitive capabilities in humanoid robots.

The iCub platform itself has as its aim the replication of the physical and cognitive abilities of a child. This 'child' robot will act in a cognitive scenario, performing the tasks useful to learning and interacting with the environment and humans.

The high integration density of the developed SEA is due to the novel mechanical compliant module. To minimize dimensions while achieving high rotary stiffness, a mechanical structure with a three-spoke output component, a circular input pulley and six linear springs has been designed and fabricated (Figure 3.3).

The circular component forms the input of the compliant module and is fixed to the output of the reduction drive. The three-spoke element rotates on bearings with respect to the circular base and is coupled with it by means of six springs (Figure 3.3). The three-spoke component forms the output of the compliant module and the mounting base of the output link. The six linear springs when inserted in the arrangement shown in the figure experience a pre-contraction equal to half of the maximum acceptable deflection. Deflections

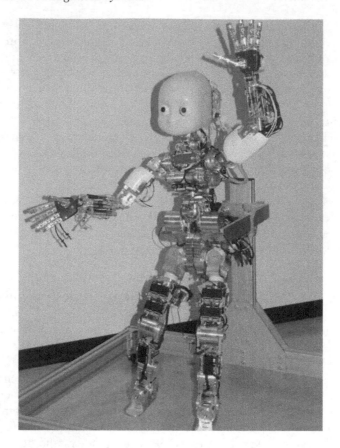

*Figure 3.2    The 'iCub' child humanoid robot*

larger than the maximum allowable are not permitted by means of mechanical pin-based locks.

Two 12 bit absolute position sensors are integrated within the actuation group measuring respectively the mechanical angle of the motor after the reduction drive and the deflection angle of the compliant module. These sensors allow not only the monitoring of the link position but also the evaluation of the joint torque. Because of the compliance introduced, it is possible to use the sensor measuring the compliant module deflection to estimate the torque.

## 3.4    The passive stiffness module

In this section, the stiffness model of the three-spoke spring structure is presented. The deflection of the compliant module results in torques through compression of the spring elements along their main axis (Figure 3.4).

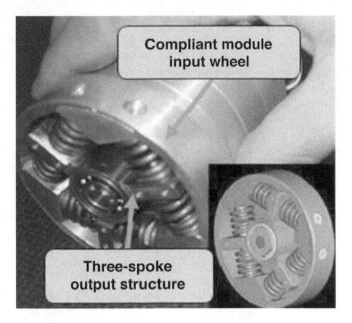

*Figure 3.3   The prototype of the compact SEA module*

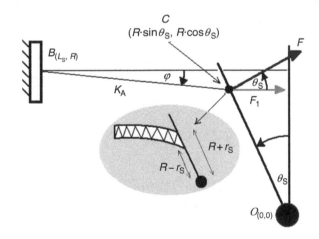

*Figure 3.4   Compression of the spring as a result of the module deflection*

Considering one of the antagonist linear spring pairs (Figure 3.4), the axial forces generated by each of the springs when the compliant three-spoke module is deflected from the equilibrium position by an angle $\theta_S$ are given by

$$F_1 = K_A \cdot (x_p + x(\theta_S)), \quad F_2 = K_A \cdot (x_p - x(\theta_S)) \tag{3.1}$$

where $F_1$ and $F_2$ are the forces generated by the two springs and $x_p$ is the spring pre-contraction, $x(\theta_S) = R \cdot \sin \theta_S$ the resulting deflection of the two springs along their main axis, $K_A$ the spring axial stiffness and $R$ is the length of the spoke arm. The combined axial force applied in any spoke is therefore

$$F = F_1 - F_2 = 2 \cdot K_A \cdot R \cdot \sin \theta_S \tag{3.2}$$

The corresponding torque generated at the joint because of the axial forces of one antagonistic pair of springs is equal to

$$T = F \cdot R \cdot \cos \theta_S = 2 \cdot K_A \cdot R^2 \cdot \sin \theta_S \cdot \cos \theta_S \tag{3.3}$$

So far we have considered that the axial spring force is concentrated at one point. Taking into account that the spring has an external radius $r_S$, the axial spring compression is not equal for the whole surface area in contact with the spoke. The areas farthest from the centre of rotation are subject to larger deflections creating higher forces (Figure 3.4).

As a result, the torque generated by the axial deflection of the antagonistic pair of springs can be computed by

$$T = \frac{1}{2 \cdot r_S} \int_{R-r_S}^{R+r_S} 2 \cdot K_A \cdot R^2 \cdot \sin \theta_S \cdot \cos \theta_S dR$$

$$= 2 \cdot K_A \cdot \left( R^2 + \frac{r_S^2}{3} \right) \cdot \sin \theta_S \cdot \cos \theta_S \tag{3.4}$$

Thus, the combined torque at the joint considering the axial forces from all three pairs (Figure 3.5) is

$$T_{\text{total}} = 3 \cdot T = 6 \cdot K_A \cdot \left( R^2 + \frac{r_S^2}{3} \right) \cdot \sin \theta_S \cdot \cos \theta_S \tag{3.5}$$

By differentiating (3.5), the rotary stiffness of the three-spoke module, due to the axial deflection of the springs, is

$$K_S = \frac{dT_{\text{total}}}{d\theta_S} = 6 \cdot K_A \cdot \left( R^2 + \frac{r_S^2}{3} \right) \cdot (2 \cdot \cos \theta_S^2 - 1) \tag{3.6}$$

Figure 3.6 shows the theoretical stiffness of the module within the range of the deflection angle, for the first prototype module with the following parameters: $K_A = 62$ kN/m, $R = 20.5$ mm, $r_S = 6.3$ mm. It can be seen that the stiffness is slightly reduced as the deflection angle increases (notice that the $y$-axis does not start from zero).

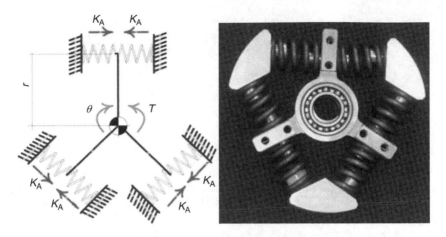

*Figure 3.5    The three-spoke spring coupling arrangement*

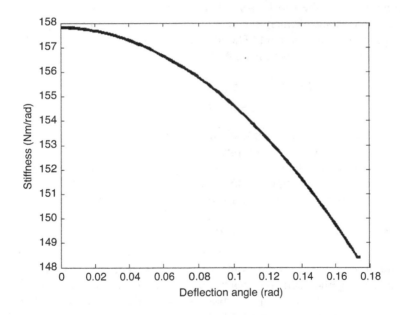

*Figure 3.6    The stiffness profile of the compliant module*

## 3.5    Actuator model and control

The compact actuator consists of three main components: a brushless DC motor, a harmonic reduction drive and the rotary compliant module outlined earlier in text. These components can be represented by the mechanical model depicted in Figure 3.7.

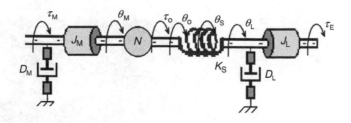

Figure 3.7 Compact SEA mechanical model diagram

The model is composed of the rotary inertia $J_M$ and viscous damping of the motor $D_M$, the gear drive with the reduction ratio $N$, the elastic module with an equivalent spring constant $K_S$ (3.6), and the output link inertia $J_L$ and axial damping coefficient $D_L$.

In addition, $\theta_M$ and $\theta_O$ are the motor mechanical angles before and after the reduction drive, $\theta_L$ the angle of the output link and $\theta_S$ is the rotary defection of the elastic module with respect to $\theta_O$ such that $\theta_L = \theta_O + \theta_S$.

Finally, $\tau_M$ is the torque provided by the actuator, while $\tau_O$ is the input torque of the elastic element and $\tau_E$ is the torque imposed to the system by the load and/or the environment. The above-mentioned system can be described by the following set of dynamic equations:

$$(J_M \cdot N^2 \cdot s^2 + D_M \cdot N^2 \cdot s + K_S) \cdot \theta_O - K_S \cdot \theta_L = \tau_O \qquad (3.7)$$

$$(J_L \cdot s^2 + D_L \cdot s + K_S) \cdot \theta_L - K_S \cdot \theta_O = \tau_E \qquad (3.8)$$

where s denotes the Laplace transform variable.

In the absence of active compliance control (i.e. a stiff position-controlled actuator), we can consider $\dot{\theta}_O = 0$ and the output impedance of the system can be obtained by differentiating (3.8) and dividing both sides by $\dot{\theta}_L$. Thus, the impedance of the system at the output can be shown to be equal to

$$Z_E = \frac{d\tau_E}{d\theta_L} = (J_L \cdot s^2 + D_L \cdot s + K_S) \qquad (3.9)$$

Equation (3.9) implies that what we will feel is a combination of the passive compliance together with the inertia and damping properties of the system. For fast contacts/impacts, the inertial forces dominate and are the ones that a user interacting with the output link will feel more. To regulate the above-mentioned preset impedance, in higher or lower values, as may be required by particular interaction properties, we use an active compliance control scheme to regulate the overall link deflection produced by the applied load. A velocity-based controller is used to achieve this by regulating $\dot{\theta}_O$.

Incorporating $\dot{\theta}_O$, the impedance expression (3.9) becomes

$$Z_E = \frac{d\tau_E}{d\theta_L} = (J_L \cdot s^2 + D_L \cdot s + K_S) - K_S \cdot \frac{d\theta_O}{d\theta_L} \qquad (3.10)$$

To make the actuator output follow a desired simulated impedance, (3.11) must be applied.

$$Z_D = \frac{d\tau_E}{d\theta_L} \tag{3.11}$$

where $Z_D$ is the desired simulated impedance of the actuator. The above equation defines the desired characteristics of the actuator motion/torque to replicate a specific compliant behaviour. Having specified this behaviour, the control law for the desired $\dot{\theta}_O$ can be derived by equating (3.10) and (3.11) and solving with respect to $\dot{\theta}_O$.

$$Z_D = (J_L \cdot s^2 + D_L \cdot s + K_S) - K_S \cdot \frac{d\theta_O}{d\theta_L} \tag{3.12}$$

$$Z_D \cdot \dot{\theta}_L = (J_L \cdot s^2 + D_L \cdot s + K_S) \cdot \dot{\theta}_L - K_S \cdot \dot{\theta}_O \tag{3.13}$$

By substituting $\theta_L = \theta_O + \theta_S$ and solving for $\dot{\theta}_O$, the final control law for the motor velocity as a function of the deflection velocity of the compliant element $\dot{\theta}_S$ can be derived.

$$\dot{\theta}_O = \left( \frac{K_S}{-J_L \cdot s^2 - D_L \cdot s + Z_D} - 1 \right) \cdot \dot{\theta}_S \tag{3.14}$$

The above equation describes only the portion of the desired velocity trajectory for $\theta_O$ that is responsible for the replication of a specific compliant behaviour at the output of the actuation unit. In addition, the term $\dot{\theta}_{TD}$ can be superimposed in Equation (3.14) to represent the desired velocity trajectory of the output motion. Therefore, the overall velocity trajectory $\dot{\theta}_{OD}$ can be written as

$$\dot{\theta}_{OD} = \dot{\theta}_{TD} + \left( \frac{K_S}{-J_L \cdot s^2 - D_L \cdot s + Z_D} - 1 \right) \cdot \dot{\theta}_S \tag{3.15}$$

A block diagram of the active control scheme expressed by (3.15) is shown in Figure 3.8, with the 'compliance regulation filter' being used to implement (3.14).

## 3.6 System evaluation

Experiments were conducted to evaluate the ability of the velocity-based control scheme and the actuation unit to regulate the output impedance for values different from the preset mechanical values. In these experiments, only the tuning of the stiffness component of the output impedance is considered, leaving the other inertial and damping components unaffected. The experiments were performed using a prototype actuation unit (Figure 3.9).

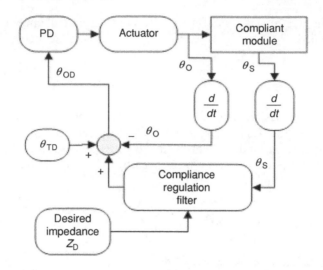

*Figure 3.8    Schematic of the active compliance control scheme*

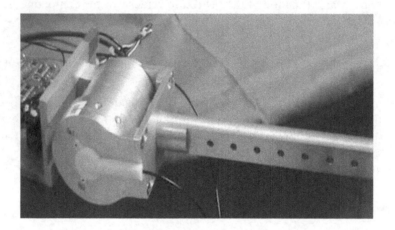

*Figure 3.9    The prototype of the compact compliant unit*

For the first prototype, six linear springs made of 2.2 mm $^{85}$C carbon steel wire were used. The springs have a stiffness of 62 kN/m, a free length of 18 mm and a maximum allowable deflection of 6 mm. When inserted as shown in Figure 3.3, the springs experience a pre-contraction of 3 mm, which is half of the acceptable deflection.

Mechanical pin-based locks prevent the spring overloading by limiting the maximum deflection to 6 mm. This gives a total rotary deflection range of $\pm 0.18$ rad. Given the stiffness of the compliant module, this deflection is considered sufficient for the motor groups used in the humanoid robot 'iCub' [1]. These motors produce a peak torque ranging from 20 to 40 Nm.

The actuator of the unit was formed by the combination of a Kollmorgen RBE1211 frameless brushless motor capable of providing 0.8 Nm of torque, with a harmonic drive CSD 17 having a reduction ratio of $N = 100$ and a peak rated torque of 40 Nm. The inertia and viscous damping parameters of the motor are $J_M = 8.47 \times 10^{-6}$ kg m$^2$, $D_M = 19.68 \times 10^{-16}$ Nm s/rad.

The output of the unit was connected to a link with a mass and inertia of $M_L = 1.6$ kg and $J_L = 5 \times 10^{-4}$ kg m$^2$, respectively. The overall dimensions of the compact prototype compliant unit together with its technical specifications are illustrated in Table 3.1.

*Table 3.1    Specifications of the compliant actuator*

| | |
|---|---|
| Diameter | 70 mm |
| Length | 80 mm |
| Power | 190 W |
| Gear ratio | 100:1 |
| Peak torque | 40 Nm |
| Max passive deflection | $\pm 0.18$ rad |
| Weight | 0.52 kg |

The unit controller and power driver are custom control boards based on the Motorola DSP 56F8000 chip with a CAN communication interface.

In the first experiment, different desired stiffness values were used to evaluate the ability of the system to realize compliances of varying amplitude. In these trials, the equilibrium position of the motor was fixed to a certain angle ($\theta_{TD} = 0$). Data for two cases are presented later in text. In the first case, Figure 3.10, the unit was commanded to retain a certain position when unperturbed and to work as an impedance source with a simulated stiffness component of 30 Nm/rad when a load is applied to the output link. This is six times smaller than the maximum mechanical passive stiffness of the joint (Figure 3.5).

The top graph in Figure 3.9 depicts the stiffness regulation performance. Note on the third graph from the top, the motor motion shows the motor working towards the direction of the applied torque. This generates the required overall link deflection in order to satisfy the commanded simulated stiffness.

In the second case, Figure 3.11, the desired simulated stiffness was set equal to 600 Nm/rad (four times the maximum mechanical preset stiffness). The third graph shows that the motion generated by the motor works in antagonism with the applied load to produce a reverse displacement that compensates the deflection of the passive spring element. This reduces the overall link displacement (Joint total) as seen by the user that applies the load.

For both the low and high stiffness cases, the spikes on the simulated compliance (top graph in Figures 3.10 and 3.11) reveal the tendency of the actuator to display the mechanical preset stiffness during high-rate perturbations. These actions cannot normally be addressed by the bandwidth of the controller.

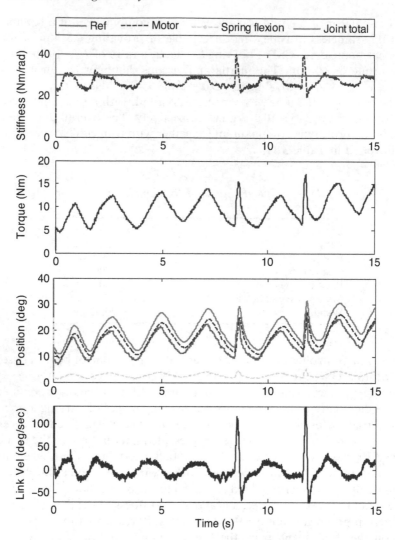

*Figure 3.10    Simulated spring component of 30 Nm/rad*

Following the above-mentioned actions, the behaviour of the actuator and its control under accidental impacts for different values of the simulated stiffness were observed. In these trials, the motor was commanded to follow a sinusoidal trajectory with a frequency of 0.8 Hz, while accidental collisions were randomly generated within the range of motion of the link. Two desired stiffness values were examined and were set equal to those in the stiffness regulation experiment ($K_S = 30$ Nm/rad and $K_S = 600$ Nm/rad).

Results of the first case ($K_S = 30$ Nm/rad) show the ability of the system to absorb the impact by effectively modulating the reference sinusoidal trajectory of the motor in Figure 3.12.

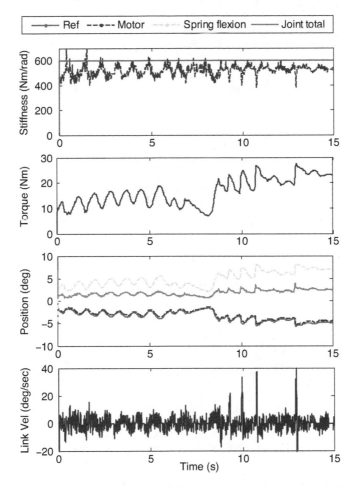

*Figure 3.11    Simulated spring component of 600 Nm/rad*

The high-simulated stiffness (Figure 3.13) does not allow the regulation of the sinusoidal trajectory when the collisions occur. The stiffness modulation term of (3.15) is not large enough to adapt the sinusoidal reference motion trajectory term $\dot{\theta}_{TD}$ during the collision phase. The result of this high actuator stiffness is the higher torques measured during the impact compared to the ones for the low virtual stiffness.

## 3.7    Conclusions

Looking on the diverse range of mechatronic actuation techniques, it can be seen that the natural world appears to have developed a much more ubiquitous

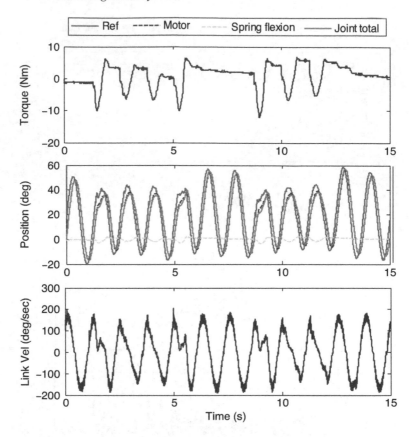

*Figure 3.12   Collisions of the link for a simulated stiffness of 30 Nm/rad*

design with the organic muscle providing power for animals ranging from the largest whales to microbes with adaptation to cope with environmental extremes. One intrinsic characteristic of the biological systems is the compliance in the actuation system. Compliance allows the biological systems to exhibit high tolerance during interactions that are either planned or even accidentally imposed. This ability to interact with the environment contributes significantly to the development of cognitive capabilities of the system. In contrast to this, the majority of the existing robot systems demonstrate very poor interaction performance, which is mainly due to their stiff body and actuation properties. Acceleration of the cognitive developments can be achieved by turning these stiff actuated machines into systems that will attempt to replicate the compliance of the biological systems even at the macro level. This requires the development of new mechatronic actuation solutions that will form the motion units of these new systems.

   In this chapter, the design of a new compact soft actuator that forms the actuation means for a small, multi-degree of freedom robot (iCub) was

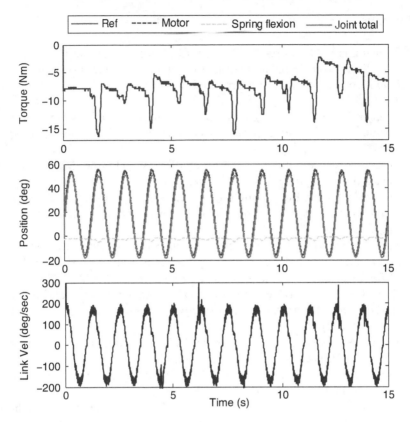

*Figure 3.13   Collisions of the link for a simulated stiffness of 600 Nm/rad*

presented. The miniaturization of the high performance unit was achieved using a novel rotary spring module realized by six linear springs arranged to constrain the motion of a three-spoke structure that rotates relatively to the reduction drive output and acts as a mounting basement for the output link.

The model and control scheme for the actuator were analysed and a velocity-based controller proposed that generates velocity commands based on the desired virtual stiffness using the spring deflection state. The overall system was evaluated with experimental trials performed using a prototype unit.

The preliminary results obtained from these experiments show that the unit and the proposed control scheme are capable of replicating simulated impedances within a wide range with good fidelity. In addition, under accidental random interactions, the system demonstrated good tolerance, effectively modulating the reference motion trajectory to reduce interaction impact forces. Additional studies will include further development of the control scheme to incorporate a stiffness adaptation agent to allow self-tuning of the stiffness according to the safety and interaction needs.

# References

1.  Tsagarakis N.G., Metta G., Sandini G., Vernon D., Beira R., Becchi F., *et al.* 'iCub – the design and realization of an open humanoid platform for cognitive and neuroscience research'. *Advanced Robotics.* 2007;**21**(10): 1151–1175.
2.  Vischer D., Kathib O. 'Design and development of high-performance torque-controlled joints'. *IEEE Transactions on Robotics and Automation.* 1995, vol. 11, pp. 537–544.
3.  Hirzinger G., Sporer N., Albu-Schaffer A., Hahnle M., Krenn R., Pascucci A., *et al.* 'DLR's torque-controlled light weight robot III – are we reaching the technological limits now?'. *Proceedings of IEEE International Conference on Robotics and Automation.* 2002, vol. 2, pp. 1710–1716.
4.  Sang-Ho H., Hale J.G., Cheng G. 'Full-body compliant human–humanoid interaction: balancing in the presence of unknown external forces'. *IEEE Transactions on Robotics and Automation.* 2007;**23**(5):884–898.
5.  Pratt J., Pratt G. 'Intuitive control of a planar bipedal walking robot'. *International Conference on Intelligent Robots and Systems.* 1998, pp. 2014–2021.
6.  Brooks R., Breazeal C., Marjanovic M., Scassellati B., Williamson M. 'The Cog Project: Building a Humanoid Robot'. In: *Computation for Metaphors, Analogy and Agents,* Vol. 1562 of Springer Lecture Notes in Artificial Intelligence, Springer-Verlag, 1998, pp. 8–13.
7.  Kazuo H., Masato H., Yuji H., Toru T. 'The development of honda humanoid robot'. *International Conference on Robotics and Automation.* 1998, pp. 1321–1326.
8.  Pratt G., Williamson M. 'Series elastic actuators'. *Proceedings of IEEE/ RSJ International Conference on Intelligent Robots and Systems.* 1995, vol. 1, pp. 399–406.
9.  Pratt G. 'Low impedance walking robots'. *Integrative and Comparative Biology.* 2002;**42**(1):174–181.
10. Sugar T.G. 'A novel selective compliant actuator'. *Mechatronics Journal.* 2002;**12**(9–10):1157–1171.
11. Hurst J.W., Chestnutt J., Rizzi A. 'An actuator with physically variable stiffness for highly dynamic legged locomotion'. *Proceedings of the International Conference on Robotics and Automation.* 2004, vol. 5, pp. 4662–4667.
12. Migliore S.A., Brown E.A., DeWeerth S.P. 'Biologically inspired joint stiffness control'. *Proceedings of the IEEE International Conference on Robotics and Automation.* 2005, pp. 4508–4513.
13. Tonietti G., Schiavi R., Bicchi A. 'Design and control of a variable stiffness actuator for safe and fast physical human/robot interaction'. *Proceedings of the IEEE International Conference on Robotics and Automation.* 2005, pp. 526–531.

14. Van Ham R., Vanderborght B., Van Damme M., Verrelst B., Lefeber D. 'MACCEPA, the mechanically adjustable compliance and controllable equilibrium position actuator: design and implementation in a biped robot'. *Robotics and Autonomous Systems*. 2005;**55**(10):761–768.
15. Koganezawa K., Nakazawa T., Inaba T. 'Antagonistic control of multi-DOF joint by using the actuator with non-linear elasticity'. *Proceedings of the IEEE International Conference on Robotics and Automation*. 2006, pp. 2201–2207.
16. Hollander K.W., Sugar T.G., Herring D.E. 'A robotic "Jack Spring" for ankle gait assistance'. *Proceedings CIE 2005, ASME 2005 International Design Engineering Technical Conferences*. 2005, California, USA, pp. 24–28.
17. Wolf S., Hirzinger G. 'A new variable stiffness design: matching requirements of the next robot generation'. *Proceedings of the International Conference on Robotics and Automation*. 2008, pp. 1741–1746.
18. Schiavi R., Grioli G., Sen S., Bicchi A. 'VSA-II: a novel prototype of variable stiffness actuator for safe and performing robots interacting with humans'. *Proceedings of the International Conference on Robotics and Automation*. 2008, pp. 2171–2176.
19. Morita T., Sugano S. 'Development of 4-DOF manipulator using mechanical impedance adjuster'. *Proceedings of IEEE International Conference on Robotics and Automation*. 1996, vol. 4, pp. 2902–2907.
20. Zinn M., Khatib O., Roth B., Salisbury J.K. 'A new actuation approach for human friendly robot design'. *International Journal of Robotics Research*. 2004;**23**(4–5):379–398.
21. Sup F., Bohara A., Goldfarb M. 'Design and control of a powered knee and ankle prosthesis'. *Proceedings of the IEEE International Conference on Control Applications*. 2006, pp. 2498–2503.
22. Schulte H.F. 'The characteristics of the McKibben artificial muscle', In: *The Application of External Power in Prosthetics and Orthotics*, Publication 874, National Academy of Sciences – National Research Council, Washington DC, Appendix H, pp. 94–115.
23. Daerden F., Lefeber D. 'Concept and design of pleated pneumatic artificial muscles'. *International Journal of Fluid Power*. 2001;**2**(3):41–50.
24. Saga N., Nakamura T. 'Development of artificial muscle actuator reinforced by Kevlar Fiber'. *Proceedings of IEEE ICIT*. 2002, pp. 950–954.
25. Mihajlov M., Hubner M., Ivlev O., Graser A. 'Modeling and control of fluidic robotic joints with natural compliance'. *IEEE International Conference on Control Applications*. 2006, pp. 2498–2503.
26. Verrelst B., Van Ham R., Vanderborght B., Daerden F., Lefeber D. 'The Pneumatic Biped "LUCY" actuated with pleated pneumatic artificial muscles'. *Autonomous Robots*. 2005, vol. 18, pp. 201–213.
27. Tondu B., Ippolito S., Guiochet J., Daidie A. 'A seven-degrees-of-freedom robot-arm driven by pneumatic artificial muscles for humanoid robots'. *The International Journal of Robotics Research*. 2005;**24**(4–5): 257–274.

28.   Mills J.K. 'Hybrid actuator for robot manipulators: design, control and performance'. *Proceedings of IEEE International Conference on Robotics and Automation*. 1990, pp. 1872–1878.
29.   Shin D., Sardellitti I., Khatib O. 'A hybrid actuation approach for human-friendly robot design'. *IEEE International Conference on Robotics and Automation*. 2008.

*Chapter 4*
# Toward a dynamical enactive cognitive architecture

*Boris A. Durán*

## Abstract

The increasing complexity of humanoid robots and their expected performance in dynamic and unconstrained environments demand an equally complex, autonomous, adaptable, and dynamic solution. In this article, motivated by the enactive paradigm of cognitive systems, we develop the first steps toward the creation of such a solution. Enactive approaches assert that the primary basis of cognition is anticipative skill construction by embodied systems. The system's ontogenetic development is central to this approach since it is the system's experience – its dynamics and interaction with its environment – that defines its cognitive understanding of the world in which it is embedded rather than any *a priori* rules or representations of the system designer. Enactive systems are typically realized by dynamical systems theory, connectionism, and self-organization. In this article, we survey briefly the main attributes of nonlinear dynamical systems and then introduce Dynamic Field Theory (DFT) – a mathematical framework grounded in dynamical systems and neurophysiology – as a plausible way of combining the power of nonlinear dynamical systems theory and the autonomous developmental aspects of enactive systems. The potential of this approach is demonstrated by replicating in an anthropomorphic robotic platform the responses found in 7- to 12-month-old infants when exhibiting the A-not-B error, a classic simple cognitive behavior in developmental psychology.

## 4.1  Introduction

Research in humanoid robotics dates back approximately 30 years and was founded on a strong dualist view of human nature. On one hand, the body of an agent[1] has been controlled by using a 50-year-old tradition of control theory

---

[1] The term *agent* will represent in this book an artificial entity used for the simulation of human behavior.

that started with industrial automation at the beginning of the 1960s. And on the other, the mind of an agent has been treated independently of its body by using *cognitivist* approaches, which was a very promising area of research that gave birth to what it is known as *artificial intelligence.*

*Cognitivism* is one of the two main paradigms of cognition and has been relatively successful in solving task-specific problems. This approach sees cognition as a set of computations defined over symbols or representations of the world; these representations are predesigned by a human programmer [1]. The main problems this approach has been and is still facing are due to its dependency on the programmer's knowledge of the world: the symbol grounding problem, the frame problem, and the combinatorial problem [2].

The focus of the present chapter is on the *enactive* paradigm of cognition. This paradigm sees cognition as a process of co-determination between the environment and the agent. In other words, the agent constructs its reality through the interaction with the world [2]. Therefore, in contrast with the *cognitivist* approach, the mind cannot be independent of the body and the knowledge about the world is not predefined by the designer.[2]

Developmental psychology is one of the main areas of scientific research from where cognitive robotics has been taking inspiration during the past decade. It studies the psychological changes that occur in human beings through their lives, being the first stages of infant development the most interesting for cognitive robotics. Among the most important works within this area, the one followed by Esther Thelen and Linda Smith [3] has been of great interest not only for psychologists but mathematicians and roboticists as well. The main reason for this multidisciplinary interest in their approach is based on their choice of nonlinear dynamical systems theory to explain the different paradigms found in infant development. According to Thelen and Smith, the different behaviors found in adult people are the result of actions that are not influenced by the dynamic responses to the environment only but have a strong component of previous experiences as well. They see human development as a landscape with an infinite number of valleys that represent different behaviors being created and reshaped continuously (Figure 4.1). As time passes by, the previous 'history' in the formation of these valleys influences strongly their current tendencies and creates in some cases deep and stable behaviors.

The formation of cognitive behaviors could be described in most cases by what is known as *habituation*. Habituation is the process by which the responses to certain stimulus vary less and less with each repetition of the experience during time. The mathematical model adopted by Thelen and colleagues for replicating this kind of behavior is called *Dynamic Field Theory* (DFT) [4].

This chapter describes the main characteristics of the theory behind dynamical systems, and the results of its implementation in a classical infant paradigm known as the *A-not-B error paradigm*. A previous implementation of

---

[2] See Reference 1 for a thorough study on cognitive systems.

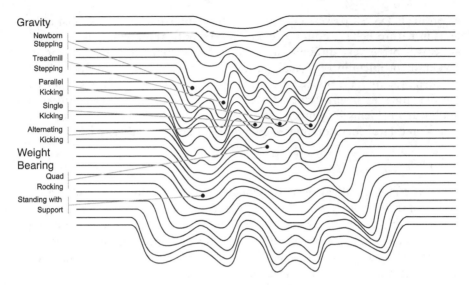

Gravity
Newborn Stepping
Treadmill Stepping
Parallel Kicking
Single Kicking
Alternating Kicking
Weight Bearing
Quad Rocking
Standing with Support

*Figure 4.1   Schematics of an ontogenetic landscape for locomotion
[After Reference 3]*

this paradigm was done by Gregor Schöner and colleagues using a khepera robot. In our case, the implementation was done on a more human-like robotic platform, iCub's upper body [5]; making use of its stereo vision and arms movements to implement the reaching motion performed by infants.

## 4.2   Nonlinear dynamical systems

The theory of nonlinear dynamical systems tries to study and model all physical phenomena that go through spatial and temporal evolution. The range of phenomena studied with this theory goes from simple pendulum analysis to the motion of planets and celestial bodies, from chemical reactions to lasers and telecommunications. During the past decades the spectrum of scientific communities interested in nonlinear dynamical systems theory has become wider, e.g., neurosciences, economics, sociology, biology, psychology; and with them the theory has also grown significantly. The amount of knowledge accumulated during the years is enormous and would require several books to describe the whole scope of this theory. In this section will try to describe the basic concepts and tools of the theory by studying the different *attractors* used to represent the qualitative nature of nonlinear dynamical systems.

A dynamical system is said to be 'nonlinear' when it cannot be mathematically described as the linear sum of its components, i.e., it includes terms such as products ($x_1x_2$), powers ($x^3$), and functions of $x$ (sin $x$, $e^x$). Almost all systems in nature behave nonlinearly and mathematical modeling of this type of systems is often very difficult and in some cases impossible. One option is to divide the problem in very small parts that behave linearly and find solutions for them, the major drawback of this approach is that it cannot see changes of the dynamics at macro scales. Instead of trying to find exact solutions for nonlinear problems it is much easier to study the long-term behavior of the system and later focus on interesting points of changing dynamics.

## 4.2.1   Fixed points

In a strict mathematical sense we would say that a value $x^*$ is called a fixed point if it satisfies $f(x^*) = x^*$. In other words, a system remains *fixed* (constant) in some specific value regardless of the initial value of $x$. In a cognitive system a single static memory like the color of an object or its position in space could be represented by a dynamic variable that enters into a fixed point attractor since they describe states that stay unchanged over time.

Finding fixed points and their stabilities by means of an analytical analysis gives us certain degree of confidence on our results. There is, however, a much simpler and faster way of finding fixed points and their behaviors: graphically. The different plots in Figure 4.2 are called *cobweb* plots and respond to a very simple principle of graphical iteration. The studied function is plotted together with a bisector and starting with $x(0)$ on the $x$-axis we draw a vertical line from $x(0)$ until hitting the function. Then a horizontal line is drawn until reaching the bisector, a new vertical to the function, and a new horizontal to the bisector, and so on. This kind of plot depicts the progress in time of the iterations within the function, thus showing how stable or unstable our fixed points are. Plots of activations versus time are also shown in each one of the examples.

To understand a bit more what the different *attractors* are and how they can be analyzed, we will use a special function from the family of quadratic maps (4.1a). The logistic map, (4.1b), was studied by the biologist Robert May as a model of population growth and is one of the simplest and most used examples of one-dimensional discrete maps. This second-order difference equation shows very rich and complex dynamics through the parameter $\alpha$ that controls the nonlinearity of the system. In order to keep the system bounded with normalized values, i.e., between 0 and 1, $\alpha$ takes values between 0 and 4. If $\alpha > 4$ the system just blows up to infinity very rapidly, if it is less than zero the system starts oscillating around zero including negative numbers in the overall dynamics until a point where it also explode to infinity.

$$x = a_2x^2 + a_1x + a_0; \qquad a_i, x \in \Re \qquad\qquad (4.1a)$$

$$x = \alpha x(1 - x); \qquad 0 < \alpha < 4, \qquad 0 < x < 1 \qquad\qquad (4.1b)$$

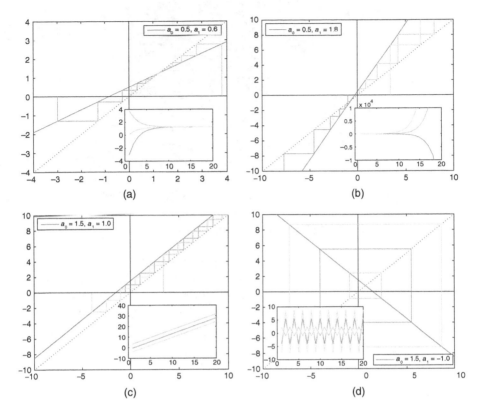

Figure 4.2  Cobweb plots for the general linear case for $x(k + 1) = a_0 + a_1 x(k)$; $x, a_i \in \mathfrak{R}, k \in \mathbb{N}$. (a) $|a_1| < 1$; (b) $|a_1| > 1$; (c) $a_1 = 1$; (d) $a_1 = -1$

The first step in our analysis is to find the fixed points of (4.1b). To do this we have to find those values where $f(x) = x$

$$f(x) = \alpha x(1 - x) = x \rightarrow x(\alpha(1 - x) - 1) = 0$$

$$yx^* = \begin{cases} 0 \\ 1 - \frac{1}{\alpha} \end{cases}$$

Then we need to know how stable these points are. Fixed points come in two different forms, and in order to see their differences we can compare them to being either a hill or a valley. If a ball is exactly at the top of the hill it will remain there, but we know that even the slightest force in any direction will make the ball roll over the hill and away from that steady point where it was (Figure 4.3, right). This kind of fixed point is called a *repellor* or an unstable fixed point. On the other hand a ball resting at the bottom of a valley will remain there or will go back there even if a temporary force takes it out of its

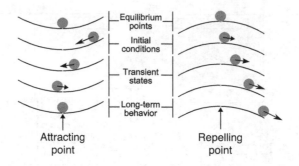

*Figure 4.3    Fixed points representations: stable (left) and unstable (right)*

steady state (Figure 4.3, left). This kind of point is called an *attractor* or a stable fixed point.

The stability of a point in a discrete system is found analytically by evaluating the first derivative of the function on that point, i.e., $f(x^*)$. If the absolute value of this evaluation is less than one, i.e., $|f(x^*)| < 1$ we say the fixed point is stable, and if it is greater than one, i.e., $|f(x^*)| > 1$ we say it is unstable.

$$f'(x) = \alpha - 2\alpha x \tag{4.2}$$

$$\rightarrow |f'(x^*)| = \begin{cases} \alpha; & x^* = 0 \\ 2 - \alpha; & x^* = 1 - \frac{1}{\alpha} \end{cases} \tag{4.3}$$

As we see, the stability of our map depends on the parameter $\alpha$, and since we have constrained $\alpha$ to be greater than 0 but less than 4 we can start by saying that $x^* = 0$ is an attractor when $0 < \alpha < 1$ and a repellor otherwise. On the other hand, the point $x^* = 1 - 1/\alpha$ will always be a negative value for $0 < \alpha < 1$, hence not considered as part of the solution. In summary for $0 < \alpha < 1$ we have one single fixed point ($x^* = 0$) and it is an attractor.

The second-order function of (4.1b) depicts a parabola with its vertex located at $x = 0.5$ and height proportional to $\alpha$: $h = \alpha/4$ (Figure 4.4a). A cobweb plot is used to show the stability of attractor $x^* = 0$ by plotting three different initial conditions (Figure 4.4b).

A *bifurcation*[3] occurs at $\alpha = 1$ where $x^* = 0$ becomes a repellor and $x^* = 1 - 1/\alpha$ an attractor. Figure 4.5 gives some examples of the new points of intersection between the logistic function and the bisector. This type of bifurcation is called a *tangent* or *saddle node* bifurcation since it occurs when the tangent to the function crosses the bisector. Three different initial conditions are used to

---

[3] Bifurcation is the term given to sudden changes of the dynamics of a system when varying its control parameters. It is not within the scope of this article to analyze dynamical systems from the bifurcations point of view but to make use of stable states found around these points. A whole body of research has been developed around the different types of bifurcations found in nonlinear dynamical systems.

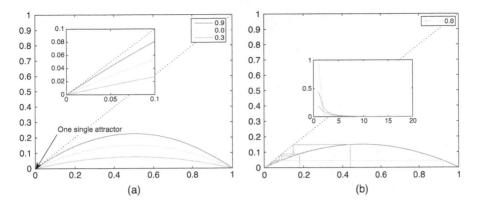

*Figure 4.4    A graphical method for finding fixed points and their stabilities.*
*(a) $0 < \alpha < 1$; (b) $\alpha > 1$*

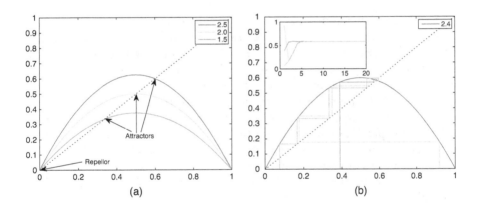

*Figure 4.5    A graphical method for finding fixed points and their stabilities.*
*Examples for $1 < \alpha < 3$. (a) $0 < \alpha < 1$; (b) $\alpha > 1$*

find the stability of the new fixed points (Figure 4.5b). It is possible to see that values close to $x^* = 0$ are driven away with time making it a repellor, all initial conditions are attracted to the newly created fixed point at the crossing of both curves.

## 4.2.2    Periodicity

The previous section gave us an initial analysis of fixed points found in the logistic map and their stabilities for values below and above $\alpha = 1$. However the condition for having stability changes once again as $\alpha$ goes beyond 3 since $|f(x^*)| > 1$.

$$f'(x^*)|_{x^*=1-(1/\alpha)} = |2 - \alpha| > 1, \qquad \text{for } \alpha > 3 \qquad (4.4)$$

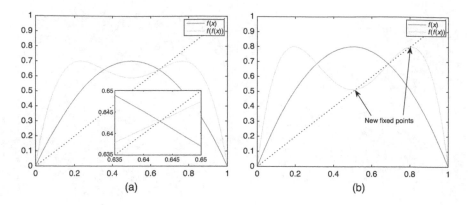

*Figure 4.6    First- and second-iteration plots of logistic map. (a) $\alpha = 2.8$;*
*(b) $\alpha = 3.2$*

To see better what happens at this new bifurcation we will plot the second iteration of our function. Figure 4.6 shows two plots for $\alpha = 2.8$ (Figure 4.6a) and for $\alpha = 3.2$ (Figure 4.6b). For $1 < \alpha < 3$, the second-iteration curve has the same stable fixed point as the first-iteration curve. Once $\alpha$ goes beyond 3, this fixed point loses stability and two new fixed points are created when the tangent of the fourth-order function, second-iteration curve, crosses the bisector curve. The main difference between these new fixed points and the one studied in the previous section is that the function does not settle down to just one of them as time increases but it alternates between both on each time step. By doing so, an entirely new behavior is created known as *periodicity* and the name given to this type of bifurcation is *period-doubling* or *pitchfork* bifurcation.

In order to find the location of the new fixed points we need to follow the same methodology as we did in the previous section, but in this case we will work with a second-iteration function, i.e., $f(f(x))$, since the second fixed point is located two steps ahead of our present state.

$$f(x) = \alpha x(1 - x)$$

$$f(f(x)) = \alpha(f(x))(1 - f(x)) = \alpha(\alpha x(1 - x))(1 - \alpha x(1 - x))$$

$$= \alpha^2 x(1 - x)(1 - \alpha x(1 - x))$$

$$= -\alpha^3 x^4 + 2\alpha^3 x^3 - (\alpha^2 + \alpha^3)x^2 + (\alpha^2 - 1)x$$

(4.5)

Once again we use the condition to find the fixed points of a discrete map, $f(x) = x$. Equation (4.5) should give us four roots or fixed points, however we already know two solutions to this equation from the roots of the first-iteration function, namely $x^* = 0$ and $x^* = 1 - 1/\alpha$. We use these two already known roots to further simplify (4.5):

$$f(f(x)) = x$$
$$-\alpha^3 x^4 + 2\alpha^3 x^3 - (\alpha^2 + \alpha^3)x^2 + (\alpha^2 - 1)x = x$$
$$(-\alpha^3 x^3 + 2\alpha^3 x^2 - (\alpha^2 + \alpha^3)x + (\alpha^2 - 1))x = 0 \qquad (4.6)$$
$$-\alpha^3 x^3 + 2\alpha^3 x^2 - (\alpha^2 + \alpha^3)x + (\alpha^2 - 1) = 0$$

Dividing (4.6) by $x - (1 - 1/\alpha)$ we obtain:

$$-\alpha^3 x^2 + (\alpha^2 + \alpha^3)x - (\alpha^2 + \alpha) = 0$$
$$\rightarrow x^2 - \frac{\alpha + 1}{\alpha}x - \frac{\alpha + 1}{\alpha^2} = 0 \qquad (4.7)$$
$$yx^* \begin{cases} x_h^* = \frac{\alpha + 1 + \sqrt{\alpha^2 - 2\alpha - 3}}{2\alpha} \\ x_l^* = \frac{\alpha + 1 - \sqrt{\alpha^2 - 2\alpha - 3}}{2\alpha} \end{cases}$$

These new solutions are defined only for $\alpha \geq 3$ and their stabilities could be found by the same methodology as before, i.e., by evaluating the derivative of the function on each solution:

$$|f'(f(x))|_{x^*} = -4\alpha^3 x^3 + 6\alpha^3 x^2 - 2(\alpha^2 + \alpha^3)x + (\alpha^2 - 1) \qquad (4.8)$$

where

$$x^* = \begin{cases} x_1^* = 0 \\ x_2^* = 1 - \frac{1}{\alpha} \\ x_3^* = \frac{\alpha + 1 + \sqrt{\alpha^2 - 2\alpha - 3}}{2\alpha} \\ x_4^* = \frac{\alpha + 1 - \sqrt{\alpha^2 - 2\alpha - 3}}{2\alpha} \end{cases}$$

Finding the stability of these points by algebraic means becomes very difficult already for a second-iterate function, not mentioning finding fixed points and their stabilities for higher-order periods. The principle of graphical iteration and the power of simulations becomes very helpful at this point allowing us to explore a wider space of the control parameter. By plotting a limited number of initial conditions and their iterative progress within the map, it is possible to find out how stable or unstable certain points are. Figure 4.7 shows the behavior of our logistic map for three different initial conditions when $\alpha > 3$. The new fixed points pull any initial condition toward them in an alternating pattern, thus becoming a limit cycle attractor (Figure 4.7a). Plotting our function versus time gives a better picture of what happens after a transient period (Figure 4.7b). On the other hand, fixed point $x_1^* = 0$ is still a repellor since all orbits starting close to zero are still driven away from it.

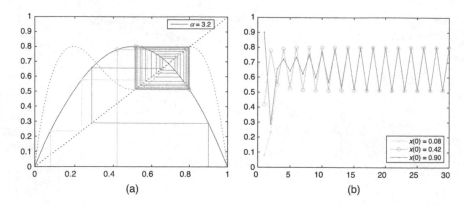

*Figure 4.7    Logistic map plots as a cobweb and in time for* $\alpha = 3.2$*. (a) Graphical
iterations; (b) activations versus time*

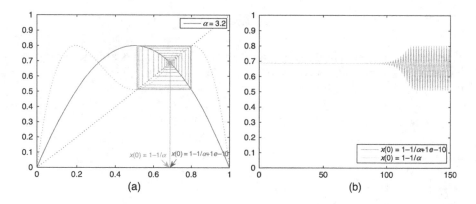

*Figure 4.8    Logistic map plots for* $\alpha = 3.2$*, showing the lack of stability of fixed
point* $x_2^* = 1 - 1/\alpha$*. (a) Graphical iterations; (b) activations versus
time*

Finally, Figure 4.8 shows two very close initial conditions and their long-term behavior at fixed point $x_2^* = 1 - 1/\alpha$. When the initial point is exactly at $1 - 1/\alpha$ the system stays fixed at that point, however any infinitesimal variation from this point will make it lose stability and ultimately end up in the limit cycle create by the attractor points $x_3^*$ and $x_4^*$. Hence we conclude that fixed point $x_2^*$ has become a repellor for $\alpha > 3$.

To study what happens for values of $\alpha > 3$ we will introduce a new graphical tool that will simplify the rest of the analysis for any nonlinear map. Figure 4.9 is called a *bifurcation diagram* and consists of, after getting rid of transients, plotting as many random initial conditions as possible for each value of the control parameter (Figure 4.9a). This plot gives a much more comprehensible picture of the whole dynamic space of the map being studied.

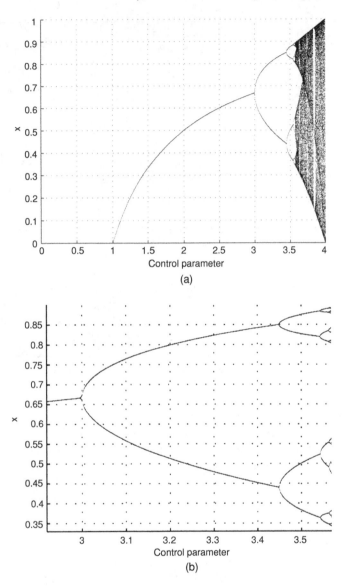

*Figure 4.9 Bifurcation diagram for the logistic map. (a)* $0 < \alpha < 4$;
*(b)* $2.92 < \alpha < 3.57$

Figure 4.9b shows the following period-doubling bifurcation points for the logistic map, giving us an immediate look into the dynamics of this function in the periodic regime.

From these plots it is easy to see the values that any initial condition will take on the long term depending only on the value of $\alpha$. Previously we chose three different initial conditions to see the behavior of the logistic map in

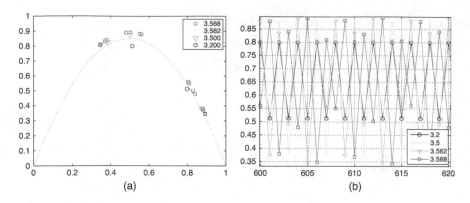

*Figure 4.10   Bifurcation diagram for the logistic map. (a) $0 < \alpha < 4$;
(b) $2.92 < \alpha < 3.57$*

Figure 4.7, we can see the two-cycles obtained in those figures represented in the bifurcation map at $\alpha = 3.2$. In the same way as it happened before, the increase of $\alpha$ will bring us new period-doubling bifurcations, thus creating limit cycles of periods 2, 4, 8, 16, etc. Figure 4.10a depicts the *return map* of our function, it shows the activations points only and not the cobweb lines as before, in this way it is much easier to see the different periods formed after the transients. Figure 4.10b shows a window of activations in time for limit cycles of periods 2, 4, 8, and 16.

The property of periodicity found in nonlinear dynamical systems give us the possibility to represent repetitive actions. Motor behaviors such as walking, crawling, or jumping can be the embodied version of limit cycles of different periods and frequencies. In the same way, periodicity can be used to characterize imitation tasks by considering them as a sequence of states that need to be repeated.

### 4.2.3   Chaos

The previous sections showed us how rich dynamics can be obtained from a very simple equation such as the logistic map, (4.1b). Starting from any point in our space, and depending on the value of our control parameter, we were able to decrease to zero, stay at constant values, or oscillate with different periods at a steady state. However, Figure 4.9a shows a much more complex area to the right of our period-doubling tree, starting from somewhere around $x \approx 3.569945...$, the *Feigenbaum point*.[4] The Feigenbaum point marks the end

---

[4] The *Feigenbaum constant*, $\delta = 4.6692...$, was discovered by physicist Mitchel Feigenbaum in October 1975 and is obtained as the relationship between the distance between two successive bifurcation points. This constant can be found not only in primitive mathematical models but also in real physical phenomena, a property known as *universality* [6].

of the period-doubling tree and the beginning of *Chaos*, the richest and most difficult of all regimes in a nonlinear dynamical system.

After decades of research in chaos theory it is still difficult to find a universal definition of *chaos* or methodology to study and describe a chaotic system. Nevertheless, most authors agree in some key properties of chaotic systems grouped in the following sentence. *Chaos* is the part of Nonlinear Dynamical Systems Theory that studies the *aperiodic long-term behavior* of systems governed by *deterministic rules* that exhibit *sensitive dependence on initial conditions.*

To better understand these and other properties of chaotic systems we will make use of simulations of the logistic map for values of $\alpha$ greater than the Feigenbaum point. Aperiodic long-term behavior means that there are trajectories that do not settle down to constant, periodic, or quasi-periodic values as time goes to infinity (Figure 4.11). Deterministic rules are those free of noisy or random inputs or parameters; the nonlinearity of the system is created by the internal dynamics alone. Finally, sensitivity to initial conditions refers to the property of a system where any arbitrarily small interval of initial values will be enlarged significantly with each iteration (Figure 4.12). In other words, the error of two nearby trajectories will have the same magnitude as the signal itself after few iterations (Figure 4.12b).

Chaos has two other central properties which are usually not mentioned in most textbooks: *mixing* and *dense periodic orbits.* A system is said to show *mixing behavior* when any point within an arbitrarily small subinterval of the state space, eventually reaches points in another arbitrarily small subinterval of the same space; in other words, we can get everywhere from anywhere. Besides sensitivity to initial conditions and mixing behavior, the existence of a dense set of *unstable periodic orbits* (UPO) is the other necessary condition for validating the presence of chaos. Within the chaotic regime, trajectories will not settle

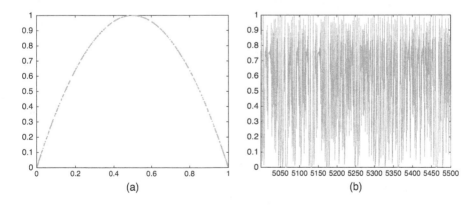

(a)                                          (b)

*Figure 4.11     Simulation of the logistic map for $\alpha = 4$. Plotting 500 activations after 5000 time steps, no periodic dynamics are found. (a) Return map; (b) time series*

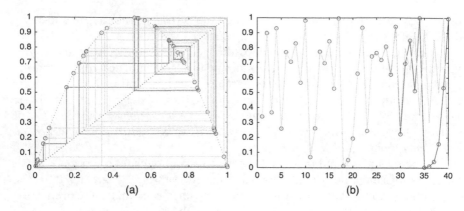

*Figure 4.12    Simulation of the logistic map for $\alpha = 4$. Plotting two initial conditions separated by $1\mathrm{e} - 10$. In a chaotic system this difference is amplified exponentially fast. (a) Return map; (b) time series*

down to a single periodic orbit but will wander in a sequence of close approaches to these orbits.

So far we have been using the logistic map (4.9a) to show that although simple this equation is rich with a great variety of dynamics. Other examples of one-dimensional discrete maps that present the kind of behaviors we have been studying so far, including chaotic dynamics are: shift map, (4.9c); tent map, (4.9d); sine map, (4.9b). Examples of two-dimensional chaotic maps are: the tent map, (4.9e); the tent map, (4.9f).

$$f(x) = \alpha x(1 - x) \tag{4.9a}$$

$$f(x) = \alpha \sin(\pi x) \tag{4.9b}$$

$$f(x) = 2x, \quad mod(1) \tag{4.9c}$$

$$f(x) = \begin{cases} \alpha x, & 0.0 \le x \le 0.5 \\ \alpha(1 - x), & 0.5 < x \le 1.0 \end{cases} \tag{4.9d}$$

$$f(x) = f\begin{pmatrix} x_1 \\ x_2 \end{pmatrix} = \begin{pmatrix} x_2 + 1 - \alpha x_1^2 \\ \beta x_1 \end{pmatrix} \tag{4.9e}$$

$$f(x) = f\begin{pmatrix} x_1 \\ x_2 \end{pmatrix} = \begin{pmatrix} -\beta x_2 + \alpha x_1 - x_1^3 \\ x_1 \end{pmatrix} \tag{4.9f}$$

### 4.2.3.1    Strange attractors

In the previous sections simple kinds of attractors representing constant values (fixed points), and periodic oscillations (limits cycles) were studied. Within

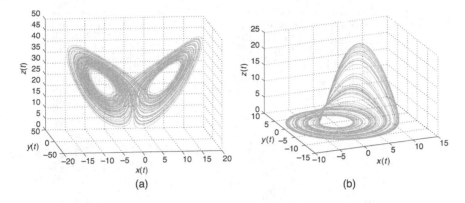

*Figure 4.13    Strange attractors. (a) The Lorenz attractor; (b) the Rössler attractor*

chaotic regimes some dynamical systems form very distinctive long-term patterns called *strange attractors*. The shape of these kinds of attractors repeats over and over regardless of the initial conditions of the system. This is a surprising condition since we already saw how sensitive to initial conditions a chaotic system is (Figure 4.12b). We would think that two different initial conditions would create completely different patterns in the long term since already during the first few steps they diverge exponentially, but the surprise comes when analyzing the long-term behavior of these orbits: both stay bounded within a very distinctive *basin of attraction*. Figure 4.13 shows two of the most known strange attractors, both of them are continuous three-dimensional functions: Lorenz (Figure 4.13a); and Rössler (Figure 4.13b).

Strange attractors do not only challenge the notion of sensitivity to initial conditions but present other interesting properties like those of fractals and mixing behavior. Both, strange attractors and chaos have received a lot of attention during the past decades but they are still in their infancy when trying to find a definite mathematical description. Even so, their popularity is not only among mathematicians and physic scientists but for people in neural sciences, sociology, or psychology to name a few.

### 4.2.3.2    Intermittency

Another very interesting property of chaotic systems that we will make use in this thesis is *intermittency*. *Intermittency* could be seen as the alternation of almost periodic dynamics and chaotic dynamics. When looking more carefully into some regions of the chaotic regime it is possible to find periodic windows that interrupt the chaotic behavior of the system for certain values of the control parameter (Figure 4.14a). At the beginning of these windows it is possible to find values of the control parameter for which the system seems undecided between behaving periodically or chaotically (Figure 4.14b).

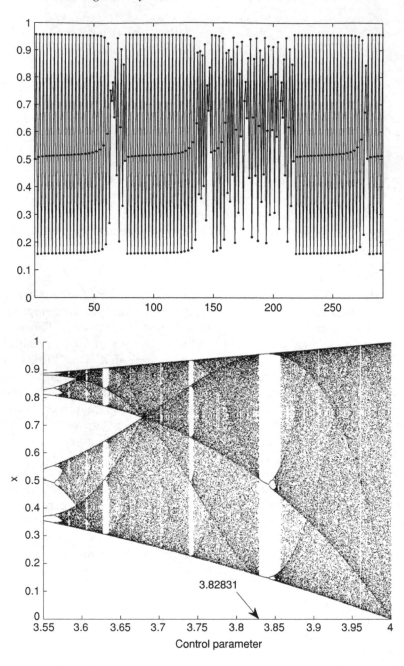

*Figure 4.14    Periodic windows within the chaotic regime of the logistic map contain points where chaos and periodic behavior interrupt each other. (a) Zoom of chaotic regime for the logistic map; (b) time series for α = 3.83821*

## 4.3  Dynamic field theory

DFT is a mathematical framework based on the concepts of Dynamical Systems and the guidelines from Neurophysiology. In this theory, fields could be seen as populations of neurons and their activations as the continuous responses to external stimuli. A field has the same structure of a recurrent neural network and were studied first by Amari [7] as a class of bi-stable neural networks. They could be compared to globally coupled maps since all units influence each other in some degree, *global inhibition*. But they could also be seen as coupled map lattices since close neighbor units have a stronger influence than far neighbors, *local excitation*. Global inhibition and local excitation are the two types of interactions among field sites $x$ and having them embedded in the dynamics of the system allow the generation of single peaks of activation (Figure 4.15).

The units of representation in DFT are the peaks of activation along the *dimension* being studied [8]. A *dimension* represents any perceptual feature, movement or cognitive decision, e.g., position, orientation, color, speed. Besides global and local interactions, the dynamics of a field depend also on inputs coming from external sources. In the absence of external inputs a stable 'off' state is usually found as the resting position of a field. External inputs add lifting forces to this 'off' state, and if sufficiently strong for rising above the zero-point threshold, they bring the system into an unstable condition where local excitation and global inhibition will decide the new final stable state.

Local and global interactions are an implicit mechanism for sensor fusion and decision making. Local excitation makes neighboring field sites gain force once the input has become larger than the resting level of the field, moreover stabilizes the newly formed peaks against gradual decay. On the other hand, global inhibition makes that a strong input in a field site suppresses the activation of weaker input in another site, the weaker the input the less inhibitory

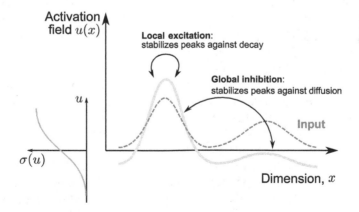

*Figure 4.15   Typical activations in a dynamics field [After Reference 8]*

influence will have on other field sites. As a result, the competition that exists among different inputs is dynamically computed with the help of these two types of interactions [8].

In DFT it is assumed that the different changes in the activations of the field occur continuously in time. Therefore, the different models that follow this theory are defined by differential equations, e.g., $\dot{u}(x,t) = f[u(x',t)]$. Where $u$ represents the activation field and the brackets defines a function of any field site $x'$ and not just on field site $x$ [9]. The mathematical formulation of a single layer neural field is defined by a differential equation that describes the rate of change of each unit in the field, (4.10a). The four main components of this equation are a linear decay term $-u$, a constant resting level $h < 0$ that fixes the overall level of activation, the inputs to the field $S$ that may contain both task and/or specific inputs shaped by a Gaussian filter (4.10b), and finally a convolution term between the interaction kernel (4.10d) and a sigmoidal transfer function (4.10c) integrated over the whole field.

$$\tau \dot{u}(x,t) = -u(x,t) + h + S(x,t) + \int w(x-x')f[u(x',t)]dx' \qquad (4.10a)$$

$$S(x) = C \exp\left[\frac{-(x-x')^2}{2\sigma_s^2}\right] \qquad (4.10b)$$

$$f(u) = \frac{1}{1 + \exp[-\beta(u-u_0)]} \qquad (4.10c)$$

$$w(x-x') = w_e \exp\left[\frac{-(x-x')^2}{2\sigma_w^2}\right] - w_i \qquad (4.10d)$$

A single layer dynamic neural field contains eight different parameters that need to be tuned up, (4.10). First, parameter $\tau$ specifies the timescale over which the field gradually builds up or decay. As mentioned before, the resting level $h$ defines the threshold required by the inputs to activate the field. External inputs $S$ are shaped as Gaussian bells therefore we need to specify their strength $C$ and width $\sigma_s$. The sigmoidal transfer function $f(.)$ is defined by a single parameter $\beta$ that specifies its steepness. Finally, interaction kernel $w$ works as the connection matrix of a recurrent neural network which weights are defined by a Gaussian function of width $\sigma_w$, a local excitatory strength given by $w_e$ and a global inhibitory level given by $w_i$.

The parameter space is able to generate a rich range of dynamics going from mono-stable and oscillatory behaviors to saturated activation fields. However, mono-stability and bi-stability are the most useful scenarios when trying to replicate human behavior. A system is said to be mono-stable when the only attractor is the subzero resting state in the absence or after the application of an input. A system may also present a bi-stability condition when, besides the subzero attractor, a self-stabilized peak remains after the input has vanished. The property of dynamic neural fields to sustain an activation peak in the absence of input is used as a working memory.

DFT also studies the different configurations that could be created by grouping several fields and the possibility of having multidimensional fields. Fields may represent excitatory or inhibitory neural populations as well as long-term memory layers; for example, Simmering *et al.* [10] created a five-layer structure to study spatial-cognition. Bi-dimensional fields have been created to study visuo-spatial cognition where DFT provides an elegant solution to the 'binding problem' [11]. Other applications of DFT include motor planning [9], saccadic eye movements [12, 13], infant perseverative reaching [14].

## 4.4    The 'A-not-B' paradigm

The 'A-not-B' error is an infant paradigm first studied by Jean Piaget in 1954 [15]. It could be described as a game designed to study the emergence of the concept 'object' in infants between 7 and 12 months of age. Infants at this age present a partial knowledge of the location of a toy hidden at one of two locations [14]. They search repetitively for an object at a previously visited location, even though they see the desired object vanishing at another location.

Since Piaget's first description, this experiment has been repeated countless times and in different variations. Many different explanations have been proposed among developmental psychologists, thus reflecting the little consensus about the origin of this error. Thelen *et al.* [14] proposed an elegant modeling of this paradigm based on DFT, this approach will be the guideline for the implementation in our robotic platform.

The classic version of the A-not-B hiding task goes as follows [16]. Two marked locations (distinctive from the background) are presented to the infant. A few training trials consist of showing and hiding a small and attractive toy behind location 'A', after a delay the infant is allowed to reach and grab the toy from any of the two locations being 'A' the usual choice. However it is common to find experiments where the infant goes to location 'B' during these training trials. Following this training process the toy is shown and hidden behind location 'B' and again after the respective delay the infant is allowed to reach for one of the two locations. The 'A-not-B' error occurs when the reaching goes to location 'A' instead of 'B' (Figure 4.16).

## 4.5    Methodology

The hiding task deals with the location of objects in front of an agent, therefore a single dimension representing orientation was used. A one-dimensional neural field (4.11a) together with a memory trace field, (4.11b) were used as the core of the model for the 'A-not-B' error experiments, (4.11). The memory layer was active only when two conditions were fulfilled: the main layer was active ($u_x > 0$), and a 'go signal' was set to 'on', (4.11d). This layer includes

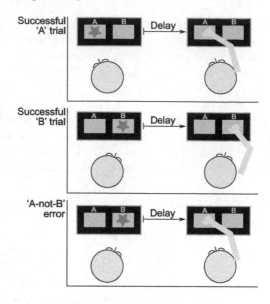

*Figure 4.16    Graphical representation of the hiding task experiment and its results*

two parameters in charge of controlling how fast memories are built ($\lambda_{\text{build}}$) and how fast these memories decay ($\lambda_{\text{decay}}$). The interaction kernel for the main field $w_{xx}(x-x')$ was given by (4.10d) described in Section 4.3. The connection weights from the memory layer onto the main layer $w_{xm}(x-x')$ were also computed through an interaction kernel of Gaussian type, (4.11c).

Layers:

$$\tau \dot{u}_x(x,t) = -u_x(x,t) + h + S(x,t) + \int w_{xx}(x-x')f[u_x(x',t)]dx'$$

$$+ \int w_{xm}(x-x')f[u_m(x',t)]dx' \tag{4.11a}$$

$$\tau \dot{u}_m(x,t) = [-u_m(x,t) + f[u_x(x,t)]]$$

$$\times [\lambda_{\text{build}}\Theta(u_x(x,t)) + \lambda_{\text{decay}}(1 - \Theta(u_x(x,t)))] \tag{4.11b}$$

Kernels:

$$w_{xx}(x-x') = C_e \exp\left[\frac{-(x-x')^2}{2\sigma_e^2}\right] - C_i \exp\left[\frac{-(x-x')^2}{2\sigma_e^2}\right] - G_i$$

$$w_{xm}(x-x') = C_m \exp\left[\frac{-(x-x')^2}{2\sigma_m^2}\right]$$

$$w_\phi(x) = -C_\phi x \exp[-2\sigma_\phi^2(x-x_i')^2] \tag{4.11c}$$

Activation function for memory layer:

$$\Theta(u(x)) = \int \theta(u(x'))dx'; \qquad \theta(u(x')) = \begin{cases} 1, & u(x') > 0 \\ 0, & \text{otherwise} \end{cases} \qquad (4.11d)$$

Inputs:

$$S(x,t) = S_{\text{task}}(x) + S_{\text{spec}}(x,t) \rightarrow$$

$$S_{\text{task}}(x) = C_{AB}\left[\exp\left[\frac{-(x-x'_A)^2}{2\sigma_s^2}\right] + \exp\left[\frac{-(x-x'_B)^2}{2\sigma_s^2}\right]\right] \qquad (4.11e)$$

$$S_{\text{spec}}(x) = C_i \exp\left[\frac{-(x-x'_i)^2}{2\sigma_s^2}\right]; \qquad i \in \{A, B\}$$

Output:

$$\dot{\phi}(x) = \sum u_x(x,t) \odot w_\phi(x) \qquad (4.11f)$$

The input of the system is given by the sum of two terms representing the task and the specific input, (4.11e). The task input has two constant Gaussian bells through all the experiment whereas the specific input changes its amplitude and position according to the trial, i.e., location 'A' or 'B'. Finally, the output to the motors was taken as the sum of the point-to-point multiplication between the activation layer and an attracting kernel, (4.11f). The output kernel $w_\phi$ dynamically computes the location and amplitude of the signal that activates the motor; a positive activation of $\phi$ performs an action in one direction, a negative value performs the action in the opposite direction. The speed of motion is controlled by the steepness of the attracting area in the output kernel $\Sigma_\phi$.

### 4.5.1 Hardware

The iCub platform [5] was the physical layer where all of our experiments on DFT were implemented. Seven degrees-of-freedom (DOF) were actuated: 5DOF in the head (yaw and pitch for the neck, a single pitch motion for both eyes, and independent yaw motors for each eye) and 2DOF in the shoulders (Figure 4.17). The other DOF were blocked for these experiments. Three different Blade units were used in order to handle image analysis, motor control, and dynamic fields, respectively.

The task panel itself was simulated inside a computer screen in front of the robot. Two small rectangles representing the lids (task input) behind which the object would be hidden (specific input). Since the task deals only with the location of an object and not with the differentiation of lids and toys, the object itself was simulated by a rectangle of larger size than the lid but with the same color. The major advantage of simulating the 'A-not-B' panel and 'toy' in this

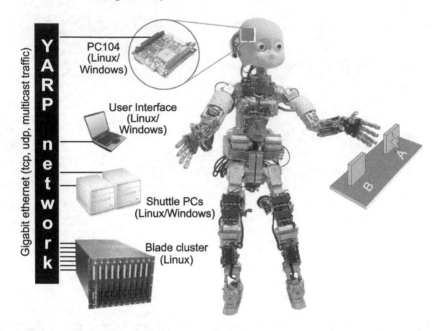

*Figure 4.17    Schematics of the experimental setup for the hiding task*

way was the possibility of adapting the 'objects' to those sizes and colors better detected by the color segmentation algorithm depending on the room's luminance.

### 4.5.2    Software

The set of equations described in Section 4.5 contains around 20 different parameters to be tuned. Most of these parameters remain constant and are easy to define, however some others produce important changes in the overall dynamics of the system. In order to have a 'live' feeling of the behavior of the fields when changing these parameters a graphical interface was implemented using Matlab.[5] Figure 4.18 shows screen shots of the two interfaces designed for an interactive simulation of the 'A-not-B' error task. Besides the para- meters described previously, a local level of noise was added to the dynamics of the system. Two different sets of kernels representing the behavior of young and older infants were preset in the interface, however the user is also able to specify his/her own parameters.

For the implementation in the robotic platform, the same open-source fra- mework used in the previous chapter was used in these experiments. The algo- rithms were implemented in C++ with the same parameters used in the simulation. Whenever a 'go-signal' was present, the activations of the main layer

---

[5] This code was an adaptation of the *Interactive Dynamic Field Simulator* written by Sebastian Schneegans, Ruhr-Universitaet Bochum, 2008.

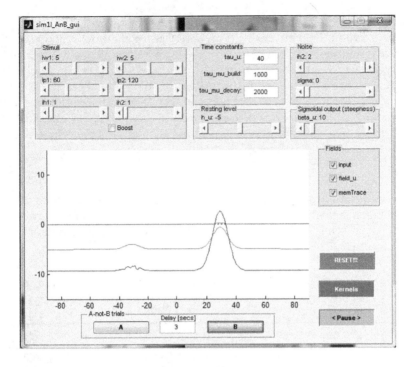

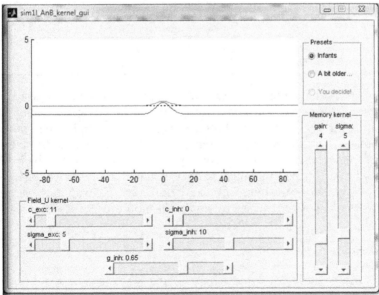

*Figure 4.18   Screen shot of the field's simulator for the 'A-not-B' task*

actuated the motors in the shoulders to simulate a reaching behavior. A color tracking algorithm was implemented in YARP using the Intel® Integrated Performance Primitive Libraries version 5.3. The eyes were at all times focused in the middle point between locations 'A' and 'B', in this way the task input to the field was given by two colored rectangles in front of the robot at each side of the middle point. Specific inputs were created by increasing the size of the rectangles whenever an 'A' or 'B' trial was performed. Thus, the values of (4.11e): $C_i \in A, B$ were given by the number of pixels at position 'A' or 'B', and the values for $x_i$, $i \in \{A, B\}$ were given by the center of the blobs in the image.

### 4.5.3    Model parameters

For comparison purposes two different sets of parameters were used in these experiments: the kind of response found in young infants when performing the hiding task was compared to the 'normal' behavior of adults or older children in the same task (Table 4.1).

*Table 4.1    Parameters for the DFT model of the 'A-not-B' error paradigm found in young infants and for the control case of adults responses*

| Parameters | $C_e$ | $\sigma_e$ | $C_i$ | $\sigma_i$ | $C_m$ | $\sigma_m$ | $C_s$ | $\sigma_s$ | $C_q$ | $\sigma_q$ |
|---|---|---|---|---|---|---|---|---|---|---|
| Infants | 11.0 | 5.0 | 0.0 | 10.0 | 4.0 | 5.0 | 7.0 | 5.0 | 0.04 | 1.0 |
| Adults | 21.0 | 5.0 | 10.0 | 10.0 | 4.0 | 5.0 | 0.7 | 5.0 | 0.04 | 1.0 |

| Parameters | $C_\phi$ | $\sigma_\phi$ | $\tau$ | $\lambda_{build}$ | $\lambda_{decay}$ | $\beta$ | $h$ | $G_i$ |
|---|---|---|---|---|---|---|---|---|
| Infants | 1.0 | 0.03 | 20 | $10^4$ | $40^4$ | 10.0 | $-5.0$ | $-0.65$ |
| Adults | 1.0 | 0.03 | 20 | $10^4$ | $40^4$ | 10.0 | $-5.0$ | $-0.95$ |

The interaction kernels $w_{xx}$ for both infants and adults was designed following the *spatial precision hypothesis* [17]. This hypothesis states that the spatial precision of neural interactions becomes more precise and more stable over development. This development is represented by two factors: the increasing sharpness and strength of local excitatory interactions controlled by $\sigma_e$ and $C_e$ respectively, and by the strengthening of lateral inhibitions controlled by $C_i$ and $\sigma_i$. Figure 4.19a shows the interaction kernels $w_{xx}$ created with these parameters as well as the output kernel used to obtain the motor response (Figure 4.19b).

## 4.6    Results

In this section we present the results of the implementation of the model described by (4.11) with the conditions presented in Table 4.1.

### 4.6.1    Young infants

First, the kind of behavior found in infants between 7 and 12 months of age was simulated and implemented. Infants at this age show repeatedly the

'A-not-B' type of error described in Section 4.4. Figure 4.20 shows the evolution of both the main activation layer and the memory traces left when the reaching behavior is allowed. The model is able to simulate the same kind of

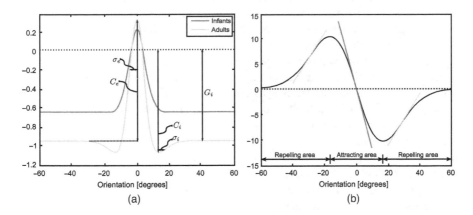

Figure 4.19    Plots of the kernels used in the DFT model of the 'A-not-B' error paradigm, (4.11c). (a) Interaction kernels $w_{xx}$ for infants and adults; (b) output kernel $w_\phi$

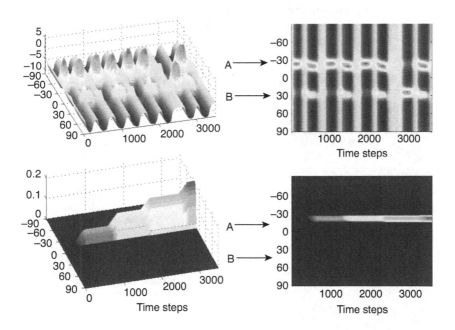

Figure 4.20    The 'A-not-B' task performed by an 'infant' agent. Top row: 3D and top views of the main activation layer. Bottom row: 3D and top views of the memory layer

results found in human infants. When presenting the visual stimulus at location 'B', the pointing/reaching action goes back to location 'A'. Memory traces are created only when a motor activation is present, in turn the main layer will get an extra influence in those locations where the memory layer has values larger than zero. In this way a primitive way of *expectation* is created.

A single experiment consists on three training trials with the purpose of building a motor habituation at the 'A' location, and one testing trial. Each trial is composed of two parts separated by a 3 s delay: a visual stimulation where the 'toy' is showed and a motor activation where the robot is allowed to 'point/reach'. Further 'B' trials will end up activating location 'A' with the consequence of creating a higher memory trace at this location. Thus, a *motor habituation* is created at one location. The only way of breaking this *habit* is to wait for an internal decrease of these memory traces due to the effect of the memory decay term controlled by $\lambda_{\text{decay}}$. Depending on the experiment this value can be changed to induce a faster or slower 'forgetting' effect.

The model is able to replicate the spontaneous errors reported by 'A-not-B' experimenters. Spontaneous errors are described as the selection of 'B' location when showing an object at 'A' in the training stage. These type of errors are replicated by the DFT model when adding noise to the dynamics of the system.

### 4.6.2    Adult control

Several trials in both locations were performed continuously in order to test the response of 'adult' conditions (Figure 4.21). Three 'A' trials were performed in

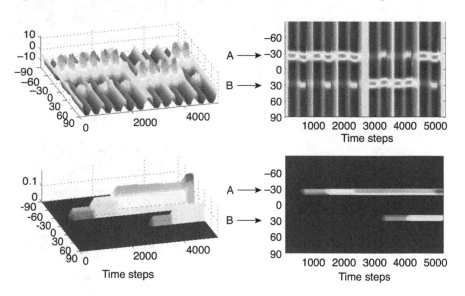

*Figure 4.21    The 'A-not-B' task performed by an 'older' agent. Top row: 3D and top views of the main activation layer. Bottom row: 3D and top views of the memory layer*

a row before testing the response of 'B' locations. Then two trials at location 'B' gave successful responses and the creation of the respective memory traces. Trials at locations 'A' and 'B' were performed in single or multiple groups each to test the performance of the system. Only one 'A' trial is presented here due to the lack of space, however all trials went to their correct locations.

The plots depicting the memory layer in this experiment allow us to have a better look at the effect of $\lambda_{decay}$. This term represents the rate of decay of memories through time and the designer should decide how fast an old experience remains present. During a 'B' trial, the memory trace left by 'A' experiences starts its decay, this effect is better seen in the bottom left plot of Figure 4.21.

## 4.7 Conclusions

Nonlinear dynamical systems theory and the *enactive* paradigm of cognitive systems are the basis of our approach to cognition. Next the most important concepts of dynamical systems and DFT were introduced. And finally the theory was implemented on a robotic platform by using an infant paradigm from developmental psychology; thus demonstrating the feasibility of our approach.

We would like to argue that traditional approaches constrain the dynamics of agents to modular versions of limited solutions based on the point of view of the designers. Consequently, unexpected circumstances at the moment of solving a task make these systems to either halt or continue with their programs without considering the new information. This is a crucial point for our proposal of a new dynamic, flexible, and autonomous way of understanding and implementing human behaviors in robotic platforms.

Dynamical systems theory was our answer for a dynamic world. Dynamical systems theory let us see the body of an agent and the environment as two parts of single dynamic and continuous flow of information in a dynamic world. The body of an agent represents the apex of this world where information is condensed in their simplest forms (stable states, fixed points) through the use of sensors and motors. The further we go into the inner world of an agent or out into the environment the more complex dynamics will be found. With this in mind we could create the concept of a chaotic cognitive architecture for artificial agents where more complex attractors are created through the fusion of simpler stable states.

The strategic use of tools provided by dynamical systems theory showed that neither the agent nor the environment require to be modeled since these two parts are seen as nonlinear systems; moreover, the way of solving a task is not specified in advance, as compared with the traditional control approach where the agent is told the steps to follow for solving a task or for overcoming certain problems. In short, if both approaches still need to know what to do, nonlinear dynamical systems theory free us from knowing how to solve a task;

thus overcoming the slow or none reactions to unexpected circumstances in dynamic environments found in traditional control models.

DFT has proved to have the potential to link neurophysiological studies with real human behaviors in a dynamic way. Even though it is still under development, this theory models in very elegant ways human paradigms that have been difficult to replicate with other approaches. This mathematical model embed in its dynamics very important and useful properties. The hysteresis property of bi-stable regimes implement a short-term memory, and long-term memories are created as field traces that feed-back past experiences. The way of seeing the world and its variables in terms of dimensions let us have implicit mechanisms for sensor fusion and decision-making problems. The most attractive properties of this approach are its implicit mechanisms for sensor fusion, decision making, and dynamic way of building short- and long-term memories.

The 'A-not-B' error paradigm has been studied by developmental psychologists as a proof of a very elementary form of cognition. This game designed to test the emergence of the concept 'object' in infants between 7 and 12 months of age was modeled by DFT and implemented in the iCub platform. The model replicates the same responses found in infants and adults when faced with this kind of task. This approach showed us a new novel, powerful, and dynamic way of implementing a human behavior in a robotic platform: motor habituation.

Further research is needed for DFT to replicate more complex human behaviors. In this approach fields represent static spaces that could take different shapes depending on the different experiences with the environment. However, most of the experiences in a time–space world are sequences of actions that can or cannot be periodic. Periodic sequences are used in many human activities such as walking, breathing, pattern recognition, etc. Moreover the large number of parameters becomes as usual one of the bigger challenges for most mathematical models. A single layer neural field with memory traces has already more than 20 parameters to be tuned. However in DFT, a few experiences with the way fields work give the designer the feeling of which parameters create the most important dynamics. This reduces enormously the size of the parameter space.

## Acknowledgments

I would like to express my gratitude to Prof. David Vernon for the important comments and suggestions in the creation of this document. I really cannot thank him enough for his motivating words and continuous faith in my work. I must acknowledge the wonderful work of Prof. Gregor Schöner and his students. I owe a special note of gratitude to Christian Faubel for the helpful hand with the understanding and implementation of the DFT model, without him I could not have been able to complete my work.

The work presented in this document has been partially supported by the RobotCub project (IST-2004-004370), funded by the European Commission through the Unit E5-Cognitive Systems.

## References

1. D. Vernon and D. Furlong. 'Philosophical foundations of AI'. In *50 Years of Artificial Intelligence*, vol. 4850 of *Lecture Notes in Artificial Intelligence*, pp. 53–62. Springer-Verlag, Heidelberg, Germany, 2007.

2. D. Vernon. 'Cognitive vision: The case for embodied perception'. *Image and Vision Computing*, 26(1):127–140, 2008.

3. E. Thelen and L. Smith. *A Dynamic Systems Approach to the Development of Cognition and Action*. The MIT Press Inc., Cambridge, MA, 1994.

4. G. Schöner and E. Thelen. 'Using dynamic field theory to rethink infant habituation'. *Psychological Review*, 113(2):273–299, 2006.

5. G. Sandini, G. Metta and D. Vernon. 'The iCub cognitive humanoid robot: An open-system research platform for enactive cognition'. *Journal of Bionics Engineering*, 1(3):191–198, 2004.

6. M. J. Feigenbaum. 'Quantitative universality for a class of nonlinear transformations'. *Journal of Statistical Physics*, 19:25–52, 1978.

7. S.-I. Amari. 'Dynamics of pattern formation in lateral-inhibition type neural fields'. *Biological Cybernetics*, 27:77–87, 1977.

8. G. Schöner. 'Development as change of system dynamics: Stability, instability, and emergence'. Chapter 2. In *Toward a New Grand Theory of Development? Connectionism and Dynamic Systems Theory Re-Considered*, vol. 1, pp. 20–50. Oxford University Press, New York, 2009.

9. W. Erlhagen and G. Schöner. 'Dynamic field theory of movement preparation'. *Psychological Review*, 109(3):545–572, 2002.

10. V. Simmering, A. Schutte, and J. P. Spencer. 'Generalizing the dynamic field theory of spatial cognition across real and developmental time scales'. *Computational Cognitive Neuroscience [special issue]. Brain research*, vol. 1202, pp. 68–86, April 2008.

11. J. Johnson, J. P. Spencer, and G. Schöner. 'Moving to higher ground: The dynamic field theory and the dynamics of visual cognition. *Dynamics and Psychology [special issue]'. New Ideas in Psychology*, 26:227–251, 2008.

12. K. Kopecz and G. Schöner. 'Saccadic motor planning by integrating visual information and pre-information on neural dynamic fields'. *Biological Cybernetics*, 73(1):49–60, 1995.

13. C. Wilimzig, S. Schneider, and G. Schöner. 'The time course of saccadic decision making: Dynamic field theory'. *Neural Networks*, 19(8):1059–1074, 2006.

14. E. Thelen, G. Schöner, C. Scheier, and L. Smith. 'The dynamics of embodiment: A field theory of infant perseverative reaching'. *Behavioral and Brain Sciences*, 24:1–86, 2001.

15. J. Piaget. *The Construction of Reality in the Child*. Basic Books, Inc., New York, 1954.
16. G. Schöner and E. Dineva. 'Dynamic instabilities as mechanisms for emergence'. *Developmental Sciences*, 10(1):69–74, 2007.
17. A. Schutte, J. Spencer, and G. Schöner. 'Testing the dynamic field theory: Working memory for locations becomes more spatially precise over development'. *Child Development*, 74(5):1393–1417, 2003.

*Chapter 5*

# A computational model of object affordances

*Luis Montesano, Manuel Lopes, Francisco Melo,*
*Alexandre Bernardino and José Santos-Victor*

## Abstract

The concept of object affordances describes the possible ways whereby an agent (either biological or artificial) can act upon an object. By observing the effects of actions on objects with certain properties, the agent can acquire an internal representation of the way the world functions with respect to its own motor and perceptual skills. Thus, affordances encode knowledge about the relationships between action and effects lying at the core of high-level cognitive skills such as planning, recognition, prediction and imitation. Humans learn and exploit object affordances through their entire lifespan, by either autonomous exploration of the world or social interaction.

Building on a biological motivation and aiming at the development of adaptive robotic systems, we propose a computational model capable of encoding object affordances during exploratory learning trials. We represent this knowledge as a Bayesian network and rely on statistical learning and inference methods to generate and explore the network, efficiently dealing with uncertainty, redundancy and irrelevant information. The affordance model serves as base for an imitation learning framework, which exploits the recognition and planning capabilities to learn new tasks from demonstrations. We show the application of our model in a real-world task in which a humanoid robot interacts with objects, uses the acquired knowledge and learns from demonstrations. Results illustrate the success of our approach in learning object affordances and generating complex cognitive behavior.

## 5.1 Introduction

Humans routinely solve tasks that require manipulation and interaction with different types of objects. From simple everyday actions such as pulling up a chair and sitting down, to complex skilled operations such as driving or repairing a car, humans purposely exploit the objects to achieve their goals. Even when an object is not designed for a particular application, humans are creative enough to give it new usages. For example, it is not uncommon to use a chair to hang a jacket, even though a chair is made for sitting. In fact what matters in obtaining a certain desired effect are the properties of the objects rather than the objects themselves.

Objects have different usages depending on the agents' motor and perceptual skills. For example, a chair is 'sitable' only by an individual of a certain height; a tree is 'climbable' only by animals with specific capabilities. James J. Gibson denominated these 'agent-dependent object usages' as *affordances* [1]. In that seminal work, all object affordances were defined as 'action possibilities' with reference to the actor's motor and sensing capabilities. In a nutshell, affordances represent the link between an agent and the environment and depend on the agent's specific motor and perceptual skills.

Our dependence on objects to achieve even the simplest goals in daily tasks emphasizes the importance of the concept of affordances in human cognition. The knowledge of object affordances is exploited in most of our decisions:

- to choose the most appropriate way of acting upon an object for a certain purpose
- to search and select objects that best suit the execution of a task
- to predict the effects of actions on objects
- to recognize ambiguous objects or actions, etc.

Affordances are at the core of high-level cognitive skills such as planning, recognition, prediction and imitation.

A key question that this chapter addresses, from a computational point of view, is how affordance knowledge can be acquired during an agent's lifespan. The importance of experience in mastering object-related skills suggests that affordance knowledge is acquired as the agent autonomously interacts with objects. In fact, infants interact with objects since early childhood and gradually learn how to use them in order to solve complex tasks. For instance, in the task of inserting shaped blocks into apertures [2], 14-month-old children demonstrate manipulation skills that are more exploratory than functional while 18-month-old children seem to understand the rules for fitting the blocks but cannot implement them yet. It is only at 22 months that children are able to have some success on the task. Finally, at 26 months they are able to solve the problem systematically. The capability to manipulate and exploit object properties is the result of a sophisticated ontogenetic development. Skills are acquired incrementally according to a genetic program conditioned by the surrounding environment, i.e., through the interaction with the world and other people.

## 5.1.1   *Robotic affordances*

In this chapter, we discuss object affordances within the context of the long-term goal of building (humanoid) robots capable of acting in a complex world and interacting with humans and objects in a flexible way. In particular, we address the following key questions:

- What knowledge representation and cognitive architecture should such a system require to be able to act in a complex and generally unpredictable environment?
- How can the system acquire task- and domain-specific knowledge to be used in novel situations?

From a robotics perspective, affordances are an extremely powerful concept that captures the essential world and object properties in terms of the actions the robot is able to perform. They can be used to predict the effects of an action, to plan actions leading to a specific goal or to select the best object to produce a given effect (see Figure 5.1).

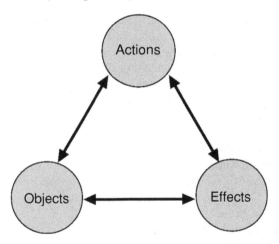

*Figure 5.1    Affordances as relations between (A)ctions, (O)bjects and (E)ffects, that can be used for different purposes: predict the outcome of an action, plan actions to achieve a goal or recognize objects/actions*

Extending the concept a bit further, affordances also play an important role when interacting with other agents.

An artificial system can gain a tremendous amount of information by observing another human or robotic agent performing actions on objects [3]. Affordance knowledge allows for action recognition in terms of the robot's motor capabilities and can be used, for example, in imitation [4]. Learning by imitation is one of the motivations behind our approach: we are interested in using generic affordances to learn a demonstrated task. We show that affordance knowledge can successfully be used to obtain imitation-like behaviors, e.g., allowing a robot to learn from a teacher a sequence of actions leading to a particular outcome.

Finally, it is important to emphasize that it is a tremendous task for a robot to learn, from scratch, a model of its interaction with the environment. It involves complex perceptual and motor skills, such as identifying objects and acting upon them. One possible way to deal with such complexity is by considering a bottom-up developmental perspective: basic sensory-motor skills are learned in initial stages upon which more complex cognitive capabilities can then be built. From the start, the robot should be able to individuate objects in the environment and execute directed exploratory motor actions upon them. In the same way, it is only after having learned a reasonable model of the world that imitation mechanisms will be operative and will enable the robot to learn from other agents.

## 5.1.2   Related work

Gibson's affordances [1] represent what the elements present in the environment afford to the agent. This very general concept was originally applied to entities such as surfaces (ground, air and water) or their frontiers. In psychology, there has been a lot of discussion to establish a definition or model of affordances [5]. Several authors have shown the presence of some type of affordance knowledge by comparing percepts among different people [6], measuring response times to tasks elicited by specific object orientations [7] or perceiving heaviness [8]. Unfortunately, there is little evidence on how humans learn affordances.

From the robotics standpoint, affordances have been mainly used to relate actions to objects. Several works use affordances as prior information. A computational cognitive model for the learning of grasping actions by infants was proposed in Reference 9. The affordance layer in this model provides information that helps the agent to perform the action. Affordances have also been used as prior distributions for action recognition in a Bayesian framework [10] or to perform selective attention in obstacle avoidance tasks [11].

Several works have investigated the problem of learning affordances and their subsequent application to different tasks. In Reference 12, a robot learned the direction of motion of different objects when poked and used this information at a later stage to recognize actions performed by others. The robot used the learned maps to push objects so as to reproduce the observed motion. A similar approach was proposed in Reference 13, where the imitation is also driven by observed effects. However, this work focuses on the interaction aspects and do not consider a general model for learning and using affordances. The biologically inspired behavior selection mechanism of Reference 14 uses clustering and self-organizing feature maps to relate object invariants to the success or failure of an action.

All previous approaches learn specific types of affordances using the relevant information extracted from sensor inputs. A more complete solution has been recently proposed in Reference 15, where the learning procedure also selects the appropriate features from a set of visual SIFT descriptors. The work in Reference 16 focuses on the importance of sequences of actions and invariant perceptions to discover affordances in a behavioral framework. Finally, based on the formalism of Reference 17, a goal-oriented affordance-based control for mobile robots has been presented in Reference 18: previously learned behaviors such as *traverse* or *approach* are combined to achieve goal-oriented navigation.

Regarding imitation, one must consider two fundamental problems: description of the observed motion in terms of the imitator's own motor capabilities (body correspondence) and selection of the goal of imitation (imitation metric). The former has been addressed in different ways in the literature. Possible approaches include hand coding the correspondence between the teacher and the imitator actions [19] or describing world state transitions at the trajectory level [20].

Recent research in neuroscience has triggered some alternative ways of addressing the correspondence problem. In particular, area F5 in the primate's premotor cortex is dominated by action coding neurons, but several of them also respond to visual stimuli (e.g., canonical neurons [21] and mirror neurons [22,23]). Canonical neurons show object-related visual responses that are, in the majority of cases, selective for objects of certain size, shape and orientation. On the other hand, mirror neurons respond to the observation of actions upon objects. They respond to actions executed by the observer but also to similar actions executed by another individual. This suggests that the same neuronal circuitry may be involved in both the recognition and generation of object-directed actions.

The discovery of mirror neurons triggered a large interest in the study of action recognition and imitation behaviors. Mirror neurons constitute an observation/execution matching system that maps observed actions to the observer's internal motor representations. This facilitates the recognition of actions because, unlike visual data, internal motor representations are invariant to viewpoint and other visual distortions [10]. Mapping actions to internal representations allows the imitation of actions even when individuals are not morphologically identical. The imitator chooses from its own motor repertoire the action that best matches the observations. It may choose to match only the effect of the action (emulation) [12,13] or also part of the action itself [24].

In summary, imitation can be interpreted at different levels from trajectory mimicking to effect emulation. There is no single solution for this problem and even humans change imitation strategies taking into account contextual information [25,26,27]. In artificial systems several authors have proposed different imitation metrics that result in behaviors distinct from pure imitation [19,24,28].

## 5.1.3 Our approach

Learning affordances from scratch without assuming any previous knowledge can be overwhelming. On one hand, it involves learning relations between motor and perceptual skills, resulting in an extremely high-dimensional search problem. On the other hand, affordances can be defined more appropriately once the robot has already learned a suitable set of elementary actions to explore the world.

We adopt a developmental approach [29,30], in which the robot acquires skills of increasing difficulty on top of previous ones. Similarly to newborn children, the robot 'starts' with a minimal subset of core (phylogenetic) capabilities [31] to bootstrap learning mechanisms that progressively lead to the acquisition of new skills by means of self-experimentation and interaction with the environment and other agents.

We follow the developmental roadmap proposed in Reference 32 and extend it to include the learning and usage of affordances in the world interaction phase. This framework considers three main stages in a possible developmental architecture for humanoid robots: (i) sensory-motor coordination, (ii) world interaction and (iii) imitation (see Table 5.1). In the sensory-motor coordination stage, the robot learns how to use its motor degrees of freedom and to encode the relationships between motor actions and perception. In the world interaction phase, the robot learns by exploring the effects of its own actions upon elements of the environment. In the imitation phase, the robot learns by observing and imitating other agents.

*Table 5.1    Learning stages of the developmental approach*

| Sensorimotor Coordination | 1. Learn basic skills |
| | 2. Develop visual perception of objects |
| World Interaction | 3. Perception of effects and categorization |
| | 4. Improve motor skills |
| | 5. Learn object affordances |
| | 6. Prediction and planning skills |
| Imitation | 7. Perform imitation games – Task inference from observation |

Affordances are central in the world interaction stage. In this stage, the robot has already developed a set of perceptual and motor skills required to interact with the world. In this chapter, we propose a general model to represent knowledge about affordances, i.e., relationships between the agent's actions, object properties and effects observed on these objects. The model consists of a Bayesian network (BN) [33], a probabilistic graphical model that represents dependencies between variables. In other words, a BN is a directed acyclic graph whose nodes represent (random) variables and whose connections express the correlations between them. The BN thus encodes the relation between different types of information. From the probabilistic model, the robot can use the information available from its sensory inputs to make different types of predictions about the world, infer situations from incomplete information and plan for actions depending on its goals.

In the second part of the chapter, we use the world knowledge acquired in the form of affordances to be able to imitate others. Affordances were learned in a task-independent way and so they need to be written in a way that allows sequential decision-making. Equipped with this knowledge, the robot is able to infer the goal of an observed demonstration using Bayesian Inverse Reinforcement Learning [34]. Affordances provide another source of invaluable information, i.e., the recognition of observed actions. The robot is thus able to extract information from the demonstration by recognizing the observed motions in terms of its own motor repertoire.

We used the humanoid robot Baltazar (see Figure 5.4) to validate the approach. We conducted several experiments to illustrate the capability of the system to discover affordances associated with manipulation actions (e.g., grasp, tap and touch) when applied to objects with different properties (color, size and shape). The effects of these actions consist of changes perceived in the sensor measurements, e.g., persistent tactile activation for grasp/touch actions and object motion for tap actions.

To summarize this chapter presents a model for learning and using affordances and its application to robot imitation. The main characteristics of the proposed model are: (i) it captures the relations between actions, object features and effects; (ii) it is learned through observation and interaction with the world; (iii) it identifies the object features that matter for each affordance; (iv) it provides a seamless framework for the learning and exploiting affordances; and (v) it allows social interaction by learning from others.

## 5.2  Affordance modeling and learning

In this section, we address the problem of modeling and learning object affordances. According to Table 5.1, we assume the robot has already acquired a set of skills that allows it to reason in a more abstract level than joint positions or raw perceptions. More specifically, the robot has a parameterized set of actions available that allows it to interact with the world and is able to detect and extract categorical information from the objects around it (see Section 5.4 for further details).

We pose the affordance learning problem at an abstraction level where the main entities are actions, object properties and effects. A discrete random variable $A$ taking values in the set $A = \{a_i, i = 1, \dots, n_a\}$ models the activation of the different motor actions. Each action $a_i$ is parameterized by a corresponding set of parameters $\lambda_i$. For example, when approaching an object to perform a grasp action, the height of the hand with respect to the object or the closing angles of the hand are free parameters.[1] It is important to note that, from a sensory-motor point of view, different values for these free parameters result in the same action. Hence, at this stage of development, the robot cannot distinguish between them, since the differences will only be evident when interacting with those objects.

---

[1] Refer to Section 5.4 for further details on the actual action implementation.

The object properties and effects are also modeled using discrete random variables as detected by the robot. We denote by $F_r = \{F_r(1), \ldots, F_r(n_r)\}$ and $F_o = \{F_o(1), \ldots, F_o(n_o)\}$ the sets of random variables corresponding to the descriptors (features) extracted by the perceptual modules and representing, respectively, the agent itself and object $o$. Finally, we let $E = \{E(1), \ldots, E(n_e)\}$ denote the set of random variables corresponding to the possible effects detected by the robot after executing an action. The difference between object features and effects is that the former sets can be acquired by simple observation, whereas the latter set requires interaction with the objects. Thus, clustering the effects correspond to the first step of the world interaction stage in Table 5.1 and precedes the actual learning of the affordances.

We use a BN to encode the dependencies between object features, actions exerted upon such objects and effects of those actions (see Figure 5.2). Such a representation has several advantages: it allows us to take into account the inherent uncertainty in the world; it encodes some notion of causality; and it provides a unified framework for both learning and using affordances. In continuation, we briefly survey the fundamental concepts concerning representation, inference and learning using BNs and show how to apply them to our affordance problem.

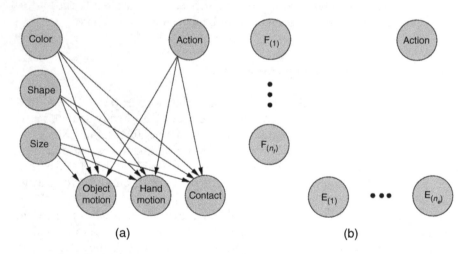

(a)                              (b)

*Figure 5.2    BN model representing affordances. (a) An example of the proposed model using: color, shape and size as object features and motion and contact information as effects. (b) Generic model where the nodes represent the action, A, the object features available to the robot, $F_{(1)}, \ldots, F_{(n_f)}$ and the effects obtained through the actions, $E_{(1)}, \ldots, E_{(n_e)}$*

A BN is a probabilistic graphical model that represents dependencies between random variables as a directed acyclic graph. Each node in the network represents a random variable $Y_i$, $i = 1, \ldots, n$ and the (lack of) arcs between two nodes $Y_i$ and $Y_j$ represents conditional independence of the corresponding variables. BNs are able to represent causal models since an arc $Y_i \to Y_j$ can be interpreted as $Y_i$ causes $Y_j$ [35]. The conditional probability distribution (CPD) of each variable $Y_i$ in the network, denoted as $\mathbf{P}[Y_i|Y_{Pa(Y_i)}, \theta_i]$, depends on the parent nodes of $Y_i$, denoted collectively as $Y_{Pa(Y_i)}$, and on a set of parameters $\theta_i$. The joint distribution of the BN thus decomposes in the following way:

$$\mathbf{P}[Y_1, ..., Y_n|\theta] = \prod_{i=1}^{n} \mathbf{P}[Y_i|Y_{Pa(Y_i)}, \theta_i] \tag{5.1}$$

where $\theta$ represents all the parameters in the different CPDs. If the conditional distributions and the priors on the parameters are conjugate, the CPDs and marginal likelihood can be computed in closed form, resulting in efficient learning and inference algorithms.

The set of nodes in the network, $Y$, comprises the discrete variable $A$ and those in $F_r$, $F_o$ and $E$, i.e.,

$$Y = \{A, F_r(1), ..., F_r(n_r), F_o(1), ..., F_o(n_o), E(1), ..., E(n_e)\}$$

Our ultimate goal is to uncover the relations between the random variables in $Y$, representing actions, object features and effects (see Figure 5.2). To this purpose, the robot performs an action on an object and observes the resulting effects. By repeating this procedure several times, the robot acquires a set of $N$ samples of the variables in $Y$, $D = \{y^{1:N}\}$.[2]

Let us assume for the moment that the dependencies between the variables in $Y$ are known, i.e., the structure of the BN representing the affordances is known. Given the discrete representation of actions, object features and effects, we can use a multinomial distribution and the corresponding conjugate – the Dirichlet distribution – to model the CPDs $\mathbf{P}[Y_i|Y_{Pa(Y_i)}, \theta_i]$ and the corresponding parameter priors $\mathbf{P}[\theta_i]$. Let $y_i$ denote the range of values of the random variable $Y_i$ and $y_{Pa(Y_i)}$ the range of values of the parents of $Y_i$. Assuming independence between the samples in $D$, the marginal likelihood for the random variable $Y_i$ and its parents given is $D$ [36].

$$\mathbf{P}[y_i^{1:N}|y_{Pa(y_i)}^{1:N}] = \int \prod_{i=1}^{N} \mathbf{P}[y_i^n|y_{Pa(y_i)}^n, \theta_i]\mathbf{P}[\theta_i]d\theta_i$$

$$= \prod_{j=1}^{|y_i|} \frac{\Gamma(\alpha_{ij})}{\Gamma(\alpha_{ij} + N_{ij})} \prod_{k=1}^{|y_{Pa(y_i)}|} \frac{\Gamma(\alpha_{ijk} + N_{ijk})}{\Gamma(\alpha_{ijk})}$$

where $\Gamma$ represents the gamma function, $N_{ijk}$ counts the number of samples in which $Y_i = j$ and $Y_{Pa(y_i)} = k$, and $N_{ij} = \sum_k N_{ijk}$. The pseudocounts $\alpha_{ijk}$ denote the Dirichlet hyperparameters of the prior distribution of $\theta_i$ and $\alpha_{ij} = \sum_k \alpha_{ijk}$.

---

[2] Capital letter $Y$ represents a random variable, lowercase $y$ represents its realizations.

The marginal likelihood of the data is simply the product of the marginal likelihood of each node,

$$\mathbf{P}[D|G] = \mathbf{P}[Y^{1:N}|G] = \prod_i \mathbf{P}[y_i^{1:N}|y_{Pa(y_i)}^{1:N}] \tag{5.2}$$

where we have made explicit the dependency on the graph structure, $G$.

### 5.2.1  Learning the structure of the network

We are now interested in learning the structure of the network, $G$, which is actually an instance of a model selection problem. In a Bayesian framework, this can be formalized as estimating the distribution over all possible network structures $G \in g$ given the data. Using the Bayes rule, we can express this distribution as the product of the marginal likelihood and the prior over graph structures,

$$\mathbf{P}[G|D] = \eta \mathbf{P}[D|G]\mathbf{P}[G] \tag{5.3}$$

where $\eta = 1/\mathbf{P}[D]$ is a normalization constant. The term $\mathbf{P}[G]$ (the prior) allows to incorporate prior knowledge on possible structures. Unfortunately, the number of possible BN structures is superexponential in the number of nodes [37]. Thus, it is infeasible to explore all the possible graph structures and one has to rely on some form of approximation of the full distribution. Markov Chain Monte Carlo (MCMC) methods have been proposed to approximate the distribution $\mathbf{P}[G|D]$ [38]. In our case, this can be important during the first iterations of the learning process, as it allows the robot to keep a set of alternative hypotheses on the possible affordance model.

As the robot performs the actions itself, it is usually able to obtain information on all the variables $Y_i$ in the BN. The model can also be applied to learning by observation, i.e., by observing other people performing the actions. However, in this situation there may be some missing information. For example, the action is not directly available and has to be inferred from visual measurements. In this case, the learning task is much harder and several algorithms have been proposed such as augmented MCMC or structural EM [39].

Finally, it is important to consider causality. The previous learning techniques are able to distinguish among equivalence classes of graph structures.[3] An equivalence class contains different causal interpretations between the nodes in the network. It is necessary to use *interventional data*, where some of the variables are held fixed to a specific value to disambiguate between graph structures in the same equivalence class, so as to be able to infer the correct causal dependency.

In the case of a robot interacting with its environment, there are several variables that are actively chosen by the robot: the action and the object. These variables are actually interventional, since they are set to their specific values

---

[3] Two directed acyclic graphs $G$ and $G$' are equivalent, if for every BN $B = (G, \Theta)$ there exists another network $B' = (G', \Theta')$ such that both define the same probability distribution.

by the robot at each experience. Interventional data is currently an important research topic within BN learning algorithms [40]. Under the assumption of a perfect intervention of node $Y_i$, the value of $Y_i$ is set to the desired value, $y_i^*$, and its CPD is just an indicator function with all the probability mass assigned to this value, i.e., $\mathbf{P}[Y_i|Y_{Pa(Y_i)}, \theta_i] = \mathbf{I}(Y_i = y_i^*)$. As a result, the variable $Y_i$ is effectively cut off from its parents $Y_{Pa(Y_i)}$.

### 5.2.2 Parameter learning and inference

Once the structure of the network has been determined, the parameter $\theta_i$ of each node is estimated using a Bayesian approach [41]. The estimated parameters can still be updated on line, allowing the incorporation of the information provided by new trials.

Since the structure of the BN encodes the relations between actions, object features and effects, we can now compute the distribution of a (group of) variable(s) given the values of others. The most common way to do this is to convert the BN into a tree and then apply the junction tree algorithm [42] to compute the distribution of interests. It is important to note that it is not necessary to know the values of all the variables to perform inference.

Based on these probabilistic queries, we are now able to use the affordance knowledge to answer the questions outlined in Figure 5.1 simply by computing the appropriate distributions. For instance, predicting the effects of an observed action $a_i$ given the observed object features $f_i$ can easily be performed from the distribution $\mathbf{P}[E|A = a_i, F = f_i]$. The query can combine features, actions and effects both as observed information and as the desired output.

## 5.3 Imitation learning

After interacting with the objects, the robot has acquired important information to support the social stage of its development. This knowledge provides the basis for allowing the robot to learn how to perform tasks by observing others and imitate their goals and actions. A general model for imitation learning is presented in the following. We adopt the formalist of Reference 24 to learn complex task descriptions from human demonstrations. Afterwards, and based on this formalism, we explain how affordance knowledge can be used to provide the required inputs for imitation learning.

### 5.3.1 A model for imitating tasks

To imitate complex tasks, the robot must first understand the tasks executed by others. The robot will observe an action/sequence of actions by a human demonstrator and then choose its own actions accordingly. At each time instant, the robot must choose an action from its action repertoire $A$, depending on the perceived state of the environment. We represent the state of the environment at time $t$ by $X_t$ and let $x$ be the finite set of possible environment states.

At this level of abstraction, action selection can be seen as a *decision process*. The state of the world evolves according to some probabilistic transition model,

$$\mathbf{P}[X_{t+1} = z | X_t = x, A_t = a] = \mathbf{P}_a(x, z) \tag{5.4}$$

where $A_t$ denotes the robot's action at time $t$. The action-dependent transition matrix $\mathbf{P}_a$ thus describes the dynamic behavior of process $\{X_t\}$.

At this stage we assume that the robot is able to recognize the actions performed during the demonstration. Later, we will show how the affordance knowledge can be used to achieve this goal. Bearing this assumption in mind, we consider that the demonstration consists of a sequence $H$ of state—action pairs.

$$H = \{(x_1, a_1), (x_2, a_2), ..., (x_i, a_i), ..., (x_n, a_n)\}$$

Each pair $(x_i, a_i)$ exemplifies to the robot the expected action $(a_i)$ in each of the states visited during the demonstration $(x_i)$. From this demonstration, the robot is then expected to perceive what the demonstrated task is and learn how to perform it optimally, possibly relying on some experimentation of its own. A *policy* is a map $\pi : X \rightarrow A$, a decision-rule that determines the action of the robot as a function of the state of the environment. The robot must then *infer the task* from the demonstration and *learn the corresponding optimal policy*.

In our formalism, the task can be defined using a function $r : X \rightarrow \mathbf{R}$ describing the 'desirability' of each particular state $x \epsilon X$. This function $r$ works as a *reward* for the robot and, once $r$ is known, the robot should choose its actions to maximize the total reward collected during its lifespan, represented as the functional:

$$J(x, \{A_t\}) = E\left[\sum_{t=1}^{\infty} \gamma^t r(X_t) | X_0 = x\right]$$

where $\gamma$ is a discount factor between 0 and 1 that assigns greater importance to those rewards received in the immediate future than to those in the distant future. We remark that, once $r$ is known, the problem falls back to the standard formulation of dynamic programming [43].

To express the recursive relation between the function $r$ describing the task and the optional behavior rule, we define the function $V_r$ as follows:

$$V_r(x) = \max_{a \in A} \left[r(x) + \gamma \sum_{z \in x} \mathbf{P}_a(x, z) V_r(z)\right] \tag{5.5}$$

The value $V_r(x)$ represents the expected (discounted) reward accumulated along a path of the process $\{X_t\}$ starting at state $x$ when the optimal behavior rule is followed. The optimal policy associated with the reward function $r$ is thus given by:

$$\pi_r(x) = \underset{a \in A}{\operatorname{argmax}} \left[ r(x) + \gamma \sum_{z \in X} \mathbf{P}_a(x, z) V_r(z) \right]$$

The computation of $\pi_r$ (or, equivalently, $V_r$) given $\mathbf{P}$ and $r$ is a standard problem and can be solved using any of several standard methods available in the literature [43].

## 5.3.2 Learning task descriptions

In the formalism just described, the fundamental imitation problem lies in the estimation of the function $r$ from the observed demonstration $H$. Notice that this is closely related to the problem of *inverse reinforcement learning* as described in Reference 44. We adopt the method described in Reference 24, which is a basic variation of the Bayesian inverse reinforcement learning (BIRL) algorithm in Reference 34.

For a given function $r$, we define the *likelihood of a pair* $(x, a) \in X \times A$ as:

$$L_r(x, a) = \mathbf{P}[(x, a)|r] = \frac{e^{\eta Q_r(x,a)}}{\sum_{b \in A} e^{\eta Q_r(x,b)}}$$

where $Q_r(x, a)$ is defined as:

$$Q_r(x, a) = r(x) + \gamma \sum_{z \in X} \mathbf{P}_a(x, z) V_r(z)$$

and $V_r$ is as in (5.5). The parameter $\eta$ is a user-defined *confidence parameter* that we describe further ahead. Notice that $L_r(x, a)$ is simply a *softmax* distribution over the possible actions, and translates the plausibility of the choice of action $a$ in state $x$ when the underlying task is described by $r$. Given a demonstration sequence

$$H = \{(x_1, a_1), (x_2, a_2), ..., (x_n, a_n)\}$$

the corresponding likelihood is:

$$L_r(H) = \prod_{i=1}^{n} L_r(x_i, a_i)$$

We use MCMC to estimate the posterior distribution over the space of possible $r$-functions (usually a compact subset of $\mathbf{R}^p$, $p > 0$) given the demonstration, as proposed in Reference 34. We then choose the $r$-function corresponding to the maximum of this distribution. Since we consider a uniform prior to the distribution over $r$-functions, the selected reward is the one whose corresponding optimal policy 'best matches' the demonstration. The confidence parameter

$\eta$ determines the 'trustworthiness' of the demonstration: it is a user-defined parameter that indicates how 'close' the demonstrated policy is to the optimal policy [34].

Some remarks are in order. First of all, to determine the likelihood of the demonstration for each function $r$, the algorithm requires the transition model in **P**. If such transition model is not available, then the robot will only be able to *replicate particular aspects of the demonstration*. However, as argued in Reference 24, the imitative behavior obtained in these situations may not correspond to actual imitation.

Second, it may happen that the transition model available is *inaccurate*. In this situation (and unless the model is significantly inaccurate) the robot should still be able to perceive the demonstrated task. Then, given the estimated $r$-function, the robot may only be able to determine a *suboptimal policy* and will need to resort to *experimentation* to improve this policy. We discuss these aspects in greater detail in the continuation.


### 5.3.3    Combining affordances with imitation learning

In this section, we discuss, in greater detail, how the information provided by the affordances described in Section 2 can be combined with the imitation learning approach described in Section 5.3.2. We discuss the advantages of this approach, as well as several interesting issues that arise from this combination.

In the methodology described in Section 5.3.2. We assumed the robot to be able to recognize the actions performed by the demonstrator. This action recognition does not need to be explicit, i.e., the agent is not required to determine the actual movements executed by the demonstrator. Instead, it needs only to *interpret* the observed action in terms of its own action repertoire. This interpretation may rely on the observed state transition or in the corresponding effects. It is important to emphasize that transitions and effects are different concepts: the same transition may occur from different actions/effects and the same effect can be observed in different transitions. To clarify this distinction, consider moving or jumping from one place to the other: the effects are different but the transition is the same. For another example, consider moving between different places with the same speed: the effect is the same (motion at a given speed) but the transitions are different.

If no action recognition/interpretation takes place, the robot will generally be able to learn only how to replicate particular elements of the observed demonstration. In our approach we want the robot to *learn the task* more than to replicate aspects of the demonstration. As seen in Section 5.2, affordances provide a functional description of the robot's interaction with its surroundings as well as the action recognition capabilities necessary to implement imitation.

Affordance-based action recognition/interpretation works as follows. For each demonstrated action, the robot observes the corresponding effect. The affordance network is then used to estimate the probability of each action in the robot's action repertoire given the observed effects, and the action with greatest probability is picked as the observed action. Clearly, there will be some uncertainty in the identification of the demonstrated action, but as will be seen in the experimental section, this does not significantly affect the performance of the learning algorithm.

On the other hand, given the demonstration – consisting on a sequence of state–action pairs – the robot should be able to *infer* the task to be learned. This means in particular that once the robot realizes the task to be learned, it should be able to learn how to perform it *even in situations that were never demonstrated.*

Choosing between two policies generally requires the robot to have a model of the world. Only with a model of the world will the robot have the necessary information to realize *what task is more suitably accomplished by the demonstrated policy.* If no model of the world is available, then the robot will generally only *repeat the observed action pattern*, with no knowledge on what the underlying task may be. Also, the absence of a model will generally prevent the robot from *generalizing* the observed action pattern to situations never demonstrated.

As argued in Section 5.2, affordances empower the robot with the ability to predict the effect of its actions in the surrounding environment. Once the adequate state space for a particular task is settled, the information embedded in the affordance network can be used to extract the dynamic model describing the state evolution for the particular task at hand. This action-dependent dynamic model consists of the transition matrix **P** described in the previous subsection.

Figure 5.3 depicts the fundamental elements in the imitation learning architecture described. It is important to notice that the affordance network is *task independent* and can be used to provide the required transition information for different tasks. Of course, the interaction model described in the affordance network could be enriched with further information concerning the state of the system for a specific task. This would make the extraction of the transition model automatic, but would render the affordance network *task dependent.* This and the very definition of affordances justify the use of a more general affordance model, although, in such a case, the transition model might have to be extracted separately for each particular task. This means that imitation can be successfully implemented in different tasks, provided that a single sufficiently general and task-independent model of interaction is available (such as the one provided by the affordances).

Another important observation is concerned with the fact that the affordance network is learned from interaction with the world. The combination of

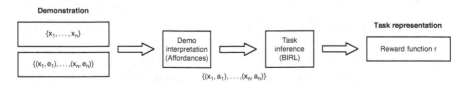

*Figure 5.3   Representation of the fundamental elements of an imitation learner*

both learning blocks (affordance learning and imitation learning) gives rise to a complete architecture that allows the acquisition of skills ranging from simple action recognition to complex sequential tasks.

In the remainder of the chapter, we describe the implementation of this combined architecture in a humanoid robot. We illustrate the learning of a sequential task that relies on the interaction model described in the affordance network. We discuss the sensitivity of the imitation learning to action recognition errors.

## 5.4   Experimental setup

In this section we present the robot used in the experiments, the experimental playground and the basic skills required to start interacting with the world and to learn the affordances. These skills include the basic motor actions and the visual perception of objects and effects.

### 5.4.1   Robot platform and playground

The experiments were done using *Baltazar*, a 14-degrees-of-freedom humanoid torso composed of a binocular head and an arm (see Figure 5.4). The robot has implemented a set of parameterized actions based on a generic controller:

$$\dot{\Theta} = m_i(\Theta^*, y, \lambda, \psi)$$

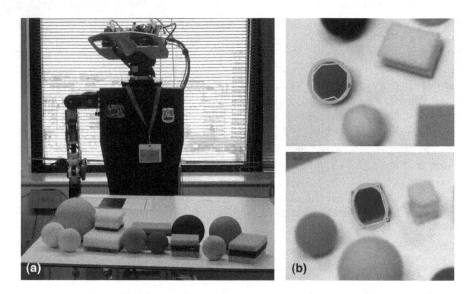

*Figure 5.4   Experimental setup. (a) The robot's workspace consists of a white table and some colored objects with different shapes, sizes and colors. (b) Objects on the table are represented and categorized according to their size, shape and color, e.g., 'ball' and 'square'*

where $\dot{\Theta}$ represents the time derivatives of the controlled variables, $\Theta^*$ is the final objective and $y$ the available proprioceptive measurements of the robot. Parameters $\psi$ describe the kinematics/dynamics of the robot. Parameters $\lambda$ can be used to shape the controller, i.e., change desired velocities, energy criteria or postures. They can be tuned during affordance learning (refer to Figure 5.8 in Section 5.8), but are frozen by the system during the initial learning stage.

In this work we focus on object manipulation actions such as grasping, tapping and touching (see Figure 5.5). Each of these actions consists of three distinct steps: (i) bring the hand to the field of view in open loop; (ii) approach the object using visual servoing; and (iii) actually grasp, tap or touch the object. The two former steps are learned by self-experience (see [32] for further details), while the latter is pre-programmed due to practical limitations of our current robotic platform. Using this approach, *Baltazar* is able to perform three different actions. In terms of our model, we have $A = \{a_1 = \mathrm{grasp}(\lambda),$ $a_2 = \mathrm{tap}(\lambda), a_3 = \mathrm{touch}(\lambda)\}$, where $\lambda$ represents the height of the hand in the 3D workspace when reaching the object in the image.

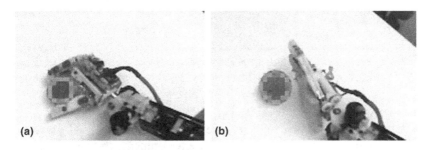

(a)                                        (b)

*Figure 5.5    Examples of actions as seen by the robot. (a) Grasping. (b) Tapping*

The robot executed its actions upon several different objects, each having one of two possible shapes ('box' and 'ball'), four possible colors and three possible sizes (see Figure 5.4). We recorded a set of 300 trials following the protocol summarized in Figure 5.6. In each trial, the robot was presented with a random object. *Baltazar* randomly selected an action from its repertoire and approximated its hand to the object. On reaching the object, it performed the selected action (grasp($\lambda$), tap($\lambda$) or touch ($\lambda$)) and returned the hand to the initial position. During action execution, the object features and effects were recorded. Note that the robot does not receive any feedback concerning the success or failure of the actions. The goal of this learning stage is to understand the causal relations between actions and effects in an unsupervised manner.

## 5.4.2    Visual perception of objects

Regarding object perception and feature extraction, we assume the system has simple segmentation and category formation capabilities already built in. For

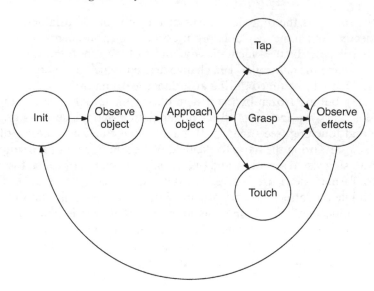

*Figure 5.6    Experiments protocol. The object to interact with is selected manually and the action is randomly selected. Object properties are recorded in the Init to Approach transition, when the hand is not occluding the object. The effects are recorded in the final Observe. Init moves the hand to a predefined position in open loop*

the sake of experimental simplicity, we have constructed the 'playground' environment shown in Figure 5.4. In this environment, the robot plays with simple colorful objects over a white table and observes people playing with the same objects. At this stage, we rely on fast techniques such as background and color segmentation to allow the robot to individuate and track objects in real time. Along time, the robot collects information regarding simple visual object properties, such as color, shape, size, etc. Figure 5.4 illustrates the robot's view of several objects, together with their color segmentation and extracted contour. After interacting with the objects for some time, the robot is able to group their properties into meaningful categories. The set of visual features used consist of color descriptors, shape descriptors and size (in terms of the image). The color descriptor is given by the hue histogram of pixels inside the segmented region (16 bins). The shape descriptor is a vector containing measurements of convexity, eccentricity, compactness, roundness and squareness.

## 5.4.3    Perception of effects

In our framework, effects are defined as salient changes in the perceptual state of the agent that can be correlated to actions. For example, upon interacting with an object, the robot may observe sudden changes in the object position or tactile information.

All effects are processed in the same way: when the action starts, the agent observes its sensory inputs during a certain time window that depends on the action execution time and the duration of the effects. It then records the corresponding information flow. We fit a linear model to the recorded temporal information associated with each observed effect and use the corresponding inclination and bias parameters as a compact representation of this effect. The regression parameters for the velocity of either an object, the hand or the 'object–hand pair' are determined from a sequence of image velocity norms. In this case, only the inclination parameter is used, since the bias parameter only reflects the absolute position in the image. Concerning contact effects, we consider only the bias parameter (offset), which gives a rough estimation of the duration of contact.

### 5.4.4   Discretization of perceptual information

This is an important step in the overall learning architecture, since it provides the basis for the discretization and categorization used in the affordance learning algorithm. In our example, we used features describing object's color, shape and size. Each feature takes values in some $n$-dimensional vector space. We applied the *X-means* algorithm [45] to detect clusters in the space of each object feature and in the effects. We also discretized the space of the free actuator parameters, $\lambda$, using a predefined resolution and the same clustering algorithm. It is important to note that our ultimate goal is to learn the affordances given a set of available motor and perceptual skills and not to make a perfect object classification. Indeed, the clustering introduces some errors, among other things, due to different illumination conditions during the trials. As such, the features of some objects were misclassified and the affordance learning had to cope with this noise.

Figure 5.7(a) shows the results of the *X-means* algorithm for the object shape feature. The two resulting clusters are able to easily separate 'balls' from 'boxes' based mostly on roundness and eccentricity descriptors. Figure 5.7(b) gives the equivalent result for colors, where the feature vector is a histogram of hue values. As the objects have uniform color, each histogram has only one salient peak. Finally, for the unidimensional size, three clusters were enough to represent five different sizes of the objects presented to the robot.

Figure 5.7(c) shows the classes of object velocities and contact patterns detected by the robot, following the procedure described in Subsection 5.4.3. Roughly speaking, a grasp action resulted in 'medium' velocity (except in one case where the ball fell down the table), a tap action produced different velocity patterns depending on the shape and size of the object and a touch action induces small velocities. Also, contact information was more pronounced for grasp and touch actions than for tap ones. The combination of the different features produced patterns in the feature space that were used to infer statistical dependencies and causality. Table 5.2 summarizes the clustering results for the different variables and provides the notation used in the remainder of this section.

This categorization was conducted after the robot interacted with different objects for several trials, during which it collected information about the effects of its actions on the objects. The obtained clustering resulted in groups that are close in the sensory space. We thus have to assume that the motor and perceptual capabilities of the robot are such that the same action applied to the same object will, in average, yield similar effects. For example, all successful grasps will have the pressure sensors persistently activated. This clustering is not restricted to observed objects, because new objects will be categorized according to their similarity to known ones.

## 5.5 Experimental results

In this section we present the experimental results obtained with our approach. We start by illustrating the affordance learning stage, carefully evaluating its

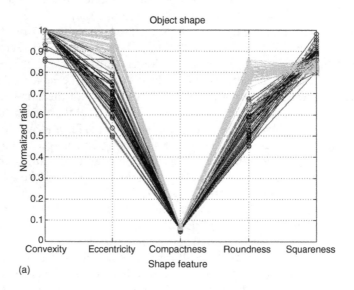

(a)

*Figure 5.7    Clustering of object features and effects. (a) Shape description of*
*the objects. Five features: convexity, eccentricity, compactness,*
*roundness and squareness describe the objects. For the objects*
*considered in the experiments, box and balls, they can be clustered*
*automatically. Different clusters are represented by circles or plus*
*signs. (b) Color histograms with the corresponding clusters.*
*Each bin relates to a given Hue value. The clusters correspond to:*
*yellow, green$_1$, green$_2$ and blue. (c) Clustering of object velocity*
*and contact. For each observation, grasp is represented by "×",*
*tap by "Δ" and touch by "O". The vertical lines show the cluster*
*boundaries for velocity and the horizontal line for contact*

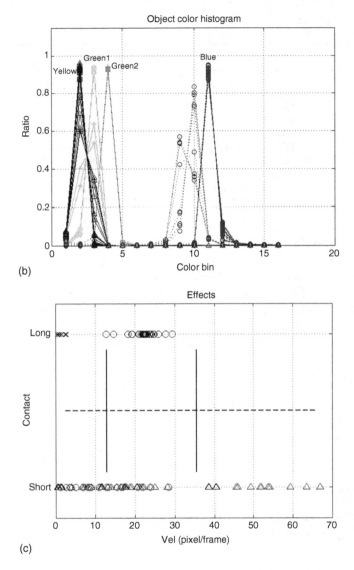

(b)

(c)

*Figure 5.7   Continued*

capabilities. We then proceed to presenting the application of the learned affordances in interaction games and in learning by demonstration of a complex task.

## 5.5.1   Affordances

We now describe the different experiments conducted to illustrate the ability of the proposed model to capture object affordances.

*Table 5.2   Summary of variables and values*

| Symbol | Description | Values |
| --- | --- | --- |
| A | Action | *Grasp, Tap, Touch* |
| H | Height | Discretized in ten values |
| C | Color | *Green₁, Green₂, Yellow, Blue* |
| Sh | Shape | *Ball, Box* |
| S | Size | *Small, Medium, Big* |
| V | Object velocity | *Small, Medium, Big* |
| HV | Hand velocity | *Small, Medium, Big* |
| Di | Object-to-hand velocity | *Small, Medium, Big* |
| Ct | Contact duration | *None, Short, Long* |

### 5.5.1.1   Controller optimization

The objective of the first experiment is to find the influence of a free actuator parameter on an action. The robot tries the action for different values of the free parameters. For a grasp, these parameters are closure of the fingers and approaching height of the hand. The former is used after reaching the object in the closing of the hand, whereas the latter is a free parameter of the sensory-motor map used to approximate the hand to the object. We computed the maximum likelihood graph with a random starting point and *BDeu* priors [36] to give uniform priors to different equivalence classes.[4]

Figure 5.8(a) shows how the resulting network captures the dependency of the effects on these parameters. Interestingly, the CPDs provide the probability of producing different effects according to the values of the free parameters. Figure 5.8(b) shows the estimated probability of different height values conditioned on the observation of a long contact (indicating a successful grasp) for medium and small objects. Since big objects cannot be grasped by the robot's hand, all heights have zero probability for this class.

Note that the distribution of Figure 5.8(b) can be used directly to adjust the height of the action for different object sizes and, as such, perform an optimization of the controller parameter based on the target object.

### 5.5.1.2   Affordance network learning

In the second experiment, we illustrate the robot's ability to distinguish the effects of different actions and simultaneously identify the object features that are relevant to this purpose. As in the previous experiment, we use *BDeu* priors and random initialization.[5] For the MCMC algorithm, we used 5 000 samples with a burn-in period of 500 steps.

---

[4] The implementation of the algorithms is based on the BNT toolbox for Matlab, http://bnt. sourceforge.net/.

[5] Although it is possible to use conditional independence tests to provide a rough initialization, in our case we got similar results using randomly generated networks.

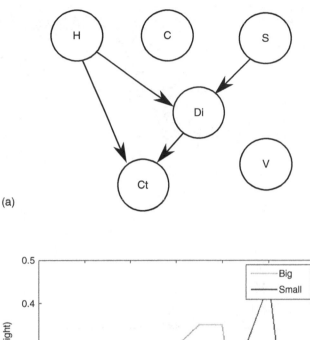

(a)

(b)

*Figure 5.8*   *Tuning the height for grasping a ball. (a) Dependencies discovered by the learning algorithm. The action and shape for this example are fixed and the color does not have an impact on the effects. Node labels can be found in Table 5.2. (b) CPD of the height value given that the robot obtained a long contact (successful grasp)*

Figures 5.9(a–c) show the three most likely networks computed by MCMC and Figure 5.9(d) shows the posterior probability distribution over all sampled models. To show the convergence of the network toward a plausible model, we have estimated a network structure using datasets of different lengths. For each length, we have randomly created 100 datasets from the complete dataset, estimated the posterior over-graph structures using MCMC and computed the

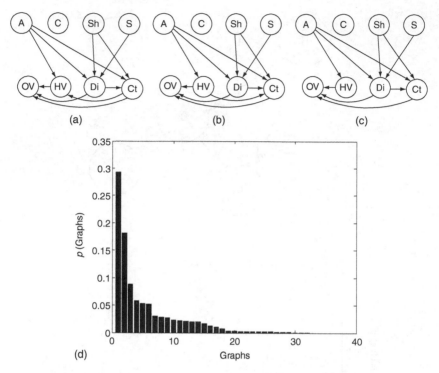

*Figure 5.9    Affordance model estimated by the MCMC algorithm. Node labels can be found in Table 5.2. (a–c) Three most likely network structures obtained by the MCMC algorithm for three different datasets of common length. (d) Posterior probability over graphs, as computed by MCMC*

likelihood of the whole data for the most likely model. Figure 5.10 shows how the marginal likelihood of the data converges as the length of the dataset increases. The figure also indicates that, after 100 trials, the improvement of the likelihood of the data given more experiments is very small, since the model was already able to capture the correct relations.

### 5.5.1.3    Affordances conditional probability distributions

To ensure that the network actually captures the correct dependencies, we computed some illustrative distributions. The actual dependencies are encoded in the multinomial CPDs of each node and, as such, we cannot rely on the typical mean squared error to validate the fitting on the training set. Although there is no ground truth to compare the estimated network structure, we see that color has been detected as irrelevant when performing any action. Shape and size are important for grasp, tap and touch since they have an impact on the observed velocities and contact.

Figure 5.11(a) depicts the predicted contact duration for a grasp action on a ball of different sizes. It basically states that successful grasps (longer contact

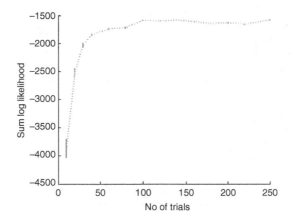

*Figure 5.10*     *Marginal likelihood of the data given the learned network structure as the number of trials included in the dataset increases. The vertical bars show the variance of the likelihood*

between the hand and the object) occur more often with smaller balls than with bigger ones. Figure 5.11(b) shows the distribution over ball sizes after a tap action for different observed velocities. According to these results, smaller balls move faster than bigger ones and medium ball velocities are somewhat unpredictable (all velocities have similar likelihood). This actually reflects the behavior of the objects during the trials. For example, the mean and variance of the ball velocity ($\mu$ [pixel/frame] and $\sigma^2$ [pixel$^2$/frame$^2$]) were (33.4, 172.3), (34.3, 524.9) and (17.5, 195.5) for a small, a medium and a big ball, respectively.

We can also verify if the robot is able to recognize the actions in the training set. To this purpose, we performed leave-one-out cross-validation. For each trial, we computed the network structure and parameters using the data from the remaining trials and the MCMC algorithm. We then estimated the probability of each action given the object features and the object velocity, hand velocity and object–hand velocity. Since contact is a proprioceptive measurement, it is not usually available when observing other actions. The most likely action was correctly identified in more than 85% of the tests. The errors were due mainly to the absence of contact information, which makes touching and tapping of boxes very similar from the point of view of observed effects. After including contact information, the ratio of correct recognition raised to 98%.

### 5.5.1.4  Summary

We have shown how the robot can tune its motor controllers through experimentation by including the effects of its actions. Once this information is available, the robot can start to establish relationships between the features of the objects and the effects of its actions. The learning of the affordance model depends on the motor and perceptual skills of the robot and was conducted in a completely unsupervised manner. There is no notion of success or failure and the network may not be able to distinguish between nonseparable objects,

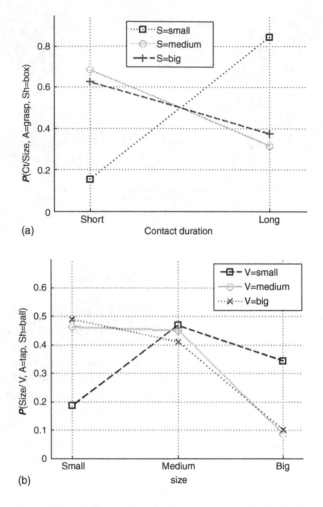

*Figure 5.11    Examples of CPDs for the learned network. (a) CPD of the contact duration when grasping a box, for different values of the size parameter: **P[Ct|S, A=grasp, Sh=box]**. (b) CPD of the size of a tapped ball, for different values of the observed velocity: **P[S|V, A=tap, Sh=ball]***

given the used descriptors. However, it is still able to construct a plausible model of the behavior of the different objects under different actions that can readily be used for prediction and planning.

### 5.5.2    Interaction games

Before delving in the problem of imitation learning, we present the results obtained in several simple interaction games using the learned affordance network. These results further validate the use of our proposed affordance model and illustrate the basic prediction, recognition and planning capabilities at the core of imitation learning.

We implemented a one-step emulation behavior. The robot observes a demonstrator perform an action on a given object. Then, using one of the functions described in Figure 5.1, it selects an action/object that is more likely to achieve the observed effect.

Figure 5.12 depicts the demonstration, the objects presented to the robot and the selected action/object for different situations. We used two different demonstrations, a tap on a small ball, resulting in high velocity and medium hand–object distance and a grasp on a small square, resulting in small velocity and small hand–object distance. Notice that contact information is not available when observing others. The goal of the robot is to replicate the observed effects.

The first situation (Figure 5.12a) is trivial, as only tap has a nonzero probability of producing high velocity. Hence, the imitation function selected a tap on the only object available. In Figure 5.12(b), the demonstrator performed the same action, but the robot had to decide between two different objects. Table 5.3 summarizes the probability of observing the desired effects given the six possible combinations of actions and objects. The robot selected the one with highest probability and thus performed a tap on the ball.

Figures 5.12(c) and 5.12(d) illustrate how the inclusion of the object features in the goal criteria may lead to different behaviors. After observing the grasp demonstration, the robot had to select among three objects: a big yellow ball, a small yellow ball and a small blue box. In the first case (Figure 5.12c), the objective was to replicate the same effects. The probability for each of the objects is 0.88, 0.92 and 0.52, respectively, and the robot grasped the small yellow ball despite the presence of the actual object used in the demonstration. Notice that this is not a failure, since it maximizes the probability of a successful grasp. This was the only requirement specified by the task goal. When the goal is modified to also include a similar shape, the robot successfully selects the blue box (Figure 5.12d). More details are provided in Reference 46.

In the continuation, we address the more complex scenario in which the robot must learn, by imitation, a full sequence of actions.

## 5.5.3 Imitation learning

To implement the imitation learning algorithm described in Section 5.3, we considered a simple recycling game in which the robot must separate different objects according to their shape (Figure 5.13). In front of the robot are two slots (Left and Right), where three types of objects can be placed: large balls, small balls and boxes. The boxes should be dropped in a corresponding container and the small balls should be tapped out of the table. The large balls should be touched upon, since the robot is not able to efficiently manipulate them. Every time a large ball is touched, it is removed from the table by an external agent. Therefore, the robot has a total of six possible actions available:

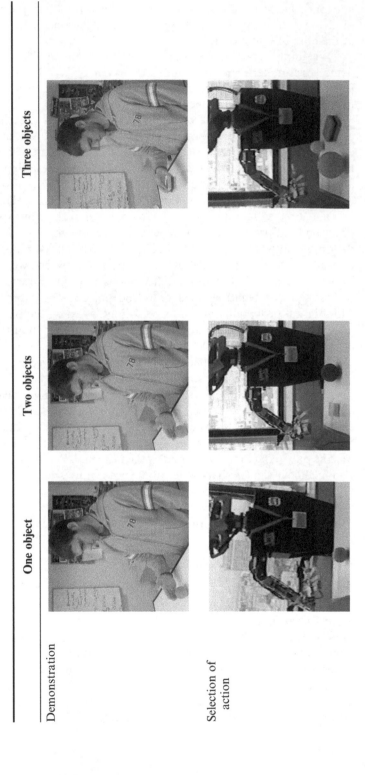

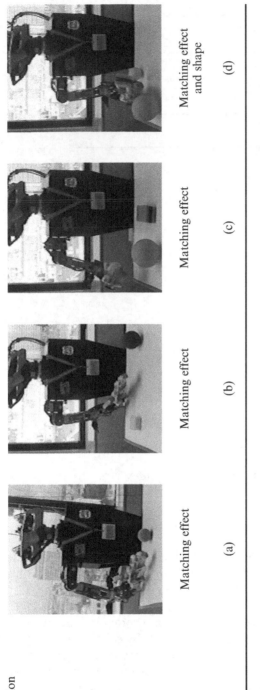

Imitation

Matching effect      Matching effect      Matching effect      Matching effect and shape

(a)      (b)      (c)      (d)

*Figure 5.12 Different imitation behaviors. (Top) Demonstration. (Middle) Set of possible objects. (Bottom) Imitation. (a–c) Imitation by matching the observed effect. (d) Imitation by matching the observed effect and shape*

*Table 5.3*   *Probability of achieving the desired effects for each possible action*
*and each object in Figure 12(b)*

| Object\Action | Grasp | Tap | Touch |
|---|---|---|---|
| Blue big ball | 0.00 | 0.20 | 0.00 |
| Yellow small box | 0.00 | 0.06 | 0.00 |

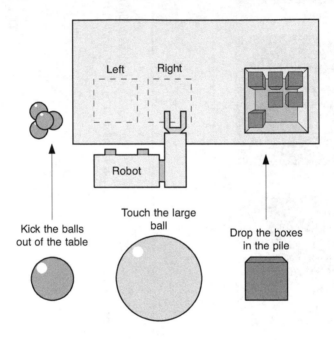

*Figure 5.13   A simple recycling game*

Touch Left (TcL), Touch Right (TcR), Tap Left (TpL), Tap Right (TpR),
Grasp Left (GrL) and Grasp Right (GrR).

To describe the process $\{X_t\}$ for the task at hand, we considered a state
space consisting of 17 possible states. Of these, 16 correspond to the possible
combinations of objects in the two slots (including empty slots). The 17th
state is an invalid state that accounts for the situations where the robot's
actions do not succeed. As described in Section 5.3, determining the dynamic
model consists of determining the transition matrix **P** by considering the possible effects of each action in each possible object. From the affordances in
Figure 5.9 the transition model for the actions on each object are shown in
Figure 5.14. Notice that, if the robot taps a ball on the right while an object
is lying on the left, the ball will most likely remain in the same spot. However,
since this behavior arises from the presence of two objects, *it is not captured in*

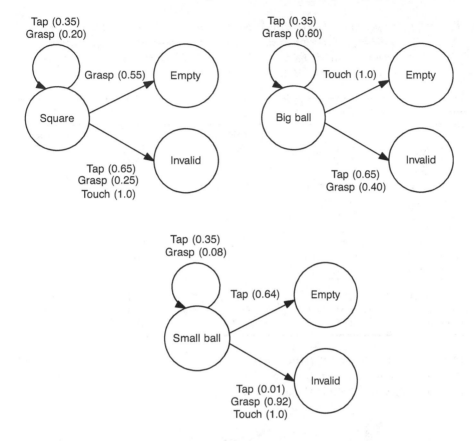

*Figure 5.14  Transition diagrams describing the transitions for each slot/object*

*the transition model* obtained from the affordances. This means that the transition model extracted from the affordances necessarily includes some inaccuracies.

To test the imitation, we provided the robot with an error-free demonstration of the desired behavior rule. As expected, the robot was successfully able to reconstruct the optimal policy. We also observed the learned behavior when the robot was provided with *two* different demonstrations, both optimal, as described in Table 5.4. Each state is represented as a pair $(S_1, S_2)$ where each state element can take one of the values 'Ball' (Big Ball), 'ball' (Small Ball), 'Box' (Box) or $\phi$ (empty). The second column of the table lists the observed actions for each state, and the third column lists the learned policy. Notice that, once again, the robot was able to reconstruct an optimal policy, by choosing one of the demonstrated actions in those states where different actions were possible.

In another experiment, we provided the robot with an *incomplete and inaccurate* demonstration. In particular, the action at state $(\phi, \text{Ball})$ was never demonstrated and the action at state (Ball, Ball) was *wrong*. Table 5.4 shows the demonstrated and learned policies. Notice that in this particular case the

*Table 5.4    Experiment 1: Error free demonstration (demonstrated and learned*
*policies). Experiment 2: Inaccurate, incomplete demonstration*
*(demonstrated and learned policies). The boxed elements in column*
*Demo2 correspond to the incomplete and inaccurate actions*

| State | Demo1 | Learned | Demo2 | Learned |
|-------|-------|---------|-------|---------|
| $(\phi,$ Ball) | TcR | TcR | ▭ | TcR |
| $(\phi,$ Box) | GrR | GrR | GrR | GrR |
| $(\phi,$ ball) | TpR | TpR | TpR | TpR |
| (Ball, $\phi$) | TcL | TcL | TcL | TcL |
| (Ball, Ball) | TcL, TcR | TcL, TcR | GrR | TcL |
| (Ball, Box) | TcL, GrR | GrR | TcL | TcL |
| (Ball, ball) | TcL | TcL | TcL | TcL |
| (Box, $\phi$) | GrL | GrL | GrL | GrL |
| (Box, Ball) | GrL, TcR | GrL | GrL | GrL |
| (Box, Box) | GrL, GrR | GrR | GrL | GrL |
| (Box, ball) | GrL | GrL | GrL | GrL |
| (ball, $\phi$) | TpL | TpL | TpL | TpL |
| (ball, ball) | TpL, TcR | TpL | TpL | TpL |
| (ball, Box) | TpL, GrR | GrR | TpL | TpL |
| (ball, ball) | TpL | TpL | TpL | TpL |

robot was able to recover the *correct policy*, even with an incomplete and
inaccurate demonstration. Figure 5.15 illustrates the execution of the optimal
learned policy for the initial state (Box, ball).[6]

*Figure 5.15    Execution of the learned policy in state (Box, ball)*

[6] For videos showing additional experiences see http://vislab.isr.ist.utl.pt/baltazar/demos/.

*Figure 5.15    Continued*

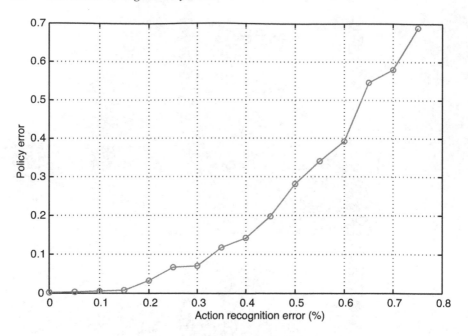

*Figure 5.16    Percentage of wrong actions in the learned policy as the action
recognition errors increase*

To assess the sensitivity of the imitation learning module to the action
recognition errors, we tested the learning algorithm for different error recognition rates. For each error rate, we ran 100 trials. Each trial consists of 45
state–action pairs, corresponding to three optimal policies. The obtained
results are depicted in Figure 5.16.

As expected, the error in the learned policy increases with the increase in
the number of wrongly interpreted actions. Notice, however, that for small
error rates (≤15%) the robot is still able to recover the demonstrated policy
with an error of only around 1%. In particular, if we consider the error rates of
the implemented action recognition method (between 10% and 15%), we
observe that the optimal policy is accurately recovered. This allows us to
conclude that action recognition using the affordances is sufficiently precise to
ensure the recovery of the demonstrated policy.

The accuracy of the recognition varied depending on the performed action
on the demonstrator and on the speed of execution, but for all actions
the recognition was successful with an error rate between 10% and 15%. The
errors in action recognition are justified by the different viewpoints during the
learning of the affordances and during the demonstration. During learning,
the robot observes the consequence of its own actions, while during recognition
the scene is observed from an external point-of-view. In terms of the image, this
difference in viewpoints translates in differences on the observed trajectories

and velocities, leading to some occasional misrecognition. Refer to Reference 28 for a more detailed discussion of this topic.

## 5.6 Conclusions

We have presented a computational model of affordances relying on BNs and, based on it, an imitation learning framework. On one hand, the model captures the relations between actions, object properties and the expected action outcomes. Using well-established learning and inference algorithms, a robot can learn these relations in a completely unsupervised way simply by interacting with the environment. Our results show how the learned network captures the structural dependencies between actions, object features and effects even in the presence of the perceptual and motor uncertainty inherent to real-world scenarios.

On the other hand, affordances provide the basic skills required for social interaction. The proposed imitation framework learns reward functions from task demonstrations. It uses the inference capabilities of the learned affordance network to recognize the demonstrated actions, predict the potential effects and plan accordingly. In this sense, imitation is not limited to mimicking the detailed human actions. Instead, it is used in a goal-directed manner (emulation), since the robot may choose a very different action when compared to that of the demonstrator, as long as its experience indicates that the desired effect can be met.

The model has a number of interesting properties that are worth pointing out: not only does it bridge the gap between sensory-motor loops and higher cognitive levels in humanoid robots but also allows us to possibly gain some insight concerning human cognition. One interesting remark is that, by using this approach, objects are represented by taking into account not only their visual attributes (e.g., shape) but also their 'behavior' when subject to actions. In other words, 'objecthood' is defined as a consequence of the robot's own actions (and embodiment) and the corresponding action outcomes. This concept of action-based object categorization is fundamental for planning but it is also relevant from the point of view of human perception and behavior.

A second point to notice in our affordance model is that the same basic knowledge and inference methods are used both for action selection and action recognition. In a sense, our model displays a 'mirror' structure as suggested by the existence of the so-called mirror neurons in the premotor cortex of macaque monkeys. The same neuronal structures simultaneously support action generation and recognition.

Finally, we have illustrated how the ability of the learned model to predict the effects of actions and recognize actions can be used to play simple interaction games and to implement learning by imitation in a robot.

We believe that modeling the interplay between actions, objects and actions outcomes is a powerful approach not only to develop goal-oriented behavior, action recognition, planning and execution in humanoid robots but also to shed some light into some of the fundamental mechanisms associated with human learning and cognition.

## Acknowledgements

This work was partially supported by the Information and Communications Technologies Institute, the Portuguese Fundação para a Ciência e Tecnologia, under the Carnegie Mellon-Portugal Program, the Programa Operacional Sociedade de Conhecimento (POS C) in the frame of QCA III and the PTDC/ EEA-ACR/70174/2006 project, and in part by the EUProjects (IST-004370) RobotCub and (EU-FP6-NEST-5010) Contact and (EU-FP7-231640) Handle.

Parts of this chapter were previously published in the IEEE Transactions on Robotics [46] and IEEE/RSF International Conference on Intelligent Robotic Systems [4].

## References

1.  J. J. Gibson, *The Ecological Approach to Visual Perception.* Boston: Houghton Mifflin, 1979.
2.  H. Ornkloo and C. von Hofsten, 'Fitting objects into holes: On the development of spatial cognition skills', *Developmental Psychology*, **43**(2): 404–416, 2007.
3.  S. Schaal, 'Is imitation learning the route to humanoid robots', *Trends in Cognitive Sciences*, **3**(6): 233–242, 1999.
4.  M. Lopes, F. S.Melo, and L.Montesano, 'Affordance-based imitation learning in robots', *in IEEE/RSJ Intelligent Robotic Systems (IROS'07)*, USA, 2007.
5.  A. Chemero, 'An outline of a theory of affordances', *Ecological Psychology*, **15**(2): 181–195, 2003.
6.  J. Konczak, H. Meeuwsen, and M. Cress, 'Changing affordances in stair climbing: The perception of maximum climbability in young and older adults', *Journal of Experimental Psychology: Human Perception & Performance*,vol. 19, pp. 691–697, 1992.
7.  E. Symes, R. Ellis, and M.Tucker, 'Visual object affordances: Object orientation', *Acta Psychologica*, **124**(2): 238–255, 2007.
8.  M. Turvey, K. Shockley, and C. Carello, 'Affordance, proper function, and the physical basis of perceived heaviness', *Cognition*, **73**(2): B17–B26, 1999.
9.  E. Oztop, N. Bradley, and M. Arlib, 'Infant grasp learning: A computational model', *Experimental Brain Research*, vol. 158, pp. 480–503, 2004.

10. M. Lopes and J. Santos-Victor, 'Visual learning by imitation with motor representations', *IEEE Transactions on Systems, Man, and Cybernetics – Part B: Cybernetics*, **35**(3), June 2005.
11. A. Slocum, D. Downey, and R. Beer, 'Further experiments in the evolution of minimally cognitive behavior: From perceiving affordances to selective attention', *in Conference on Simulation of Adaptive Behavior*, Paris, France, 2000.
12. P. Fitzpatrick, G. Metta, L. Natale, S. Rao, and G. Sandini, 'Learning about objects through action: Initial steps towards artificial cognition', *in IEEE International Conference on Robotics and Automation*, Taipei, Taiwan, 2003.
13. H. Kozima, C. Nakagawa, and H. Yano, 'Emergence of imitation mediated by objects', *in Second International Workshop on Epigenetic Robotics*, Edinburgh, Scotland, 2002.
14. I. Cos-Aguilera, L. Canamero, and G. Hayes, 'Using a SOFM to learn object affordances', *in Workshop of Physical Agents (WAF)*, Girona, Spain, 2004.
15. G. Fritz, L. Paletta, R. Breithaupt, E. Rome, and G. Dorffner, 'Learning predictive features in affordance based robotic perception systems', *in IEEE/RSJ International Conference on Intelligent Robots and Systems*, Beijing, China, 2006.
16. A. Stoytchev, 'Behavior-grounded representation of tool affordances', *in International Conference on Robotics and Automation*, Barcelona, Spain, 2005.
17. E. Sahin, M. Cakmak, M. Dogar, E. Ugur, and G. Ucoluk, 'To afford or not to afford: A new formalization of affordances towards affordance-based robot control', *Adaptive Behavior*, **124**(2): 238–255, 2007.
18. M. Dogar, M. Cakmak, E. Ugur, and E. Sahin, 'From primitive behaviors to goal-directed behavior using affordances', *in IEEE/RSJ International Conference on Intelligent Robots and Systems*, 2007.
19. A. Alissandrakis, C. L. Nehaniv, and K. Dautenhahn, 'Action, state and effect metrics for robot imitation', *in 15th IEEE International Symposium on Robot and Human Interactive Communication (RO-MAN 06)*, Hatfield, United Kingdom, 2006, pp. 232–237.
20. A. Billard, Y. Epars, S. Calinon, G. Cheng, and S. Schaal, 'Discovering optimal imitation strategies', *Robotics and Autonomous Systems*, vol. 47, pp. 2–3, 2004.
21. A. Murata, L. Fadiga, L. Fogassi, V. Gallese, V. Raos, and G. Rizzolatti, 'Object representation in the ventral premotor cortex (area f5) of the monkey', *Journal of Neurophysiology*, vol. 78, pp. 2226–2230, 1997.
22. L. Fadiga, L. Fogassi, G. Pavesi, and G. Rizzolatti, 'Motor facilitation during action observation: A magnetic stimulation study', *Journal of Neurophysiology*, **73**(73): 2608–2611, 1995.
23. V. Gallese, L. Fadiga, L. Fogassi, and G. Rizzolatti, 'Action recognition in the premotor cortex', *Brain*, vol. 119, pp. 593–609, 1996.

24. F. Melo, M. Lopes, J. Santos-Victor, and M. I. Ribeiro, 'A unified framework for imitation-like behaviors', *in 4th International Symposium in Imitation in Animals and Artifacts*, Newcastle, UK, April 2007.

25. A. Whiten, V. Horner, C. A. Litchfield, and S. Marshall-Pescini, 'How do apes ape?' *Learning & Behavior*, **32**(1): 36–52, 2004.

26. G. Gergely, H. Bekkering, and I. Kiraly, 'Rational imitation in preverbal infants', *Nature*, vol. 415, p. 755, Feb 2002.

27. H. Bekkering, A. Wohlschlager, and M. Gattis, 'Imitation of gestures in children is goal-directed'. *Quarterly Journal of Experimental Psychology*, vol. 53A, pp. 153–164, 2000.

28. M. Lopes and J. Santos-Victor, 'Visual transformations in gesture imitation: What you see is what you do', *in IEEE – International Conference on Robotics and Automation*, Taipei, Taiwan, 2003.

29. J. Weng, 'The developmental approach to intelligent robots', *in AAAI Spring Symposium Series, Integrating Robotic Research: Taking The Next Leap*, Stanford, USA, Mar 1998.

30. M. Lungarella, G. Metta, R. Pfeifer, and G. Sandini, 'Developmental robotics: A survey', *Connection Science*, **15**(40): 151–190, 2003.

31. E. Spelke, 'Core knowledge', *American Psychologist*, vol. 55, pp. 1233–1243, 2000.

32. M. Lopes and J. Santos-Victor, 'A developmental roadmap for learning by imitation in robots', *IEEE Transactions on Systems, Man, and Cybernetics – Part B: Cybernetics*, **37**(2), 2007.

33. J. Pearl, *Probabilistic Reasoning in Intelligent Systems: Networks of Plausible Inference*. San Francisco, CA: Morgan Kaufmann, 1988.

34. D. Ramachandran and E. Amir, 'Bayesian inverse reinforcement learning', *in 20th International Joint Conference on Artificial Intelligence*, India, 2007.

35. J. Pearl, *Causality: Models, Reasoning and Inference*. Cambridge: Cambridge University Press, 2000.

36. D. Heckerman, D. Geiger, and M. Chickering, 'Learning Bayesian networks: The combination of knowledge and statistical data', *Machine Learning*, **20**(3): 197–243, 1995.

37. G. Cooper and E. Herskovits, 'A Bayesian method for the induction of probabilistic networks from data', *Machine Learning*, **9**(4): 309–347, 1992.

38. D. Madigan and J. York, 'Bayesian graphical models for discrete data', *International Statistical Review*, vol. 63, pp. 215–232, 1995.

39. N. Friedman, 'The Bayesian structural EM algorithm', *in Uncertainty in Artificial Intelligence*, UAI, 1998.

40. D. Eaton and K. Murphy, 'Exact Bayesian structure learning from uncertain interventions', *in Proceedings of the Eleventh International Conference on Artificial Intelligence and Statistics*, San Juan, Puerto Rico, March 2007, vol. 2, pp. 107–114.

41. D. Heckerman, M. (eds.) 'A tutorial on learning with Bayesian networks', *Learning in Graphical Models*. Cambridge, MA: MIT Press, 1998.
42. C. Huang and A. Darwiche, 'Inference in belief networks: A procedural guide', *International Journal of Approximate Reasoning*, 15(3): 225–263, 1996.
43. M. Puterman, *Markov Decision Processes: Discrete Stochastic Dynamic Programming*. New York, NY, USA: John Wiley & Sons, Inc., 1994.
44. A. Y. Ng and S. J. Russel, 'Algorithms for inverse reinforcement learning', *in Proceedings of 17th International Conference on Machine Learning*, USA, 2000.
45. D. Pelleg and A. W. Moore, 'X-means: Extending k-means with efficient estimation of the number of clusters', *in International Conference on Machine Learning*, San Francisco, CA, 2000.
46. L. Montesano, M. Lopes, A. Bernardino, and J. Santos-Victor, 'Learning object affordances: From sensory-motor coordination to imitation', *IEEE Transactions on Robotics*, 28(1), February 2008.

## Chapter 6

# Helping robots imitate: metrics and technological solutions inspired by human behaviour

*Aris Alissandrakis, Nuno Otero, Joe Saunders,*
*Kerstin Dautenhahn and Chrystopher CL. Nehaniv*

## Abstract

In this chapter we describe three lines of research related to the issue of helping robots imitate people. These studies are based on observed human behaviour, technical metrics and implemented technical solutions. The three lines of research are: (a) a number of user studies that show how humans naturally tend to demonstrate a task for a robot to learn, (b) a formal approach to tackle the problem of what a robot should imitate and (c) a technology-driven conceptual framework and technique, inspired by social learning theories, that addresses how a robot can be taught. In this merging exercise we will try to propose a way through this problem space, towards the design of a human–robot interaction (HRI) system able to be taught by humans via demonstration.

## 6.1  Introduction

As robots move from their traditional use in industry into assisting and collaborating in public and domestic spaces, new paradigms need to be developed towards more 'natural' forms of human–robot interaction (HRI) avoiding the need, for example for technical or scientific expertise from the human.

Such robots can be expected to behave more like 'companions', i.e. (a) must be able to perform a range of useful tasks or functions and (b) must carry out tasks or functions in a manner that is socially acceptable, comfortable and effective for people sharing their environment and with whom they interact (Dautenhahn, 2007).

Social learning (in the sense of taking advantage of the skills available from the presence of other agents), compared to individual learning (which may be more prone to time consuming trial-and-error approaches), allows for task learning and skill transfer in more efficient and adaptive ways.

Imitation and social learning have been studied by psychologists and ethologists for over a century. Imitative learning of this type is defined and interpreted in many ways (Zentall and Galef, 1988; Zentall, 1996, 2001; Galef and Heyes, 1996); however, here we consider imitation in terms of the agent-based perspective (see Dautenhahn and Nehaniv, 2002), and within this focus we address mainly the question of *what* (information is transmitted by teaching), and to a lesser degree the *how* (to reproduce the task). The answers to the questions of *who* (to imitate), *when* (to imitate) and *how to evaluate the reproduction* are here assumed as given, or not directly addressed.

Imitation is a powerful learning tool when a number of agents interact in a social context. The demonstrator and imitator agents may or may not belong to the same class (different size – a parent teaching a child, different species – a human training an animal) or even be biological and artificial entities (e.g. in HRI). The latter is a very interesting paradigm explored in computer science and robotics, with researchers influenced by work on biology, ethology and psychology working to design mechanisms that would allow their robots to be programmed and learn more easily and efficiently (Demiris and Hayes, 2002; Breazeal and Scassellati, 2002; Billard, 2001; Nicolescu and Matarić, 2001; Schaal, 1999; Kuniyoshi *et al.*, 1994; Erlhagen *et al.*, 2005).

Concerning the *what* and the *how* issues within imitation, we consider three aspects: (a) the description and discussion of a number of user studies that show how humans naturally tend to demonstrate a task for a robot to learn (Section 6.2), (b) a formal approach to tackle the problem of what a robot should imitate (Section 6.3) and (c) a technology-driven conceptual framework and technique, inspired by social learning theories, that addresses how a robot can be taught (Section 6.4).

## 6.2    Investigating gestures in human to robot demonstration activities

In relation to the overall conceptualisation of HRIs and the demonstration/ learning of everyday routine home tasks by robots, we have been considering a dialogical perspective where the system and the human need to establish common ground regarding their understanding of the tasks and on-going surrounding activities (Otero *et al.*, 2008a–c; Saunders *et al.*, 2007b; Alissandrakis and Miyake, 2009a, b). For example, the robot needs to 'publicise' its abilities, in order to be able to solicit further enhancements to the demonstration based on the ambiguous events at a timely and apposite manner. These responses from the robot should be tailored according to the 'human perspective', so that the human is able to interpret them, consider the robot's difficulties and be, in turn, available to engage in further specification (or generation) of alternatives regarding the way an explanation unfolds. However, what are the specifics of this communication process? One important issue is the investigation of the gestures people display when interacting with robots. As a starting point we consider the following questions: How do humans use gestures in their communication processes? How should we proceed to find design solutions for robots that take advantage of these human abilities?

Robots may need to recognise human gestures and movements not only to infer limited intent regarding these, but also to communicate their own internal state and plans to humans (Nehaniv *et al.*, 2005). Nehaniv *et al.* (2005) address the specificities of analysing and incorporating gestures in the interaction loop between humans and robots and argue for the need to consider a broad definition of gestures including many effectors actions not necessarily associated with speech, the reason being to avoid: '... inherent limitations of approaches working with a too narrow notion of gesture, excluding entire classes of human gesture that should eventually be accessible to interactive robots able to function well in a human social environment.' (Nehaniv *et al.*, 2005, p. 372).

## 6.2.1 Human gestures and the development of computational artefacts

Within the COGNIRON project,[1] we considered two goals to set the course of a research line closely connected with the themes referred to earlier in this section: (a) we wanted to describe the frequency and duration of the different gestures people produce when performing routine home tasks, which in turn could be used to inform the development of appropriate algorithms allowing recognition of these gestures (Otero *et al.*, 2006a); (b) to understand how to conceptualise and consequently design the robot to facilitate HRIs, and as much as possible conform the robot to people's expectations and ideas of comfort and acceptance (Otero *et al.*, 2006b, 2008b).

Generally speaking, gestures in HRI are closely linked with any accompanying speech in terms of timing, meaning and communicative function (see, e.g. Cassell, 2000; Kendon, 1997; McNeill, 1992, 2005). Adam Kendon's and David McNeill's work on the study of human communicative gestures are considered to be landmarks in the field (Kendon, 1997; McNeill, 1992). Gestures can be classified into distinct types, and different classificatory systems do exist (a full review of the different classification systems is beyond the scope of this present work; but see, e.g. Cassell, 2000; Kendon, 1997; Kipp, 2004; McNeill, 1992, 2005). Research on the function of gestures has been

---

[1] See http://www.cogniron.org for an overview of the project's goals and achievements.

diverse. Gestures have been studied in relation to child development and language acquisition (Iverson and Goldin-Meadow, 2005; Ozcaliskan and Goldin-Meadow, 2005), or teaching and learning strategies (Kelly *et al.*, 2002; Roth and Lawless, 2002; Singer and Goldin-Meadow, 2005). Research in relation to problem solving has found not only that gestures can convey specific information and reveal thoughts not revealed in speech but also that observing gestures can be a useful extra to speech when trying to uncover cognitive processes (Alibali *et al.*, 1999; Garber and Goldin-Meadow, 2002). However, research investigating the function of gestures in relation to speech has produced contrary results or divergent opinions. For example some authors defend the importance of gesture semantics independent from speech (Cassell *et al.*, 1999; Kendon, 1997) while others hold that the primary function of gestures is not to convey semantic information (Krauss *et al.*, 1995). Lozano and Tversky (2006) argue that gestures might be of benefit to both communicators and recipients.

The seminal work by Bolt (1980) introduced the study of using gestures in the communication loop between humans and computer artefacts. A simplified overview of research on multimodal interfaces and HRI indicates the following approaches for the inclusion of gestures in the interaction process.

- Some research has explored the use of sets of pre-defined gestures and speech to communicate with computer artefacts (see, e.g. Ghidary *et al.*, 2002; Neal *et al.*, 1989; Oh *et al.*, 2005).
- Other researchers consider that the use of gestures for communication with computer artefacts should be explored beyond the confines of a set of pre-defined gestures (Cassell *et al.*, 1998, 2000, 2007a; Cassell and Thorisson, 1999; Kipp, 2004; Nehaniv *et al.*, 2005; Severinson-Eklundh *et al.*, 2003).

Conceptually speaking, our own research efforts take a practical stance in the sense that for some tasks a combination of the two approaches might be the way forward. Some pre-defined gestures, carefully chosen among the vast repertoire and considering its specificity, might be important to kick-start the intertwined dialogue between human and robot. In turn, the initial process might foster the understanding and acquisition of new gestures by the robot. Underlying these considerations is the assumption that on one hand people might want to teach the robot new abilities, but on the other hand they will probably feel frustrated if the robot does not exhibit useful features and communicative skills, especially when dealing with routine daily tasks (such as a domestic task at home).

### 6.2.2    *The approach followed in the user studies conducted*

We have been conducting a series of user studies to understand how people naturally demonstrate some 'basic' home tasks to a robot. Up to the present, our overall strategy in addressing this problem has taken a step-by-step approach: we chose to introduce small changes (reaction/actions of the robot towards the human demonstration) from one study to the next whilst maintaining the overall experimental instruction requests to the human participants. More specifically, in the user studies described here, we were able to highlight the nature of human demonstrations of a specific routine home task: how to lay a table. This particular task was chosen because, while relatively simple and familiar to most people, the number of tableware involved, their order and their possible configurations on the table can vary depending on personal preferences.

Due to technological issues (current state-of-the-art robots are not yet able to detect and understand *unconstrained* behaviour by humans), plus the fact that for our purposes here the robot does not need to detect or respond to the actual participant's behaviour (as responses were predetermined), in all following studies the robots were not autonomous, but controlled using the Wizard-of-Oz methodology (see Dahlbäck *et al.*, 1998).

In the first two studies, two experimental conditions were set up: the participants were asked to demonstrate how to lay the table using gestures only, or using gestures and speech (please see Otero *et al.*, 2008b, for a detailed description). The rationale concerning the two conditions for these two exploratory studies was related to our additional interest in investigating to what extent the constraint of not using speech would affect the production of gestures (following Lozano and Tversky, 2006). It could be the case that by restricting the speech people would make the gesturing more explicit, relative to the goals of the task. Both studies followed a within-subjects design, meaning that the same participants were used for the two conditions. The difference between these two studies lies in the role that the robot had to perform: in the first study the robot was a passive observer of the participants' demonstration while in the second study the robot would acknowledge the actions/demonstration of the participants.

In the third study (please see Otero *et al.*, 2008a, c), the general request to the participants was, again, similar to the previous studies – demonstrate to a robot how to lay a table. However, the robot's role changed. This time the robot not only acknowledged the actions/demonstration of the participant but, when the second object was placed on the table, the robot would state its misunderstanding (see Figure 6.1). Furthermore, we also introduced an additional task: immediately after demonstrating to the robot, the participants were asked to view their own demonstrations and try to segment them (identify in the video of their demonstrations breakpoints of the activity) into meaningful chunks.

Two follow-up studies were carried out in Japan. The first one of those (please see Alissandrakis and Miyake, 2009a) was intended as a verification and extension of the third study. Here, the participants were asked to lay the table twice, using either Japanese or non-Japanese ('western') utensils. This was carried out to contrast two versions of the task, both known, but one more frequently

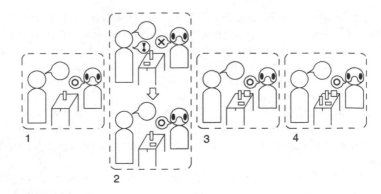

*Figure 6.1    Investigating the participant's acknowledgment and response to the robot's feedback. In the third and fourth studies, the robot would give positive feedback on the first, third and fourth instruction, but would initially state to not understand the second instruction (reverting back to positive if the participant clarified or repeated their demonstration). This also allowed to examine whether the participants remained consistent for the remaining third and fourth instructions, in the case that they changed their way of demonstrating. (Figure originally from Alissandrakis and Miyake (2009a))*

practiced. In addition, a humanoid robot (Wakamaru, Mitsubishi Heavy Industries) was used, rather than a robot with a mechanistic appearance (Peoplebot, MobileRobots Inc) that was used in all previous studies (Figure 6.2).

In the second of those studies (please see, Alissandrakis and Miyake, 2009b), the feedback from the robot was more elaborate; instead of the

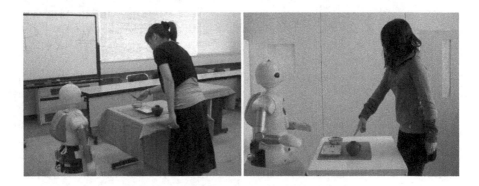

*Figure 6.2    Participants interacting with the Wakamaru robot. Pictures are from the fourth (left) and fifth (right) user-study. Both studies were conducted at Miyake Lab, Tokyo Institute of Technology, Japan*

previously used '*I do* (or *sorry, I do not*) understand [your demonstration]', the robot would in turn expound what the participant just demonstrated, using speech and gestures, in either absolute or relative terms regarding the object localisation. This explanation would be always correct regarding identifying the object, but either factually correct (in the sense of not necessarily using the same references that the participant used) or wrong, regarding the object localisation (Table 6.1).

*Table 6.1   Quick reference table for the user studies presented in Section 6.2*

| No. | Brief summary | Research issues | Related publications |
|-----|---------------|-----------------|----------------------|
| 1 | Participants instructed to use gestures only, or gestures and speech to demonstrate; robot was passive observer | Q1, Q2 | Otero *et al.* (2008b) |
| 2 | Participants instructed to use gestures only, or gestures and speech to demonstrate; robot would acknowledge the demonstration | Q1, Q2 | Otero *et al.* (2008b) |
| 3 | Participants relatively unrestrained on how to demonstrate; robot would state to understand or not the demonstration | Q3, Q4, Q5 | Otero *et al.* (2008a, c) |
| 4 | Participants relatively unrestrained on how to demonstrate; robot would state to understand or not the demonstration; two versions of the task; humanoid instead of a mechanoid robot | Q3, Q5, Q6 | Alissandrakis and Miyake (2009a) |
| 5 | Participants relatively unrestrained how to demonstrate; robot gave detailed feedback on understanding or not of demonstration, regarding object localisation; humanoid instead of a mechanoid robot | Q3, Q5, Q7 | Alissandrakis and Miyake (2009b) |

Summarising, the following set of questions gives a picture of the overall research issues being pursued in the user studies.

- *Q1*. Can we describe in a useful and efficient way the frequency and duration of different types of gestures when people are asked to explain to a robot how to perform a specific routine home task (e.g. lay a table)?
- *Q2*. Are there any differences regarding the frequency and duration of the different types of gestures produced when people are asked to use gestures only or are allowed to gesture and speak to explain this routine

home task? Can the findings be used to engender new ways of considering HRIs?

- *Q3.* Do participants clearly acknowledge the robot and change their explanations/demonstrations when the robot declares its inability to understand the participant's actions? What is the nature of this change (if any)? To what extent do people maintain the changes for the remainder of their explanation/demonstrations?
- *Q4.* What are the key moments that the participants consider to be dividing points of their own demonstrations? Does the feedback from the robot stating its misunderstanding at specific points of the demonstration alter the level of detail of the segmentation and corresponding demonstration effort?
- *Q5.* Do participants report difficulties regarding specific events on their demonstrations? Do participants consider to have been influenced by the robot's feedback?
- *Q6.* Does the level of familiarity with the task influence decisively the nature of the demonstrations produced? If yes, in what ways?
- *Q7.* How and to what extent does more elaborate feedback from the robot regarding the object localisation facilitate the interaction?

## 6.2.3   General discussion of the results from the user studies

From the two initial studies (Otero *et al.*, 2008b), we were able to reflect on the need to produce a coding scheme for the analysis of gestures in HRIs that could be readily and efficiently used to inform the development of appropriate gesture/activity recognition algorithms.

The conceptual framework presented in Nehaniv *et al.* (2005) was adapted to capture requirements for *contextual interpretation* of body postures and human activities for purposes of HRI. It defines five functional classes of gestures, which are as follows:

1. *Manipulative gestures.* These are gestures that involve the displacement of objects (e.g. picking a cup) or miming such displacements.
2. *Symbolic gestures.* These are gestures that follow a conventionalised signal. Their recognition is highly dependent on the context, both current task and cultural milieu (e.g. the thumbs-up or thumb-index finger ring to convey 'OK').
3. *Interactional gestures.* This category classifies gestures used to regulate interaction with a partner. These can be used to initiate, maintain, invite, synchronise, organise or terminate an interactional behaviour between agents (e.g. head nodding, hand gestures to encourage the communicator to continue).
4. *Referencing/pointing gestures* (Deictics). The gestures that fall into this category are gestures used to indicate objects or loci of interest.
   - *Side effects of expressive behaviour.* These are gestures that occur as side effects of people's communicative behaviour. They can be motion with

hands, arms, face, etc., but without specific interactive, communicative, symbolic or referential roles.

5. *Irrelevant gestures.* These are gestures that do not have a primary communicative or interactive function, for example adjusting one's hair or rubbing the eye.

It is important to note that certain gestures in particular situations might be multipurpose. Nehaniv *et al.* (2005) stress the importance of knowing the context in which gestures are produced since it is crucial to disambiguate their meaning. In practice, data on the interaction history and context may help the classification process. This framework is intended to compliment existing and more detailed speech-focused classification systems (see, e.g. Cassell, 2000; Cassell *et al.*, 2007b; McNeill, 1992).

The concrete utilisation within the COGNIRON project of the conceptual framework for the analysis of gestures referred to earlier in this section and its practical instantiation as a coding scheme taught us that one important challenge is to adapt the level of granularity of the coding scheme to the sensory capability of the robot, without losing the meaning of the gestures produced. Further research is needed to produce some guidelines on how to do this adaptation from a practical point of view.

In relation to the actual gestural activities observed in the user studies, one relevant issue that clearly emerged from the analysis of the data collected in these two initial studies was the low frequency of any deictic (pointing) and symbolic gestures. In fact, the expectation was that the constraint of not being allowed to use speech would make people resort to pointing (e.g. to indicate the objects and corresponding locations) and use symbolic gestures to supplement their manipulative gestures (e.g. to mark the different steps of their explanations). The unexpected result may be due to the fact that people, when performing routine daily tasks, are not naturally likely to give detailed accounts of the way in which these tasks should be performed beyond the actual simple demonstration of how to accomplish it. Our studies also clearly suggest an interaction between the type of task and the type of gestures produced. This point stresses the importance of knowing the context in which gestures are produced and the interaction history (Nehaniv *et al.*, 2005). Comparing the results from both studies, what seems more salient is the willingness of people to engage in interactions with the robot even if the trigger for such is just a 'small' acknowledgement of their actions. Moreover, people preferred the gestures and speech method of demonstrating. This choice is not surprising but it definitely supports the perspective that people might prefer to interact with robots in a 'natural' way (Dautenhahn, 1998). The design challenge for current robotic systems is to able to provide a fine balance between natural ways of interacting and the correct level of expectations generated in the final users concerning the system's capabilities.

In relation to the third study referred to previously, the actual demonstrations recorded suggest that people tried to change their demonstrations when faced with the robot's misunderstanding but the alternatives were (sometimes) not any more informative than the original ones. People need more feedback about what to address specifically. In fact, we could observe some attempts to actively engage with the robot in a more complex interaction, i.e. participants seeming to probe the robot's abilities. Furthermore, people are not particularly consistent throughout their demonstration or modifications to it when faced with negative feedback.

Considering this specific issue on a related note, it is worth mentioning that Thomaz and Breazeal (2008) make some observations about the way people tend to administer their own feedback when teaching a Reinforcement Learning agent:

1. they use the reward channel not only for feedback, but also for future-directed guidance;
2. they have a positive bias to their feedback, possibly using the signal as a motivational channel; and
3. they change their behaviour as they develop a mental model of the robotic learner.

These points about people's feedback are outside the current scope of the work presented here (focusing more on the robot's feedback), but very much of interest in the broad context.

The results regarding the participants' segmentations of their own demonstrations of the home task suggest that people might differ regarding the level of detail they spontaneously consider. To what extent this result also actually implies different levels of detail in the explanation itself is still an on-going research question. However, it is worth noting that all the participants considered breakpoints corresponding to the end of a demonstration step (marked by the positive feedback from the robot). This suggests that these breakpoints mark the higher level of detail that the participants spontaneously considered. More surprisingly, the results indicate that participants did not segment the task differently when analysing the moment when the robot gave negative feedback, although in the post-session questionnaire the feedback was considered important for the segmentation.

The participants of the fourth study also acknowledged the robot's mis-understanding of their instructions, and responded by repeating their demonstration, modifying it to some extent. Together with very infrequent deictic gestures, as previously observed, this clarification did not vary significantly from the initial instruction (e.g. most simply used a louder voice, as they thought the robot simply did not hear them). However, in the few cases where they actually modified their object localisation (from absolute to relative reference and vice versa) to resolve the misunderstanding, they kept the same reference style for the remainder of that task; this consistency was not observed in the previous third study. Receiving only limited feedback from the robot (regarding its understanding of the demonstration or not), in general the participants used a mix of localisation reference styles throughout their two tasks, which would make the detection and classification of their demonstrations by the robot a rather hard problem. Any influence by the task familiarity (Japanese and non-Japanese style) was not observed in this particular study in terms of frequency of gestures, or object localisation reference style; however, it was noted that participants would comment on the particular function of the objects only for the Japanese utensils (e.g. 'This is a soup bowl [. . .] it is used to eat *miso* soup').[2]

In the fifth study, the name of the object (always correct) and its placement (sometimes wrong, both in terms of pointing to the table and using either absolute or relative referencing) were part of the feedback, effectively allowing the robot to 'publicise' its abilities of understanding the participant's demonstration. At this time, only initial trial runs have been conducted, but they seem to indicate that the participants were favourable to this more detailed robot feedback, which in some cases made them adapt their own demonstrations accordingly. The object localisation is seen (from the 'system perspective') as an aspect of the participant's demonstration, related to the choice of effect metric(s) of similarity for imitation, which should be communicated from the robot to the participant, in such a way as to be easily interpreted in terms of the 'human perspective'.

Summarising, the user studies reported in this section highlighted some methodological, conceptual and practical challenges. More research is clearly needed to provide guidelines, efficient methods and tools that can be easily utilised by practitioners when facing the need to conduct observational studies. Furthermore, the research community probably needs to discern a useful dissemination strategy for the results from these type of observational studies to build up from past experiences. The analysis of observational studies is costly and sharing insights might be crucial to foster re-use of solutions and creativity for alternatives. In relation to the results themselves, our work suggests that people show a wide variety of behaviours/actions/gestures when faced with the task of demonstrating to a robot. The dialogical perspective that we are advocating seems to be the way forward but the details of the whole process are still very open. Nevertheless, it appears that the initial publicising by the robot of its basic skills and abilities, might facilitate the human user to adapt their own efforts. In turn, the robot in order to build up its abilities should effectively use this human effort.

Section 6.3 presents a formal approach to what the robot should imitate, including discussion on the problem of the choice among possible metrics, regarding effects, but also regarding actions and states.

---

[2] Also, note that different participants assigned a different function to the same object, e.g. the same bowl was both identified as a 'rice bowl' and as a 'soup bowl'.

## 6.3     A formal approach to what the robot should imitate

Alissandrakis' work on the correspondence problem (see Nehaniv and Dautenhahn, 2002) and metrics for the robot's successful imitation of actions, states and effects (Alissandrakis *et al.*, 2006, 2007) establishes an initial formal stance (and partial solution) for the possibility of HRIs within the context of human teaching of robots. The correspondence problem effectively states that there can be multiple definitions and interpretations of what is 'similar' when trying to reproduce a behaviour, along the aspects of actions (e.g. body motions), states (e.g. body postures) and effects (e.g. positions, orientations and states of external objects, but also changes to the body–world relationship of the agent). There is essentially not a single 'goal', and the agent has to be able to decide, among the other four questions mentioned in the introduction section, which aspects to imitate. In an analogous fashion, an imitator/student has to be able to determine what aspect of the model/teacher's demonstration is important (and therefore constitutes the task knowledge).

In the table setting task used in Section 6.2, the most important aspect would be the effects, omitting in this case the teacher's and student's particular embodiments (which influence more the consideration of action and state metrics). Even so, there are a number of possible effect metrics, taking into account (among other features) the displacement, orientation and absolute or relative reference frame. The demonstrated task knowledge could be considered as any combination of these sub-aspects. However, if the initial configuration of the objects involved is dissimilar between the demonstration and the reproduction, different choices can lead to *qualitatively* dissimilar results.[3] In the context of task knowledge transfer, these possibly undesired results need to be avoided by having the teacher provide (possibly multiple) examples that satisfy only a small set of metrics, which in turn defines what is important.

### 6.3.1     The correspondence problem in imitation

A fundamental problem when learning how to imitate is to create an appropriate (partial) mapping between the actions afforded by particular embodiments to achieve corresponding states and effects by the model and imitator agents (solving a correspondence problem) (Nehaniv and Dautenhahn, 1998).

The solutions to the correspondence problem will depend on the sub-goal granularity and the metrics used to evaluate the similarity between actions, states and/or effects, resulting in qualitatively different imitative behaviours (Alissandrakis *et al.*, 2002, 2004). The related problem of *what* to imitate addresses the choice of metrics and sub-goal granularity that should be used for imitating, depending on the context. (See Billard *et al.* (2004); Scassellati

---

[3] The issue here is that the chosen metrics may well be minimised, therefore resulting *quantitatively* in a successful imitation; however, this choice of metrics by the system might not be the same as what the human intended while demonstrating their desired way to achieve the task.

(1999) for robotic examples and Butterworth (2003); Call and Carpenter (2002); Bekkering and Prinz (2002) for ethological and psychological aspects.)

For similar embodiments, addressing the correspondence problem seems to be straightforward (although it actually involves deep issues of perception and motor control). But once the assumption that the agents belong to the same 'species', i.e. have sufficiently similar bodies and an equivalent set of actions, is dropped, as with a robot imitating a human, the problem becomes more difficult and complex. Even among biological agents, individual differences in issues of perception, anatomy, neurophysiology and ontogeny can create effectively dissimilar embodiments between members of the same species. A close inspection of seemingly similar artificial agent embodiments can yield similar conclusions due to issues such as individual sensor and actuator differences (hardware) or the particular representations and processing that these agents employ (software). In HRI, it is desirable to have different kinds of agents in the learning process, i.e. humans and robots interacting socially.

The following statement of the *correspondence problem* (Nehaniv and Dautenhahn, 2000–2002) draws attention to the fact that the model and imitator agents may not necessarily share the same morphology or may not have the same affordances:

> *Given an observed behaviour of the model, which from a given starting state leads the model through a sequence (or hierarchy [or program]) of sub-goals in states, action and/or effects, one must find and execute a sequence of actions using one's own (possibly dissimilar) embodiment, which from a corresponding starting state, leads through corresponding sub-goals – in corresponding states, actions, and/or effects, while possibly responding to corresponding events.*

In this approach, an imitator can map observed actions of the model agent to its own repertoire of actions using the correspondence found by solving the correspondence problem, as constrained by its own embodiment and by context (Nehaniv and Dautenhahn, 2000–2002). Qualitatively different kinds of social learning result from matching different combinations of matching actions, states and/or effects at different levels of granularity (Nehaniv, 2003).

Artificial agents that have the ability to imitate may use (perhaps more than one) metric to compare the imitator agent's own actions, states and effects with the model's actions, states and effects, to evaluate the imitation attempts and discover corresponding actions that they can perform to achieve a similar behaviour. The choice of metrics used is therefore very important, as it will have an impact on the quality and character of the imitation. Many interesting and important aspects of the model behaviour need to be considered, as the metrics capture the notion of the salient differences between performed and desired actions and also the difference between attained and desired states and effects (Nehaniv and Dautenhahn, 2001, 2002). The choice of metric determines, in part, *what* will be imitated, whereas solving the correspondence problem concerns *how* to imitate (Dautenhahn and Nehaniv, 2002). In general, aspects of action, state and effect as well as the level of granularity (what to imitate) do all play roles in the choice of metric for solving the problem of how to imitate (Nehaniv and Dautenhahn, 2001; Alissandrakis *et al.*, 2002; Billard *et al.*, 2004). On-going research is thus addressing the complementary problem of how to *extract sub-goals* and *derive suitable metrics* automatically from observation (Nehaniv and Dautenhahn, 2001; Nehaniv, 2003; Billard *et al.*, 2004; Calinon and Billard, 2004).

Related to work on solving the correspondence problem for imitation learning in robotics is the ALICE generic imitation framework (Alissandrakis *et al.*, 2002, 2004). The ALICE framework builds up a library of actions from the repertoire of an imitator agent that can be executed to achieve corresponding actions, states and effects to those of a model agent (according to given metrics and granularity); it provides a functional architecture that informs the design of robotic systems that can learn socially from a human demonstrator.

## 6.3.2   Action and state metrics

This subsection presents a novel generic approach to the correspondence problem, via body-mapping for the cases of state and/or action matching. In particular, partial, relative and mirror matching all arise as special cases of such correspondence mappings. Moreover, an infinite set of metrics (parameterised by correspondence matrices) for imitation performance are induced via such body correspondences.[4]

### 6.3.2.1   Different bodies

Different agent bodies can be described as simplified kinematic models, comprising of a rooted acyclic connected graph of *segments* (see Figures 6.3–6.7).[5] Each segment has a base and a tip end, and is described by the quintuple $(i, l_i, p_i, \theta_i, \phi_i)$, where:

- $i$ is the index number of the segment,
- $l_i$ is segment length,
- $p_i$ is the index number of the *parent* segment,
- and $\theta_i$ and $\phi_i$ are the *azimuth* and *polar* angles for the spherical coordinates $(l_i, \theta_i, \phi_i)$ that indicate how the segment is positioned in 3D space (relative to the end of its parent segment). *Note*: In general the range of the angles $\theta_i$ and $\phi_i$ may be constrained within given respective ranges.

---

[4] That is to say that a correspondence mapping 'induces' a metric means exactly that it mathematically determines the metric.

[5] See modelling agents as simple open kinematic chains in (Amit and Matarić, 2004).

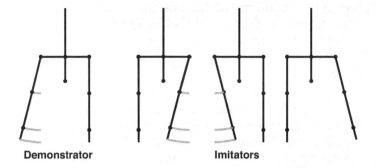

**Demonstrator**                              **Imitators**

*Figure 6.3   Examples of symmetry via correspondence mapping. The figure
shows a demonstrator (left) and three imitators, facing the reader,
each with an upper human torso embodiment. The demonstrator is
moving its right arm to its left. Each of the three imitators is using
different correspondence mappings: mapping the demonstrator's
right arm to the left arm of the imitator (second from the left), using
a weight of minus one, but maintaining the same arm mapping
(second from the right) and finally both mapping the demonstrator's
right arm to the left arm of the imitator and using a weight of minus
one (right). The grey lines trace the movement of the arms*

The values of $\theta_i$ and $\phi_i$ are *relative* for each segment, but *absolute* angles
for segment $i$, $\Theta_i$ and $\Phi_i$, can be obtained inductively starting from the segment
after the root:[6]

$$\Theta_i = \theta_i + \Theta_{p_i}$$
$$\Phi_i = \phi_i + \Phi_{p_i}$$

The *state* of such a kinematic model can be defined as the vector containing the
values of the degrees of freedom (DOF), i.e. the values of the azimuth and
polar angles for each segment in the acyclic graph.

Depending on whether the relative or the absolute angle values are used,
for an embodiment with $n$ segments, two different state vectors can be considered:

$$S_{\text{relative}} = [\theta_1 \, \phi_1 \, \theta_2 \, \phi_2 \cdots \theta_n \, \phi_n]$$
$$S_{\text{absolute}} = [\Theta_1 \, \Phi_1 \, \Theta_2 \, \Phi_2 \cdots \Theta_n \, \Phi_n]$$

Here, complying with the particular kinematic models that will be used, the
state vector is composed of the alternating azimuth and polar angles for each of

---

[6] For mathematical convenience, the root node is treated as a segment of length $l_0 = 0$, but $\theta_0$ and $\phi_0$
can have non-zero values, to orient the entire model. For expository purposes, without loss of
generality, in this chapter we ignore the latter possibilities ($\Theta_0 = \theta_0 = 0$, $\Phi_0 = \phi_0 = 0$).

the body segments.[7] For the rest of the subsection, the notation $S_j$ will be used to refer to the state value of the $j^{\text{th}}$ DOF of an agent.[8]

An *action* can be defined as the difference between two consecutive state vectors $S$ and $S'$:

$$A = S' - S$$

Using either the relative or absolute representation of state vectors for calculating an action vector produces mathematically equivalent results. Note that depending on the embodiment, a change in the relative values of the $j^{\text{th}}$ DOF can influence the absolute values of subsequent DOFs. For this subsection, the actions will be defined using the relative state vectors.

### 6.3.2.2   Some first state and action metrics

To evaluate the similarity of behaviour, with respect to states and actions, between an agent $\beta$ imitating another agent $\alpha$, we define and use appropriate *metrics*. For the moment let us assume that both agent embodiments have the same number of DOFs, $n$.

A first global *state metric* can be defined as

$$\mu_{\text{state}}(S^{\alpha}, S^{\beta}) = \sum_{j=1}^{n} |S_j^{\alpha} - S_j^{\beta}|, \tag{6.1}$$

where $S_j^{\alpha}$ *and* $S_j^{\beta}$ are the values of the state vectors for the two agents. Depending on whether the relative or absolute state vectors are used, we call the state metric *relative or absolute,* respectively.

A first global *action metric* can be defined as

$$\mu_{\text{action}}(A^{\alpha}, A^{\beta}) = \sum_{j=1}^{n} |A_j^{\alpha} - A_j^{\beta}|, \tag{6.2}$$

where $A_j^{\alpha}$ and $A_j^{\beta}$ are the values of the action vectors for the two agents. Note that instead of absolute value, one could alternatively use any $L^p$-norm, the choice of which might have consequences for optimisation in different applications.

An agent performing actions so as to minimise one (or a weighted combination) of these two metrics would successfully imitate a demonstrator in respect to states and/or actions. Note that it is not necessary, and in general will not be possible, to bring the value of the metric to zero with a matching

---

[7] Depending on the particular body representations used, the contents of the vectors (and the ordering of the elements) can of course vary. For example, if Euler angles ($\phi$, $\theta$, $\psi$) were used instead of spherical coordinates, the state vector for an embodiment with $n$ segments could be defined as $S = [\phi_1\ \theta_1\ \psi_1 \phi_2\ \theta_2\ \psi_2 \ldots \phi_n\ \theta_n\ \psi_n]$.

[8] Here, the number of DOFs for each agent will in general be twice the number of segments, depending on embodiment restrictions. For example for $S = [\theta_1\ \phi_1\ \theta_2\ \phi_2]$, $S_3 = \theta_2$.

behaviour, especially in the case of dissimilar embodiments. Instead, finding minima is the goal.[9]

### 6.3.2.3   Correspondence mapping

For two agents, demonstrator $\alpha$ and imitator $\beta$ with $n$ and $m$ DOFs, respectively, an $n \times m$ *correspondence matrix* can be defined as

$$\mathscr{C} = \begin{bmatrix} w_{1,1} & w_{1,2} & \cdots & w_{1,m} \\ w_{2,1} & w_{2,2} & \cdots & w_{2,m} \\ \vdots & \vdots & \ddots & \vdots \\ w_{n,1} & w_{n,2} & \cdots & w_{n,m} \end{bmatrix}$$

where the $w_{i,j}$ values are real-valued weights, determining how the $j^{th}$ DOF of the imitator $\beta$ depends on the $i^{th}$ DOF of the demonstrator $\alpha$. The $j^{th}$ column of the matrix can be thought as a vector indicating how the DOFs of the demonstrator influence the $j^{th}$ DOF of the imitator. Depending on how many of the weights have a non-zero value, this correspondence mapping can be *one-to-one*, *one-to-many* (or *many-to-one*) or *many-to-many*. If *partial* body imitation is desired, some DOF of the imitator (and/or the demonstrator) can be omitted by setting an entire column (respective row) to zero in the correspondence matrix.

The choice of the correspondence mapping will in general depend on the particular task. Assuming both agents share the same embodiment (and as a result have the same number of DOFs), a simple example of a one-to-one correspondence mapping would be using the identity matrix as a correspondence matrix. Alternatively, if some *mirror symmetry* is wanted, then the DOFs for the right arm and leg of the demonstrator (see example in Figure 6.3, left) could be mapped to the DOFs for the left arm and leg of the imitator, and vice versa (Figure 6.3, second from the left). Another possible form of symmetry results from mapping some of the demonstrator's DOF using a weight of minus one (e.g. if the demonstrator raises its hand, the imitator should lower its hand, or if the demonstrator turns its head to the left the imitator should turn to the right, see example in Figure 6.3, second from right).

If the agents do not have the same number of DOFs (or depending on their particular morphology), it may be useful to map a single DOF to many DOFs. For example consider correspondences between a human body as model to a dolphin-like imitator: a dolphin using its mouth corresponding to either human hand (grasping an object), or using its tail to toss a ball back to a human that used both arms comprise real-world examples of many-to-one mappings.[10] These two examples also illustrate that the correspondence need not be static – the human hand(s) are mapped to different dolphin body parts in each case – but can be adapted depending on the context and the tasks involved.

---

[9]   Of course, one can replace $\mu$ by $\mu' = \mu - m$, where $m = \inf \mu$ and then seek to solve $\mu' = 0$.
[10]  Different mappings do appear to be employed by real-life dolphins in imitating humans (Herman, 2002).

### 6.3.2.4   Induced state and action metrics

The metric definitions in Section 6.3.2.2 are appropriate for the most simple one-to-one mapping (the identity mapping), with both agents sharing the same number of DOFs (and probably a very similar morphology). But in general, using a correspondence matrix, other metric definitions can be induced.

First, the state vector $S^\alpha$ and the action vector $A^\alpha$ of the demonstrator can be multiplied with the correspondence matrix

$$\mathscr{S} = S^\alpha \times \mathscr{C} \tag{6.3}$$

$$\mathscr{A} = A^\alpha \times \mathscr{C} \tag{6.4}$$

producing two new vectors in imitator coordinates.

Combining (6.1) and (6.3) for the state metric gives

$$\mu_{\text{state}}^{\mathscr{C}}(\mathscr{S}, S^\beta) = \sum_{j=1}^{m} |\mathscr{S}_i - S_j^\beta| \varepsilon_j, \tag{6.5}$$

where $S^\beta$ is the imitator's attempted matching state and the corrective term

$$\varepsilon_j = \begin{cases} 0, & \text{if } \sum_{i=1}^{n} w_{i,j}^2 = 0 \\ 1, & \text{otherwise} \end{cases}, \tag{6.6}$$

takes the value zero if the $j^{\text{th}}$ column of the correspondence matrix contains only zeros (effectively omitting the imitator's $j^{\text{th}}$ DOF). Intuitively, the components of $\mathscr{S}$ and $\mathscr{A}$ (for such $\epsilon_j \neq 0$) can be thought as current sub-goal state and action target values. The imitator can match the state $\mathscr{S}$ by assuming state $S^\beta$ so as to minimise the metric $\mu_{\text{state}}^{\mathscr{C}}$. As in the previous definition, this state metric is called *relative* or *absolute* depending on whether the relative or absolute state vectors are used, respectively.

Finally, combining (6.2), (6.4) and (6.6) for the action metric gives:

$$\mu_{\text{action}}^{\mathscr{C}}(\mathscr{A}, A^\beta) = \sum_{j=1}^{m} |\mathscr{A}_j - A_j^\beta| \varepsilon_j, \tag{6.7}$$

where $A^\beta$ is the imitator's attempted matching action.

These $\mu_{\text{state}}^{\mathscr{C}}$ and $\mu_{\text{action}}^{\mathscr{C}}$ metrics are called the *induced state* and *action metrics for the linear correspondence* $\mathscr{C}$.

Depending on the correspondence mapping used, a plethora of new complex metrics (also allowing for dissimilar embodiments) can be induced considering state or action aspects.

### 6.3.2.5  Mapping across dissimilar bodies

For a given demonstrator and imitator embodiment pair, the imitator attempts to match the behaviour of the demonstrator by minimising a given metric (or a combination of metrics). This can be done continuously (*immediate imitation*) or after the completion of the demonstration (*deferred imitation*) (Nadel *et al.*, 1999). Moreover, the *granularity* or 'fineness' of the matching of actions, states and/or effects determines a sequence of sub-goals for the imitator to achieve, and the appropriate level of granularity may be different depending on context and task. Different correspondence mappings can be defined between the two agents, yielding qualitatively different types of matching behaviours.

In Alissandrakis *et al.* (2007) a series of simulation runs were conducted, using a variety of agent embodiments and correspondence mappings. The demonstrator performs a series of actions and the imitator tries to minimise the correspondence induced *relative state* metric. Continuously using the components of $\mathscr{S}$, for which $\epsilon_j \neq 0$, as the current sub-goals for each DOF $j$, the imitator performs actions which attempt to reduce the contribution of error in each such component. Here, the rate of change was restricted to half the component-wise error. Of course, many other selection mechanisms are possible for both immediate or deferred imitation.

*Identity and mirror symmetry mappings*

Two examples of imitation across similar embodiments are shown in Figure 6.4. Both demonstrator and imitator are humanoid. In the first example the identity correspondence mapping is used. In the second example, using the same

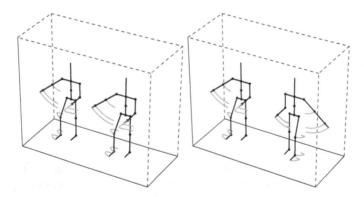

*Figure 6.4*    *Two examples of imitation across similar embodiments (humanoid).*
*Both demonstrator (left in both examples) and imitator (right*
*in both examples) share the same humanoid embodiment. In*
*the example on the left, the identity mapping is used as the*
*correspondence mapping. In the example on the right, the left arm and*
*leg of the demonstrator are mapped on the right arm and leg of the*
*imitator (and vice versa) with a weight of minus one, resulting in a*
*mirror symmetry. The grey traces visualise the body part trajectories*

demonstration, symmetry is achieved by mapping the left body parts of the demonstrator to the right body parts of the imitator and vice versa (see also examples in Figure 6.3).

### Multiple mappings between dissimilar bodies

The model of an AIBO robot is used as an imitator in the examples shown in Figure 6.5. In the first example, the right arm of the demonstrator is mapped on the right front leg of the robot, the left arm to the left front leg, the right leg to the back right leg and the left leg to the back left leg. As each of the arms and legs of the demonstrator consists of three segments, and the imitator's legs consist of two segments, only the first two segments are mapped. In the second example, the imitator's head and tail are controlled 'puppeteer-like', by mapping the first two segments of the right arm and the first segment of the left arm of the demonstrator to them, respectively. The latter can be also thought as an example of using a body part of the demonstrator (the left arm) to refer to a body part of the imitator (the tail) that does not have a direct equivalent on the demonstrator's (human) body. Also, although here both bodies have 'heads', it might be the case that it is more 'expressive' (i.e. the motions/posture easier to be perceived by the imitator and/or performed by the demonstrator) to use the right arm to indicate the head movements.

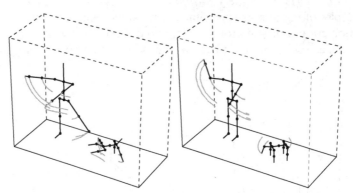

Figure 6.5    *Two examples of imitation across dissimilar embodiments*
*(humanoid and dog-like). The demonstrator (left in both examples)*
*is embodied as a humanoid, while the imitator (right in both*
*examples) is embodied as a dog-like AIBO robot. In the example on*
*the left, a simple one-to-one correspondence mapping is used,*
*mapping the first two (out of three) segments of the demonstrator's*
*arms and legs to the two segments of the four imitator's legs.*
*In the second example on the right, the demonstrator's first segment of*
*the left arm is mapped on the imitator's tail, and the demonstrator's*
*first two right arm segments to the neck and head of the imitator.*
*This results in the demonstrator controlling 'puppeteer-like' the*
*head and tail of the robot. The grey traces visualise the body part*
*trajectories*

*Partial mappings*
An example of partial mapping is shown in Figure 6.6 (left). As the imitator is an upper torso humanoid, the DOFs in the lower body parts of the (whole humanoid) demonstrator are ignored (via zero rows in the matrix), with a unity one-to-one mapping used for the upper body.

*One-to-many mappings*
When the perception of the demonstrator by the imitator is limited, a complicated one-to-many correspondence mapping could be used. Assuming a human acting as a demonstrator, but providing the system and the imitator with only the coordinates of three motion sensors, one attached to her waist and one in each hand. Filtering perception through this sensory apparatus yields a reduced representation of the demonstrator embodiment that can be modelled as a 'V' letter shape kinematic model.[11] The $\theta$ and $\varphi$ of each arm

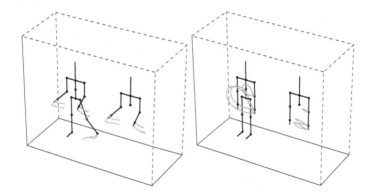

*Figure 6.6    Two examples of imitation across dissimilar embodiments (whole and upper torso only humanoids). The demonstrator (left in both examples) is embodied as a humanoid, while the imitator (right in both examples) is embodied as an upper torso humanoid robot. In the first example on the left, the arms of the demonstrator are mapped using a weight of one to the arms of the imitator. Note that the movement of the demonstrator's left leg is ignored as these demonstrator's DOFs are omitted (via a zero row in the correspondence matrix). In the example on the right, the same mapping is used, but the rate of movement of the imitator is severely limited, resulting in impersistence (see further discussion in Section 6.3.4). The grey traces visualise the body part trajectories*

---

[11] Note that as the humans move their arms around, the lengths of the two segments of the 'V' will change accordingly and not remain constant. But this can be ignored since, for the correspondence mapping, the important parameters are the azimuth and polar angles. These can be found from the (relative to the waist sensor) Cartesian coordinates of each arm sensor.

segment of the 'V' embodiment can be mapped on each of the segments of the corresponding arms of a humanoid imitator, with different weights (see example in Figure 6.7).

### 6.3.3    Effect metrics

*Effects* can be defined as changes to the body–world relationship (e.g. location) of the agent and/or to positions, orientations and states of external objects.

   Towards a characterisation of the *space of effect metrics*, i.e. those that relate to the manipulation of objects (rather than, say, body postures or limb movements), we have explored absolute/relative angle and displacement aspects and focused on overall arrangement and the trajectory of manipulated objects (Alissandrakis *et al.*, 2005). Focusing on aspects of orientation and displacement of the manipulated objects, two types of effect metrics can be used, *displacement* and *angular*. The first type relates an object's movement and position in the workspace (see Figure 6.8), and the second type to the object's orientation (see Figure 6.9). Using these metrics, one can evaluate the similarity between the *effects* on the environment (object displacement and/or rotation) of the model and the imitator, without considering the *state* or the *actions* of the agents that caused them (see Nehaniv and Dautenhahn, 1998).

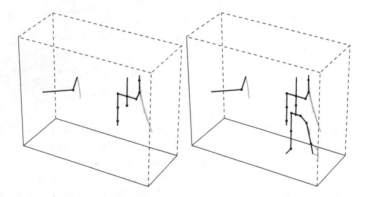

*Figure 6.7    Two examples of imitation across dissimilar embodiments using one-to-many correspondence mappings. The demonstrator (left in both examples) is embodied as an abstract 'letter V' shape, visualising the three motion sensors attached to a human (one to waist and one on each hand), while the imitator (right in both examples) is embodied as a humanoid. In the example to the left, the left segment of the demonstrator is mapped with different weight values to the imitator's left arm segments. In the second example to the right, this mapping is extended by also mapping the left segment of the demonstrator with different weight values to the imitator's left leg segments. The grey traces visualise the body part trajectories*

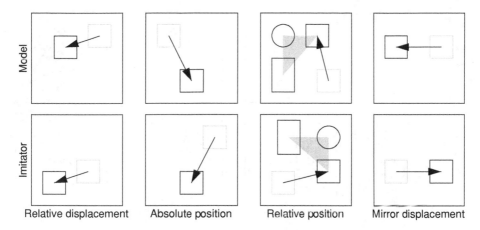

*Figure 6.8*   *A selection of displacement effect metrics. To measure the discrepancy between object displacements, the relative displacement, absolute position, relative position or mirror displacement effect metrics can be used. The first row shows four examples of effects demonstrated by the model. The second row shows the way the corresponding object (in a different workspace) needs to be moved by an imitator to match the corresponding effects according to each metric. The grey triangles are superimposed to show that for the relative position effect metric, the relative final positions of the objects are the same*

Some examples of circumstances where each of these metrics can be useful (in the context of setting up a dining table) would be placing a salad bowl or the main plate in the centre of the table (absolute position), arranging the forks and knives next to the plates (relative position and orientation) or (having placed a set of plates, silverware and glasses at each seat) repeating the dining arrangement for each person (relative displacement and rotation).

If the objects start from the same positions in the imitator's workspace as in the demonstrator's workspace, all the displacement effect metrics become equivalent (i.e. using any of them, the same trajectories will be generated); similarly if the objects start in the same orientations all the angular effect metrics become equivalent (the objects rotate in the same way as in the demonstration). But if the objects start in a dissimilar initial configuration (positions and/or orientations) to that of the demonstration, the choice of metrics affects qualitatively the character of the resulting imitative behaviour (see Figure 6.10).

See Alissandrakis *et al.* (2005) for precise mathematical definitions of these effect metrics, as well as for the capacity to use them in generalising imitative behaviours across different initial configurations and applications.

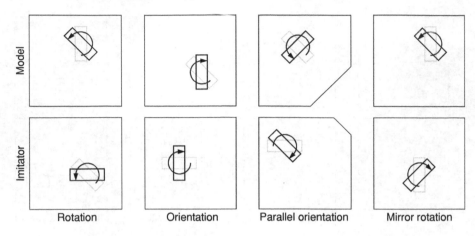

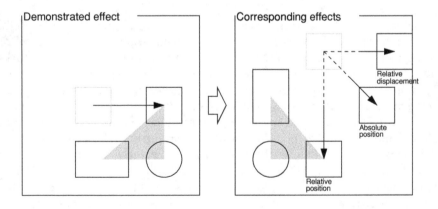

*Figure 6.9    A selection of angular effect metrics. To measure the discrepancy between object displacements, the rotation, orientation, parallel orientation or mirror rotation effect metrics can be used. The first row shows four examples of effects demonstrated by the model. The second row shows the way the corresponding object (in a different workspace) needs to be rotated by an imitator to match the corresponding effects according to each metric. Note that the shape of the two workspaces in the parallel orientation example is different, and the objects align with the highlighted diagonal edge*

*Figure 6.10    The figure illustrates three examples of using different displacement effect metrics. Depending on the effect metric used, qualitatively dissimilar imitative behaviours can result from dissimilar object configurations (here position only). The grey triangles are superimposed to highlight the relative position effect metric*

### 6.3.4    Future work directions

A robot capable of imitating a demonstrated behaviour by another agent (artificial or human) should be able to choose to match one (or a weighted combination) of different aspects of that behaviour, namely the *actions* (motions), *states* (postures) and *effects* (on the environment). To evaluate the similarity of the demonstration with the imitation attempts, appropriate metrics, such as the ones presented in Sections 6.3.2 and 6.3.3, should be used.

We have shown how partial, mirror symmetric, one-to-one, one-to-many, many-to-one and many-to-many body mappings can be characterised by (linear) *correspondence matrices*. These correspondences induce an infinite variety of absolute and relative state and action metrics that can be used to guide robotic imitation across dissimilar embodiments – even radically different ones in which neither the size of body parts, nor their type, nor number of DOFs need be preserved – enhancing existing approaches to imitation learning. The study of non-linear correspondences for achieving matching behaviour in states, actions and/or effects would extend this set of metrics. Currently the correspondence mapping is given, but finding the correspondence can be approached using reinforcement learning and an experiential history (adding memory), as in previous work with the ALICE generic imitation framework (Alissandrakis *et al.*, 2002, 2004). How such correspondences can be built is one of the hardest problems of imitation learning and comprises an important aspect of *what* to imitate problem. For related work on visuo-somatic mapping, see Asada *et al.* (2006); Cabido-Lopes and Santos-Victor (2003). Future work would naturally also address the derivation of, and switching between, appropriate correspondence mappings depending on the needs of the imitator agent in the social and task context. A developed system could eventually serve as a *correspondence engine for imitation learning*, incorporating aspects of discovering *what* to imitate, depending on context and interaction history.

The simulations presented here do not take into account joint limits, body mass distribution or detailed dynamics, that may lead to unfeasible and/or unstable solutions (see Nakaoka *et al.* (2003) for an example where the trajectory had to be corrected to ensure proper balance, after the mapping). By relying on simulation, we avoided many of the typical problems associated with the interpretation of the demonstrator's actions and with the computation of the demonstrator's state. It will in general be necessary to adapt a generated action sequence for a particular configuration of the particular physical system.

The simulations shown in Section 6.3.2 were purely reactive with no memory (and therefore no learning), which can result in certain limitations. For example in Figure 6.6 (right), the demonstrator is moving its left arm in a loop trajectory, but since the imitator is continuously reacting but with a limited rate of movement, the imitator is unable to reach the current sub-goal before it changes (until finally the demonstrator completes the entire demonstration).

This inability to sustain appropriate actions is called *impersistence* (Saunders *et al.*, 2004) and can be solved by adding memory to the system, containing the sequence of sub-goals.

The relative/absolute position and rotation of objects are important aspects of a demonstrated task to match (or not) according to effect metrics, depending on the state of the objects in the environment and the context. The exploratory characterisation of the space of effect metrics reveals that matching of results is a more sophisticated issue that generally acknowledged. This wide range of possible effect metrics illustrates that even the effect aspect of the correspondence problem for HRI by itself is already quite complex. Goal extraction in terms of effect metrics and granularity may have many different solutions that might not all be appropriate according to the desired results or context. This creates particular problems and challenges for sub-goal and metric extraction systems that can be used in programming robots by demonstration. The use of repeated demonstrations (Billard *et al.*, 2004), saliency detection (Scassellati, 1999) and goal-marking via deixis and non-verbal signalling by humans (Butterworth, 2003; Call and Carpenter, 2002; Bekkering and Prinz, 2002) may help contribute solutions to these problems.

Although useful for analysis from the 'system perspective', this research presented in this section cannot *directly* lead to a 'natural' HRI interface; we believe that the robot should be like a 'black box' willing to communicate its internal state and goals in an appropriate way, from the human perspective, with no need for the interacting human to know details about the hardware and the inner data representations and structures. To that extent, we introduced a more detailed feedback on the fifth study (discussed in Section 6.2), to examine whether the participants can adapt to the robot's chosen style (dealing with either absolute or relative effect metrics), one that is implicitly expressed by its gestures and speech during the expounding of the just observed demonstration, rather than explicitly by identifying the effect metric in technical terms.

A complementary approach to social learning, which is dealing with the correspondence problem by favouring direct interaction with the robot (rather than the robot observing the human), is presented in the Section 6.4.

## 6.4    Integrating social learning issues into a technology-driven framework

In Section 6.2 it was suggested that, first, a key design challenge was to find a balance between natural ways of interacting with a robot and managing the expectations of human users. Second, that more feedback would be required from the robot to the human to indicate the robot's success or otherwise in imitating a demonstrated task – simply giving negative feedback (e.g. 'I don't understand') would be insufficient for the human to modify the task. Third, segmentation of their own demonstrations to a robot suggested that individuals may differ as to the level of detail they spontaneously consider. In addressing these issues the robot faces many of the formalities described in

Section 6.3. Specifically these include finding a solution to the set of correspondence problems faced and deciding on the levels of granularity in the task set. Below, in Section 6.4.1, we discuss a technology-driven approach to these issues in a HRI setting.

### 6.4.1 Natural ways of interacting with a robot

A partial technological solution to the issues outlined in Section 6.4 above has been addressed in Saunders *et al.* (2007a) which describes a novel approach inspired by the notion of 'zone of proximal development' initially proposed by Vygotsky (1978, 1986), and demonstrates how teaching a robot can be achieved through active manipulation of the robot's actuators. Behavioural competencies are built step-by-step by the human, exploiting those competencies already taught. Such perspective is clearly in line with an embodied and situated perspective of cognition. The key issues that such an approach considers match those described earlier in the text. The approach has been fielded on a variety of robotic platforms including both large and small mobile robots and humanoids. Figure 6.11 shows people interacting with a humanoid robot deploying this learning architecture.

In terms of the balance of natural interaction and expectations, a mechanism that both amplifies the robots perceptions and provides a human-centred way of teaching is via the idea of 'assisted' or 'self'-imitation (Zukow-Goldring and Arbib, 2007). Here the robot's bodily configuration is physically manipulated by the human. This allows the robot to reconstruct motion paths having effectively been provided with a solution to the correspondence problem via the matching of human and robot actions. Amplification of the respective modalities can be achieved via information theoretic means whereby specific modalities which lead to the preferred outcome are favoured over those which do not. Similar mechanisms using 'assisted imitation' but employing a

*Figure 6.11    Interacting with the KASPAR II robot. The photo on the left illustrates a human physically manipulating KasparII's arms to demonstrate how to hold a box. The photo on the right shows a person teaching KasparII the name of a geometric shape. For more details on the KASPAR child-sized humanoid robot, please see http://kaspar.feis.herts.ac.uk/*

more model-based approach are also used by Calinon and Billard (Calinon *et al.*, 2006). The following subsections summarise some of the key technical aspects of this approach.

### 6.4.1.1  Learning mechanism

We use a memory based 'lazy' learning method (see Mitchell, 1997) to allow the system to learn tasks. This is a k-nearest neighbour (*k*NN) approach where the value of each feature in the system's sensorimotor state vector (see Section 6.4.1.2) is regarded as a point in *n*-dimensional space, where *n* is the number of features in the state vector. For each chosen task we collect a set of training examples (as described in Section 6.4.1.3) together with their target primitives, each primitive being chosen by the human trainer when moulding the robot's actions. We call this collection of states a *memory model*. When the task is executed, the robot continually computes its current state vector. It then computes the distance from the current state to each of the training examples held in the particular memory model.

The concepts of *scaffolding* and *putting through* can play an important part in animal learning. They support a form of self-imitation that may be the natural precursor to more complex forms of imitative learning. In our framework we use the idea of putting through directly. The human has the ability to control the robot by remotely moving it through any sequence from a set of pre-defined basic primitives. This set of primitives are basic actions available to the robot such as raise arm, turn head, etc. The human teacher has no access to the internal state of the robot. By manipulating the robot in this manner we also avoid both the problem of observation by the robot of the human actions, and of the correspondence problem between the robot and human.

During the robot 'putting through' process a snapshot of the robot's proprioceptive and exterioceptive state is recorded together with the directed primitive on each human command to the system. For each human defined task we can therefore build a memory model of state/primitive combinations, effectively a table where each row contains the robot modalities (or state vector) concatenated with the human directed primitive (or subsequent learnt behaviour).

### 6.4.1.2  Scaffolding

All of the states perceived by the system are recorded in the state vector; however, particular attributes may have more relevance to different tasks. For example during tracking of an object using the pan/tilt unit the values of the $X-Y$ object tracking may be of more relevance than the values of the sonar sensors, or the arm angles.

We use two mechanisms to ensure that the appropriate attributes are chosen. The first is based on computing the information gain (Quinlan, 1993) (effectively a measure of mutual information) to measure how well a given attribute separates the set of recorded state vectors according to the target primitive. The information gain measurement allows particular attributes in the state vector to have greater relevance by using it to weigh the appropriate dimensional axes in the $k$NN algorithm. This has the effect of either lengthening or shortening the axes in Euclidean space thus reducing the impact of irrelevant state attributes.

The second mechanism for attribute selection is the human trainer. It is assumed that the trainer already understands the task (from an external viewpoint) that the robot must carry out and therefore is able to construct the training environment appropriately so as to ensure that irrelevant features are removed. The modification of the environment allows the technical selection of relevant state features to be enhanced as the other features will now tend to have constant values and therefore a low information gain. This process of scaffolding and creating favourable conditions for learning would seem a quite natural phenomenon in social animals, and is fundamental to all forms of human teaching.

### 6.4.1.3  Learning new tasks

The trainer directs the robot using a screen-based interface which provides a number of buttons used to set operation modes such as 'execute' and 'start/ stop learning' plus an edit field to label actions and a list from which to choose existing labelled actions and primitive operations.

The robot can be operated in two modes. The first is 'execution' mode, which is its normal mode of operation where its current behaviour is executed. Alternatively, the robot can be in learning mode where the human trainer can put through, scaffold and create new activities for the robot to eventually use in execution mode.

In 'learning' mode the robot can be taught new competencies: sequences, tasks or behaviours. All three learning levels are started by pressing a 'start learning' button and terminated by pressing a 'stop learning' button. For each new competence (a behaviour, task or sequence) the trainer explicitly provides an appropriate label. When training is complete, the label is added to the set of actions available to the trainer and thus can be used immediately for further training sessions. Existing labelled actions can also be modified (or entirely deleted) with additional training episodes, as required.

The *sequence* level is where the robot can be directed through a given sequence of primitives which it records without reference to its state, i.e. sequences are entirely independent of the internal or external environment. This is similar to the ethological idea of Fixed Action Patterns where animals display certain sequences of movements that are independent of environmental stimuli (Tinbergen, 1951).

The *goal-directed task* level differs from a sequence in that during training the actions taken by the system will depend on the robot's internal and external state at that time. The trainer now has the opportunity to select not only basic primitives, but also sequences and other goal-directed tasks. The tasks are goal-directed because the trainer is able to inform the robot when the task has completed with the resulting state being recorded as a target to achieve.

The *behaviour* level allows the trainer to construct the complete behaviour for the robot from the component set of tasks, sequences and primitives. The construction of a behaviour is the same as for a task except that no goal state is required. The behaviour will run continually in execute mode and base its decision of what other behaviours, tasks, sub-task, sequences or primitives to use based on the current environmental state. A behaviour can also be used by another behaviour, task or sequence as required, the only constraint being that hierarchy must have a behaviour as its topmost node. Figures 6.12 and 6.13 show schematics of the underlying learning system.

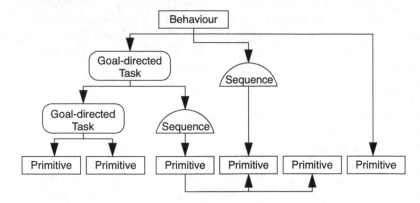

*Figure 6.12    A trained set of competencies from behaviours to primitives*

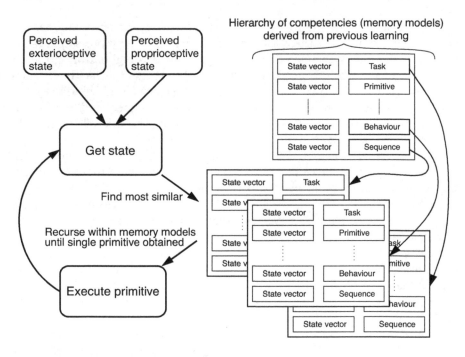

*Figure 6.13    The training/execution sequence and schematic of a memory model*

## 6.4.2 Feedback from the robot to the human

Ideas of robot feedback are also explored by Saunders *et al.* (2007a). In this instance the robot is effectively aware of what it already 'knows' in the form of a hierarchy of behavioural competencies (see Figure 6.12). By exploiting these known competencies in a predictive mode (effectively acting as a *forward model*) the robot can assess when it is being taught something that it already knows and can inform the human trainer appropriately. Based on the current state each memory model is polled using the *k*NN algorithm in section 6.4.1.1. This polling will yield a set of possible actions, one for each memory model in the system. If the trainer directs (puts through) the robot to take one of these actions, a confidence factor attached to each memory model that proposed this action is given a reward and all other tasks are penalised. As each training step is taken this cycle is repeated. A bonus is also given to a memory model that has predicted both the current action and has correctly predicted the previous action. This bonus serves to reward consistently correct predictive behaviour. Once the confidence factor of any of the memory models exceeds a global threshold (manually set between 0% and 100% of the confidence factor) the trainer is informed by the robot that it may already have knowledge of this task. Setting a low threshold (nearer to zero) will cause more of the memory models to report possible equivalence; a high threshold will require more matched actions before possible equivalence is reported. Figure 6.14 illustrates graphically an increasing confidence factor during a training episode with a mobile robot.

An extension to this idea would be to consider when the robot encounters a new and ill-described teaching event. The robot would respond with what it has understood so far in the process, for example 'I know how and where to place the knife but I have not understood the procedure after that'. Approaches to imitation have exploited combinations of *inverse* and *forward models* to model both observational and assisted imitation (Demiris and Hayes, 2002) and provide a computational mechanism which to some extent matches the mirror-neuron concept (Gallese and Goldman, 1998) in neurophysiology.

## 6.4.3 Segmentation of demonstrated tasks

In terms of spontaneous segmentation of tasks, we described in Otero *et al.* (2008c) how, in a robotic experiment (using the architecture described in Saunders *et al.* (2007a) and in section 6.4.1 above), segmenting a behaviour not only allows task re-use and reduces human training time, but can also serve to enhance the algorithmic effect in the environmental scaffolding process (in this case information gain). In that experiment, the robot was trained to visually track a coloured object by means of the human orienting its camera. The teacher trains the robot to move its pan/tilt unit so that from the teacher's viewpoint, the robot's camera is always directed at the object. The teacher is attempting to teach the robot to keep the object in view, thus as the object moves to the edge of the camera frame the teacher should instruct the robot to

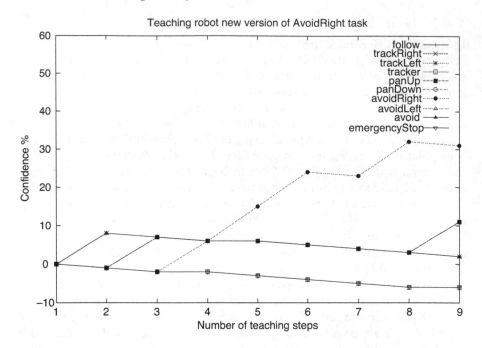

*Figure 6.14*   *An example of the predicted matching of taught and known tasks during training of a mobile robot. In this experiment the robot was taught a new behaviour to avoid obstacles on its right side. It correctly ascertained that this was similar to an existing behaviour by predicting the appropriate behaviour given the current state at each step in the training cycle. Note that some tasks in the figure above follow similar paths and cannot be graphically separated.*

carry out actions which serve to bring the object back to the centre of the image. There are a number of ways in which the teacher can achieve this:

- *Strategy 1.* Carry out a continuous training regime covering as many possibilities as possible. This is achieved by placing the tracked object in many parts of the camera frame and invoking the appropriate pan/tilt primitive action to centre the object in the frame. The trainer chooses to represent this teaching activity in two memory models, a behaviour and a goal-directed task. The behaviour is used to invoke the goal-directed task (in essence continually calling it). The goal-directed task contains all of the training steps directed by the human to the robot.
- *Strategy 2.* Train for the *X* and *Y* directions together in the camera frame. This can be carried out by giving many training examples in either the horizontal or vertical directions.
- *Strategy 3.* Train for the *X* and *Y* directions separately. In this example three memory models are created. The behavioural component being used

here to provide a scaffold which by providing a simple $X/Y$ training session giving only extreme examples in each direction invokes either of the separate training regimes shown as goal-directed tasks.

The results of this experiment (shown in Table 6.2) illustrated that the robot benefits with better performance, both on the task taught and from reduced computation when strategy 3 was employed. This strategy achieved performance very close to that of a pre-programmed control program that accurately marked every relevant point in the camera frame and adjusted the camera accordingly (marked 'control' in Table 6.2).

*Table 6.2    Performance of differing training strategies. The table shows the five memory models used for the visual tracking task. The first is a pre-generated control model using all of the camera segments. The second is the same strategy except as trained by a human. The third model shows training where only vertical and horizontal training is used. The fourth and fifth models break down the third model into its component horizontal and vertical directions. The '% accuracy' column shows predicted classification accuracy based on 10-fold cross-validation. The number of training steps is equivalent to the number of entries in each memory model. The k value shown is that which produced the optimal performance and was obtained through cross-validation*

| Memory model | % Accuracy | No. of training steps | Optimal $k$ value |
|---|---|---|---|
| No. 1 – Control | 98 | 975 | 7 |
| No. 2 – Strategy 1 | 68 | 223 | 14 |
| No. 3 – Strategy 2 | 86 | 123 | 2 |
| No. 4 – Strategy 3 (up/down) | 94 | 58 | 2 |
| No. 5 – Strategy 3 (left/right) | 95 | 65 | 4 |

The results serve to illustrate that appropriate segmentation in the robot teaching process can significantly benefit both human and robot. The human teacher benefiting as the number of training steps is significantly reduced as compared to a direct (i.e. teach everything at once – strategies 1 and 2) approach.

However, a further problematic issue (again explored via a robotic experiment using the system described in section 6.4.1 above (Saunders *et al.*, 2007b)) considers situations where humans completely disregard steps which are crucial to the robot. In this experiment a small mobile robot was taught a simple stimulus-action task. The task was to make the robot raise its grippers when it perceived a light (from a flashlight) shining on it. Otherwise it should lower its grippers. Each of the teachers taught the robot in approximately the same way; however, upon execution of the taught behaviour the teachers found that the robot did not respond as expected. This was due to the trainers omitting certain key steps in the training regime (see Table 6.3).

*Table 6.3    Results from stimulus training with missed training steps. In this
experiment, there were four actual possible states to be taught to the
robot; however, the trainers typically only trained for two states
(2 and 3, omitting 1 and 4). This is because there seems to be an
assumption that the robot will understand that the trainer wishes for it
to 'maintain' its condition, for example when the light is on,
ALWAYS keep your arm down*

| State | Arm | Light | Action that should be taught | Action actually taught |
|-------|------|-------|------------------------------|------------------------|
| 1 | Up | Off | Arm up | *Not taught* |
| 2 | Up | On | Arm down | Arm down |
| 3 | Down | Off | Arm up | Arm up |
| 4 | Down | On | Arm down | *Not taught* |

The results indicate the assumptions that were made by the teacher on how the robot learns. The training given to the robot by the teacher(s) could be due to the teacher(s) only considering the positive results of the training experience, i.e. only those instructions which achieve the desired state. Alternatively, the teacher may be assuming that there is an implicit 'while' or 'maintain' condition (which would be the case when training another human – perhaps an unspoken rule that one should do nothing in those cases not covered by the training), for example teacher says 'when light is on put your arm down'; however, what the teacher means is 'when light is on, ALWAYS keep your arm down'. Training which operates to keep the robot in a desired state does not appear to be considered in advance. Alternatively, the nature of the task (which here was highly discrete) may be problematic since a state in which no teaching has occurred but no action is required might be close to one in which an action is taught; however, this experiment was designed to be simple to illustrate situations that can arise in teaching a robot behaviours. In other more complex teaching studies the states were more numerous and far less discrete in nature. Proximity of states in which no action should be taken to those in which an action was explicitly taught could not alone allow them to be distinguished from nearby states in which generalisation of the action should occur.

Again, a possible solution to this issue would be further feedback from the robot as suggested earlier in the text.

## 6.5    Conclusions and future work

In this chapter we have argued extensively that the robot will need to publicise its own abilities and understanding of on-going tasks/activities to facilitate the interaction with a human partner. Considering this specific problem, however, one of the challenges related to teaching/learning/imitation tasks is to know what kind of initial building blocks need to be in place to make the robot's actions understandable by the human partner. At the same time, though, from the system's point of view, these building blocks have to be easily extended or adapted to different task needs. Furthermore, depending on the concrete social situations (e.g. solo human teaching or multiple instructors, resilience to interruptions and breakdowns during teaching episodes) and given the wide variability of human's behavioural repertoire, the robot's communication abilities will probably need to be tuned to the audience and/or recipient(s) (similar point made in Saunders *et al.* (2007b)).

The problem space is vast and complex and we would like to highlight that the achievements in its exploration will definitely have to consider integrative frameworks, where user studies, formal/conceptual proposals and technical solutions need to be taken into careful consideration.

Considering the expansion of robotic systems into people's everyday social places, not only some reflection needs to be made on how particular results impact on the perception of the technology to achieve concrete research goals, but also its usefulness, worth and values in society (see, e.g. Sellen *et al.* (2009)). HRI is in its infancy and, like other digital technology advancements, easily catches the public eye. However, what we should avoid is what Fernaeus *et al.* (2009) considered to be a 'robot cargo cult' phenomena, where research claims or suggestions raise expectations beyond reasonable heights, influencing the social representations people/laymen have of the field, maybe, in unexpected and undesirable ways.

This chapter aimed to illustrate state-of-the-art research in the field of imitation, outlining conceptual, experimental and technological innovations, but also raising awareness for the limitations and difficulties of this work. While the practical applications of this research for robot companions in people's homes remain a future challenge awaiting fully functioning, affordable, multipurpose, adaptable and safe robot companions, research in this field nevertheless poses many exciting research questions that can inform not only robotics and engineering disciplines, but may ultimately also contribute to our understanding of how socially interactive machines may learn from and interact with people.

# References

Alibali MW, Bassok M, Solomon KO, Syc SE, Goldin-Meadow S (1999) 'Illuminating mental representations through speech and gesture'. *Psychological Science*, 10(4):327–333.

Alissandrakis A, Miyake Y (2009a) 'Human to robot demonstrations of routine home tasks: Acknowledgment and response to the robot's feedback'. In: *New Frontiers in Human-Robot Interaction, Symposium at AISB 2009 Convention*, 6–9 April 2009, Edinburgh, Scotland, pp. 9–15.

Alissandrakis A, Miyake Y (2009b) 'Human to robot demonstrations of routine home tasks: Adaptation to the robot's preferred style of demonstration'. *18th IEEE International Symposium on Robot and Human Interactive Communication* (RO-MAN'09), pp. 135–140.

Alissandrakis A, Nehaniv CL, Dautenhahn K (2002) 'Imitation with ALICE: Learning to imitate corresponding actions across dissimilar embodiments'. *IEEE Transactions Systems, Man & Cybernetics: Part A*, 32(4):482–496.

Alissandrakis A, Nehaniv CL, Dautenhahn K (2004) 'Towards robot cultures? – Learning to imitate in a robotic arm test-bed with dissimilar embodied agents'. *Interaction Studies: Social Behaviour and Communication in Biological and Artificial Systems*, 5(1):3–44.

Alissandrakis A, Nehaniv CL, Dautenhahn K, Saunders J (2005) 'An approach for programming robots by demonstration to manipulate objects: Considerations on metrics to achieve corresponding effects'. In: *Proceedings of 6th IEEE International Symposium on Computational Intelligence in Robotics and Automation (CIRA '05)*, pp. 61–66.

Alissandrakis A, Nehaniv CL, Dautenhahn K (2006) 'Action, state and effect metrics for robot imitation'. The 15th IEEE International Workshop on Robot and Human Interactive Communication (RO-MAN 2006), pp. 232–237.

Alissandrakis A, Nehaniv CL, Dautenhahn K (2007) 'Correspondence mapping induced state and action metrics for robotic imitation'. *IEEE Transactions on Systems, Man, and Cybernetics, Part B*, 37(2):299–307.

Amit R, Matarić MJ (2004) 'A correspondence metric for imitation'. In: McGuinness DL, Ferguson G (eds.) *Proceedings of the Nineteenth National Conference on Artificial Intelligence, Sixteenth Conference on Innovative Applications of Artificial Intelligence*, July 25–29, 2004, San Jose, CA, AAAI Press/The MIT Press, pp. 944–945.

Asada M, Ogino M, Matsuyama S, Ooga J (2006) 'Imitation learning based on visuo-somatic mapping'. In: Marcelo H, Ang OK (eds.) *Experimental Robotics IX: The 9th International Symposium on Experimental Robotics, Springer Tracts in Advanced Robotics*, vol. 21, Berlin/Heidelberg, Springer, pp. 269–278.

Bekkering H, Prinz W (2002) 'Goal representations in imitative actions'. In: Dautenhahn K, Nehaniv CL (eds.) *Imitation in Animals and Artifacts*. MIT Press, pp. 555–572.

Billard A (2001) 'Learning motor skills by imitation: A biologically inspired robotic model'. *Cybernetics and Systems*, 32(1–2):155–193.

Billard A, Epars Y, Calinon S, Cheng G, Schaal S (2004) 'Discovering optimal imitation strategies'. *Robotics and Autonomous Systems*, vol. 47, pp. 2–3.

Bolt RA (1980) 'Put-that-there: Voice and gesture at the graphics interface'. In: *Proceedings of the 7th Annual Conference on Computer Graphics and Interactive Techniques*. ACM, Seattle, Washington, pp. 262–270.

Breazeal C, Scassellati B (2002) 'Robots that imitate humans'. *Trends in Cognitive Science*, vol. 6, pp. 481–487.

Butterworth G (2003) 'Pointing is the royal road to language for babies'. In: Kita S (ed.) *Pointing: Where Language, Culture, and Cognition Meet*. Lawrence Erlbaum Associate Inc, pp. 9–26.

Cabido-Lopes M, Santos-Victor J (2003) 'Visual transformations in gesture imitation: What you see is what you do'. In: *Proceedings of the 2003 IEEE International Conference on Robotics and Automation*, ICRA 2003, September 14–19, 2003, Taipei, Taiwan, IEEE, pp. 2375–2381.

Calinon S, Billard A (2004) 'Stochastic gesture production and recognition model for a humanoid robot'. In: *IEEE/RSJ Intl Conference on Intelligent Robots and Systems (IROS)*.

Calinon S, Guenter F, Billard A (2006) 'On learning, representing and generalising a task in a humanoid robot'. *IEEE Transactions on Systems, Man and Cybernetics, Part B Special Issue on Robot Learning by Observation, Demonstration and Imitation*, vol. 35, p. 5.

Call J, Carpenter M (2002) 'Three sources of information in social learning'. In: Dautenhahn K, Nehaniv CL (eds.) *Imitation in Animals and Artifacts*. MIT Press.

Cassell J (2000) 'Nudge nudge wink wink: Elements of face-to-face conversation for embodied conversational agents'. In: Cassell J, et al. (eds.) *Embodied Conversational Agents*. MIT Press, Cambridge, MA, pp. 1–27.

Cassell J, Thorisson KR (1999) 'The power of a nod and a glance: Envelope vs. emotional feedback in animated conversational agents'. *Applied Artificial Intelligence*, 13(4–5):519–538.

Cassell J, Cipolla R, Pentland A (1998) *A Framework for Gesture Generation and Interpretation*. Cambridge University Press, New York, pp. 191–215.

Cassell J, McNeill D, McCullough KE (1999) 'Speech-gesture mismatches: Evidence for one underlying representation of linguistic and non-linguistic information'. *Pragmatics and Cognition*, 7(1):1–33.

Cassell J, Cassell J, Sullivan J, Prevost S, Churchill E (2000) *Nudge Nudge Wink Wink: Elements of Face-to-Face Conversation for Embodied Conversational Agents*. The MIT Press, Cambridge, MA, pp. 1–27.

Cassell J, Koop S, Tepper P, Ferriman K, Striegnitz K (2007a) 'Trading spaces: How humans and humanoids use speech and gesture to give directions'. In: Nishida T (ed.) *Conversational Informatics*. John Wiley & Sons, New York, pp. 133–160.

Cassell J, Kopp S, Tepper P, Ferrimanand K, Striegnitz K (2007b) 'Trading spaces: How humans and humanoids use speech and gesture to give directions'. In: Nishida T (ed.) *Conversational Informatics*. John Wiley & Sons, New York.

Dahlbäck N, Jönsson A, Ahrenberg L (1998) 'Wizard of oz studies—why and how'. In: *Readings in Intelligent User Interfaces*. Morgan Kaufmann Publishers Inc, San Francisco, CA, pp. 610–619.

Dautenhahn K (1998) 'The art of designing socially intelligent agents: Science, fiction and the human in the loop'. *Applied Artificial Intelligence Journal, Special Issue on Socially Intelligent Agents*, vol. 12, pp. 12–17.

Dautenhahn K (2007) 'Socially intelligent robots: Dimensions of human-robot interaction'. *Philosophical Transactions of the Royal Society B: Biological Sciences*, 362(1480):679–704.

Dautenhahn K, Nehaniv CL (2002) 'An agent-based perspective on imitation'. In: Dautenhahn K, Nehaniv CL (eds.) *Imitation in Animals and Artifacts*. MIT Press, pp. 1–40.

Demiris J, Hayes G (2002) 'Imitation as a dual-route process featuring predictive and learning components: A biologically-plausible computational model'. In: Dautenhahn K, Nehaniv CL (eds.) *Imitation in Animals and Artifacts*. MIT Press, pp. 327–361.

Erlhagen W, Mukovskiy A, Bicho E, Panin G, Kiss C, Knoll A, van Schie H, Bekkering H (2005) 'Action understanding and imitation learning in a robot-human task'. In: Duch W, Kacprzyk J, Oja E, Zadrozny S (eds.) *Artificial Neural Networks: Biological Inspirations – ICANN 2005*. 15th International Conference, Warsaw, Poland, September 11–15, 2005, Proceedings, Part I, Springer, Lecture Notes in Computer Science, vol. 3696, pp. 261–268.

Fernaeus Y, Jacobsson M, Ljungblad S, Holmquist LE (2009) 'Are we living in a robot cargo cult?' In: *Proceedings of the 4th ACM/IEEE International Conference on Human Robot Interaction*. ACM, La Jolla, CA, pp. 279–280.

Galef BG, Heyes CM (eds.) (1996) *Social Learning in Animals: The Roots of Culture*. Academic Press.

Gallese V, Goldman A (1998) 'Mirror neurons and the simulation theory of mind-reading'. *Trends in Cognitive Sciences*, 2(12):493–501.

Garber P, Goldin-Meadow S (2002) 'Gesture offers insight into problem-solving in adults and children'. *Cognitive Science*, 26(6):817–831.

Ghidary S, Nakata Y, Saito H, Hattori M, Takamori T (2002) 'Multi-modal interaction of human and home robot in the context of room map generation'. *Autonomous Robots*, 13(2):169–184.

Herman LM (2002) 'Vocal, social, and self imitation by bottlenosed dolphins'. In: Dautenhahn K, Nehaniv CL (eds.) *Imitation in Animals and Artifacts*. MIT Press, pp. 63–108.

Iverson JM, Goldin-Meadow S (2005) 'Gesture paves the way for language development'. *Psychological Science*, 16(5):367–371.

Kelly SD, Singer M, Hicks J, Goldin-Meadow S (2002) 'A helping hand in assessing children's knowledge: Instructing adults to attend to gesture'. *Cognition and Instruction*, 20(1):1–26.

Kendon A (1997) Gesture. *Annual Review of Anthropology*, vol. 26, pp. 109–128.

Kipp M (2004) *Gesture Generation by Imitation: From Human Behaviour to Computer Character Animation*. Dissertation.com, Boca Raton, FL.

Krauss RM, Dushay RA, Chen Y, Rauscher F (1995) 'The communicative value of conversational hand gestures'. *Journal of Experimental Social Psychology*, vol. 31, pp. 533–552.

Kuniyoshi Y, Inaba M, Inoue H (1994) 'Learning by watching: Extracting reusable task knowledge from visual observations of human performance'. *IEEE Transactions of Robot Automation*, vol. 10, pp. 799–822.

Lozano S, Tversky B (2006) 'Communicative gestures facilitate problem solving for both communicators and recipients'. *Journal of Memory and Language*, 55(1):47–63.

McNeill D (1992) *Hand and Mind*. University of Chicago Press, Chicago, IL.

McNeill D (2005) *Gesture and Thought*. University of Chicago Press, Chicago, IL.

Mitchell TM (1997) *Machine Learning. McGraw-Hill International.*

Nadel J, Guerini C, Peze A, Rivet C (1999) 'The evolving nature of imitation as a format of communication'. In: Nadel J, Butterworth G (eds.) *Imitation in Infancy*. Cambridge, pp. 209–234.

Nakaoka S, Nakazawa A, Yokoi K, Hirukawa H, Ikeuchi K (2003) 'Generating whole body motions for a biped humanoid robot from captured human dances'. In: *Proceedings of the 2003 IEEE International Conference on Robotics and Automation*, ICRA 2003, September 14–19, 2003, Taipei, Taiwan, IEEE, pp. 3905–3910.

Neal JG, Thielman CY, Dobes Z, Haller SM, Shapiro S (1989) 'Natural Language with Integrated Deictic and Graphic Gestures', pp. 410–423.

Nehaniv CL, Dautenhahn K, Kubacki J, Haegele M, Parlitz C, Alami R (2005) 'A methodological approach relating the classification of gesture to identification of human intent in the context of human-robot interaction'. The 14th IEEE International Workshop on Robot and Human Interactive Communication (RO-MAN 2005), pp. 371–377.

Nehaniv CL (2003) 'Nine billion correspondence problems and some methods for solving them'. In: *Proceedings of Second International Symposium on Imitation in Animals and Artifacts* – Aberystwyth, Wales, 7–11 April 2003, Society for the Study of Artificial Intelligence and Simulation of Behaviour, pp. 93–95.

Nehaniv CL, Dautenhahn K (1998) 'Mapping between dissimilar bodies: Affordances and the algebraic foundations of imitation'. In: Demiris J, Birk A (eds.) *Proceedings of European Workshop on Learning Robots 1998 (EWLR-7)*, Edinburgh, 20 July 1998, pp. 64–72.

Nehaniv CL, Dautenhahn K (2000) 'Of hummingbirds and helicopters: An algebraic framework for interdisciplinary studies of imitation and its applications'. In: Demiris J, Birk A (eds.) *Interdisciplinary Approaches to Robot Learning*, World Scientific Series in Robotics and Intelligent Systems, pp. 136–161.

Nehaniv CL, Dautenhahn K (2001) 'Like me? – Measures of correspondence and imitation'. *Cybernetics and Systems*, 32(1–2):11–51.

Nehaniv CL, Dautenhahn K (2002) 'The correspondence problem'. In: Dautenhahn K, Nehaniv CL (eds.) *Imitation in Animals and Artifacts*. MIT Press, pp. 41–61.

Nicolescu MN, Matarić MM (2001) 'Learning and interacting in human-robot domains'. *IEEE Transactions on Systems, Man, and Cybernetics, Part A*, 31(5):419–430.

Oh JY, Lee CW, You BJ (2005) 'Gesture recognition by attention control method for intelligent humanoid robot', vol. 3681, pp. 1139–1145. Lecture Notes in Artificial Intelligence.

Otero N, Knoop S, Nehaniv CL, Syrdal DS, Dautenhahn K, Dillman R (2006a) 'Distribution and recognition of gestures in human-robot interaction'.

In: *Proceedings of the 15th IEEE International Symposium on Robot and Human Interactive Communication (RO-MAN06)*, pp. 103–110.

Otero N, Nehaniv CL, Syrdal DS, Dautenhahn K (2006b) 'Naturally occurring gestures in a human-robot teaching scenario'. In: *Proceedings of the 15th IEEE International Symposium on Robot and Human Interactive Communication (RO-MAN06)*, pp. 533–540.

Otero N, Alissandrakis A, Dautenhahn K, Nehaniv CL, Syrdal DS, Koay KL (2008a) 'Human to robot demonstrations of routine home tasks: Exploring the role of the robot's feedback'. In: *Proceedings of the 3rd ACM/IEEE International Conference on Human Robot Interaction*, ACM, Amsterdam, The Netherlands, pp. 177–184.

Otero N, Nehaniv CL, Syrdal DS, Dautenhahn K (2008b) 'Naturally occurring gestures in a human-robot teaching scenario'. *Interaction Studies*, 9(3): 519–550.

Otero N, Saunders J, Dautenhahn K, Nehaniv CL (2008c) 'Teaching robot companions: The role of scaffolding and event structuring'. *Connection Science*, vol. 20, pp. 111–134.

Ozcaliskan S, Goldin-Meadow S (2005) 'Do parents lead their children by the hand'? *Journal of Child Language*, 32(3): 481–505.

Quinlan JR (1993) *C4.5: Programs for Machine Learning*. Morgan Kaufmann, San Mateo, CA.

Roth WM, Lawless D (2002) 'Scientific investigations, metaphorical gestures, and the emergence of abstract scientific concepts'. *Learning and Instruction*, 12(3): 285–304.

Saunders J, Nehaniv CL, Dautenhahn K (2004) 'An experimental comparison of imitation paradigms used in social robotics'. In: *Proceedings of IEEE RO-MAN 2004, 13th IEEE International Workshop on Robot and Human Interactive Communication*, September 20–22, 2004 Kurashiki, Okayama, Japan, IEEE, pp. 691–696.

Saunders J, Nehaniv CL, Dautenhahn K, Alissandrakis A (2007a) 'Self-imitation and environmental scaffolding for robot teaching'. *International Journal of Advanced Robotics Systems, Special Issue on Human–Robot Interaction*, 4(1): 109–124.

Saunders J, Otero N, Nehaniv CL (2007b) 'Issues in human/robot task structuring and teaching'. The 16th IEEE International Workshop on Robot and Human Interactive Communication (RO-MAN 2007), pp. 708–713.

Scassellati B (1999) 'Imitation and mechanisms of joint attention: A developmental structure for building social skills'. In: Nehaniv CL (ed.) *Computation for Metaphors, Analogy and Agents, Springer Lecture Notes in Artificial Intelligence*, vol. 1562, pp. 176–195.

Schaal S (1999) 'Is imitation learning the route to humanoid robots'? *Trends in Cognitive Sciences*, 3(6): 233–242.

Sellen A, Rogers Y, Harper R, Rodden T (2009) 'Reflecting human values in the digital age'. *Communications of the ACM*, 52(3): 58–66.

Severinson-Eklundh K, Green A, Hüttenrauch H (2003) 'Social and collaborative aspects of interaction with a service robot'. *Robotics and Autonomous Systems*, 42(3–4): 223–234.

Singer MA, Goldin-Meadow S (2005) 'Children learn when their teacher's gestures and speech differ'. *Psychological Science*, 16(2): 85–89.

Thomaz AL, Breazeal C (2008) 'Teachable robots: Understanding human teaching behavior to build more effective robot learners'. *Artificial Intelligence*, 172(6–7): 716–737.

Tinbergen N (1951) *The Study of Instinct*. Clarendon Press.

Vygotsky LS (1978) *Mind in Society: Development of Higher Psychological Processes*. New edn. Harvard University Press.

Vygotsky LS (1986) *Thought and Language*. 2nd edn. MIT Press.

Zentall TR (1996) 'An analysis of imitative learning in animals'. In: Galef BG, Heyes CM (eds.) *Social Learning in Animals: The Roots of Culture*. Academic Press, New York, pp. 221–243.

Zentall TR (2001) 'Imitation in animals: Evidence, function and mechanisms'. *Cybernetics and Systems*, 32(1–2): 53–96.

Zentall TR, Galef, Jr. BG (eds.) (1988) *Social Learning: Psychological and Biological Perspectives*. Lawrence Erlbaum Associates, Hillsdale, NJ.

Zukow-Goldring P, Arbib MA (2007) 'Affordances, effectivities and assisted imitation: Caregivers and the directing of attention'. *Neurocomputing*, 70(13–15): 2181–2193.

## Chapter 7
# Models for cooperative decisions
# in Prisoner's Dilemma

*Maurice Grinberg, Evgenia Hristova and Emilian Lalev*

## 7.1 Introduction

The Prisoner's Dilemma (PD) game is a two-person game that is widely used as
a model of social interactions. The structure of the PD game (see Section 7.2)
represents a dilemma between the individual and the collective rationality. The
interest in studying PD game arises from the idea that many social situations
and problems such as overpopulation, pollution, energy savings, etc. have such
a dilemma structure (e.g., see Dawes, 1980). PD games are specifically used as a
tool for studying cooperative behavior.

   In our opinion, the use of games for modeling social interactions and
interpreting empirical studies should be based on the deep understanding of the
cognitive mechanisms behind game playing. This is particularly important with
regards to cooperation and coordination in human and animal behavior,
which are considered to be among the main elements of social behavior and
even a major factor for the evolution of humans and human language (e.g.,
Gärdenfors, 2006).

   From the normative point of view of formal game theory, cooperation is
not the logical solution in the PD game. At the same time, evidence from real
life and from numerous experimental studies (e.g., Rapoport and Chammah,
1965; Chater *et al.*, 2008; Sally, 1995), has shown that players cooperate even in
one-shot PD games. There are thousands of studies exploring various factors
supposed to influence the players' behavior. Such factors range from changing
the payoffs to changing the description of the game and the way it is presented
to the participants (Colman, 1995).

   Among the factors influencing cooperation in PD, the structure of the
game payoff matrix, formalized by the so-called cooperation index (Rapoport
and Chammah, 1965), occupies a central place. In most models of Iterated
PD Game (IPDG) based on reinforcement learning (RL) (e.g., Camerer *et al.*,
2002; Erev and Roth, 2001; Macy, 1991), the influence of changing payoffs
in the payoff matrix is not considered. The players in such theoretical accounts
do not consider the payoffs of the present game and base their moves on

past received payoffs or games. However, it has been shown (Hristova and Grinberg, 2005b; Grinberg *et al.*, 2006; Grinberg and Hristova, 2009; Knoepfle *et al.*, 2009) that information acquisition patterns about the payoffs are essential to understand the decision-making process during game play.

Payoffs received in the past are of course one of the main factors for strategy learning based on the positive or negative reinforcement associated with past moves in IPDG (e.g., Macy, 1991). Players can thus learn, that in the long run, cooperation might be more rewarding than defection. At the same time, it has been shown that the influence of the previous games on cooperation in the current game can be related to differences between the respective payoffs or payoff matrices, both in one-shot PD games and in IPDG (see, e.g., Vlaev and Chater, 2006; Hristova and Grinberg, 2004, 2005a). In some cases (see Chater *et al.*, 2008), the attitude of players to the payoff matrix structure can be related to preexisting biases, which can lead to some interesting implications about cooperation in one-shot PD games (leading to a Simpson's paradox situation).

Fictitious play (Brown, 1951) is a behavior in which a player evaluates reinforcement from situations that did not actually happen but were possible (Camerer and Ho, 1999). Additionally, in IPDG, players may try to guess the moves of their opponents. Therefore, it is realistic to assume that knowledge or guess about the opponent's possible moves can play an important role in the choice of strategy. It has been shown that when model players build and use models of their opponents, their behavior becomes complex (e.g., Taiji and Ikegami, 1999). Provided players are able to predict their opponent's move, they may want to maximize their own payoff by choosing the most profitable strategy given the predicted behavior. In the case when the players assume that their opponents are trying to predict their strategy, they may try to mislead them by pretending to play in a certain way (see Camerer *et al.*, 2002; Taiji and Ikegami, 1999).

Such sophisticated relations, involving theory of mind and social interactions, are related to the central role of anticipation in cognition in general (Rosen, 1985; Pezzulo *et al.*, 2008). Their influence on cooperation in IPDG is connected to forward-looking decision-making mechanisms, which have been shown to be responsible for larger cooperation (Grinberg and Lalev, 2008, 2009).

The aim of this chapter is to present the main results of our ongoing research program for investigating experimentally and theoretically all of the factors mentioned earlier in the section. Our goal is to try to understand how people play the IPDG by providing more detailed and more cognitive-science-oriented explanations of cooperation and related phenomena. To achieve these goals, we designed cognitive models that account for important cognitive aspects and constraints of game playing such as context effects and anticipation. The development of models started with a simple reinforcement model whose main features were further re-implemented in a recurrent neural network model. The latter was first validated by being applied for the description of previous experimental results and then used in multi-agent simulations of small artificial societies of agents based on IPDG.

To situate our results in the appropriate context, we first present the PD game and its characteristics used in the models and experiments (see Section 7.2). Further, we review briefly some RL models which are quite influential for modeling IPDG (see Sections 7.3). Such models are the models described in Macy (1991), Erev and Roth (1998), and Camerer and Ho (1999). The anticipation-based models of Camerer *et al.* (2002), Isaac *et al.* (1994), and Taiji and Ikegami (1999) are discussed in Section 7.4.

The model of Hristova and Grinberg (2005a), which accounts for the payoff matrix of the game and is based on subjective utility theory is introduced in Section 7.5, together with extensive comparisons with empirical data. In Section 7.6, we discuss a connectionist model, designed to use both the current game payoff matrix and predictions about future payoffs and opponent's moves (Lalev and Grinberg, 2007). Results from IPDG simulations are compared with those from experiments with human participants. In Section 7.7, simulations with the model of Lalev and Grinberg (2007) are presented, which indicate that anticipation could lead to higher cooperation and coordination levels. This is supported by the simulation of evolution in societies of agents based on the model in which anticipatory capabilities evolve when cooperation is taken to be a measure of fitness (Grinberg and Lalev, 2008, 2009). Finally, in Section 7.8, the results and insights obtained are summarized and directions for future work are discussed.

## 7.2 The Prisoner's Dilemma game

The PD game is one of the most extensively studied games. It has become popular because it is seen as fundamentally similar to real-world problems in economics, sociology, and politics.

The generalized matrix of the PD game is presented in Figure 7.1a. The players simultaneously choose their move – C (cooperate) or D (defect), without knowing their opponent's choice. R stands for the "Reward" for mutual cooperation, T for the "Temptation" to defect, P for the "Punishment" for mutual defection, and S for the "Sucker's" payoff. To represent a PD game, the payoffs must satisfy the following inequalities: $T > R > P > S$ and $2R > T + S$ (this specific choice ensures that CC outcomes are preferred to alternating CD and DC ones) (e.g., see Macy and Flache, 2002). The ordinal structure of the payoff matrix of the PD game is presented in Figure 7.1b. In this ordinal matrix the payoff 4 is the "best" (T) and the payoff 1 is the "worst" (S) possible payoff from the game.

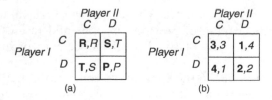

Figure 7.1    *The Prisoner's Dilemma game matrix: (a) the generalized matrix;*
*(b) the ordinal matrix. The payoffs for each game outcome in the*
*matrix are separated by commas – the first payoff is for Player I, the*
*second payoff is for Player II*

## 7.2.1    Rationality in IPDG

The PD game presents a paradox. If we turn back to Figure 7.1 and examine the restrictions for the payoffs that should be met, it is easily seen that strategy D is dominant for both players – each player gets larger payoff by choosing strategy D than by choosing strategy C, no matter the choice of the other player. Moreover, the DD outcome (both players defect) is the only Nash equilibrium of the game, i.e., neither player has any incentive to unilaterally deviate from it. However, it turns out that the payoffs for this outcome (P for each player) are lower for both players compared to the payoffs if both players choose their dominated strategies giving a CC outcome (payoff R for each player). Even if for some reason both players cooperate, and the outcome is CC, still it is not a Nash equilibrium because it is unstable due to the temptation to defect and get the largest payoff T. Because of this the PD game represents a conflict between individual and collective rationality. From individual point of view it is rational for each player to choose the dominant D strategy. Still, both players are doing better if both choose the dominated C strategy.

The argument presented earlier in the section can be applied for one-shot PD game (i.e., a PD game played just once). What would happen if the players are involved in a long run of PD games with the same opponent, as is the case in many real-life situations? In other words, what is the rational strategy choice for IPDG? In such settings, the players can adapt to their own behavior based on the observed behavior of the other player.

When there is a possibility for the two players to meet again, the current choice can influence not only the payoff received from the current game but also their future choices and payoffs (Axelrod, 1984). One could speculate that under such circumstances choosing strategy C is no longer irrational. This may be a way to signal to the other player willingness to cooperate. When the game is repeated, each player has an opportunity to "punish" the other player for previous defection. The incentive to cheat may then be overcome by the threat of punishment, leading to the possibility of a superior, cooperative outcome. However, the formal game theory analysis leads to a different solution by the use of backward induction when the number of games is known to both players.

In the finitely repeated PD game (in which the number of rounds is finite and known in advance), backward induction arguments show that the rational behavior is to defect in every move. The proof is as follows. Rational players know that the only rational strategy in the last game is to defect (to choose strategy D). Although there might have been reasons to cooperate in previous rounds, they no longer apply in the last round as there are no games to follow. Thus, it is in fact a one-shot PD game and the best strategy according to game theory is to defect. Because the outcome of the last round is predetermined and the previous round cannot influence it, the round before the last is also a one-shot PD game. If this reasoning is applied backward to the first round, it turns out that the only rational strategy is to defect on every round of the finitely iterated PD game (Colman, 2003). Some situations in which cooperation can emerge in IPDG with finite number of games in game theory involve some complexity limits of the player's strategies or uncertain knowledge about the number of games to be played (e.g., see the discussion in Neyman, 1999). An approach out of the main game-theoretical stream, proposed in Hofstadter (1985), introduced the notion of "superrational" players. Superrational players, according to Hofstadter (1985), know that their opponents will choose the same strategy as they do and thus only the CC and DD outcomes are possible. The CC must be preferred because $R > P$.

Although logical, the backward induction argument sounds counter-intuitive because there are a lot of examples in real life in which people cooperate. There are also many experiments in which participants cooperate in one-shot and finitely iterated PD games. Sally (1995) provides a meta-review of such experiments, published between 1958 and 1995, which shows that in IPDG, cooperative choices are made in 20–50% of the games (mean 47.4%) and that many players cooperate in one-shot games.

## 7.2.2　Cooperation index

There are many factors, identified experimentally, that influence the cooperation rate in playing IPDG. Among them are the framing of the game (or the way of describing the game to the participants in an experiment), players' goals and motivation, opponent's strategy, etc. (Colman, 1995; Furnham and Quilley, 1989; Sally, 1995; Komorita *et al.*, 1991). One of the important factors found to influence cooperation in PD is a quantity called Cooperation Index (CI), introduced by Rapoport and Chammah (1965). It is calculated using the equation: $CI = (R–P)/(T–S)$. CI may vary from 0 to 1 (see Figure 7.2) and it has been found to be positively correlated with the proportion of C choices. The definition of CI implies that the probability for cooperation depends not on the payoffs (T, R, P, and S) individually but rather on the ratios of their differences (Rapoport and Chammah, 1965).

Most participants are influenced by games' CI when playing IPDG and such a dependence has been observed in several experimental studies (e.g., Rapoport and Chammah, 1965; Hristova and Grinberg, 2004). Therefore, realistic models should also account for this dependency. However, some experimental studies also reveal that participants sometimes cooperate in IPDG with no dependence on the game CI (Hristova and Grinberg, 2004, 2005a). This means that the specific ratios of payoffs in the PD game matrix cannot fully predict the observed cooperation in the game and models should provide an explanation for both behaviors.

| CI=0.1 | | Player II | |
|---|---|---|---|
| | | C | D |
| Player I | C | 24, 24 | 0, 40 |
| | D | 40, 0 | 20, 20 |

| CI=0.5 | | Player II | |
|---|---|---|---|
| | | C | D |
| Player I | C | 24, 24 | 0, 40 |
| | D | 40, 0 | 4, 4 |

*Figure 7.2    Examples of PD game matrices with different CI. The first game is with CI = 0.1, the second one is with CI = 0.5*

## 7.2.3    Theories of cooperation in the PD game

A lot of theories try to explain cooperative behavior in PD games in terms of socially established values and stress the importance of social interaction and relationships as tools for achieving cooperation. Among them are theories that explain cooperation by altruism, reciprocity, or reputation building.

One of the main theories aimed to explain cooperation in IPDG is the reputation building theory (Kreps *et al.*, 1982). The reputation building theory assumes that players are self-interested (not altruists) but the repeated nature of the game creates incentives to cooperate. In this model, the player is building himself a reputation of a cooperative player and expects that the other player will also cooperate.

Reputation building is closely related to the concept of reciprocity. Many researchers point to the fact that the norm of reciprocity is widespread and is the basis of many relationships and societies (Trivers, 1972). People reciprocate cooperation with cooperation. One of the most studied strategies that are based on reciprocity is the Tit-For-Tat (TFT) strategy. A player using TFT cooperates initially, and then plays the same as her opponent did in the previous game. If previously the opponent was cooperative, the player is also cooperative. It is demonstrated analytically and in computer tournaments that the TFT strategy receives higher payoffs compared to other strategies in the long run (Axelrod, 1984; Komorita *et al.*, 1991).

Another influential theory about cooperation in PD game is based on the concept of altruism. Unlike the reputation building theory, this theory assumes that some players are not strictly self-interested and benefit from cooperation and from other's payoffs (Cooper *et al.*, 1996). For altruistic players, the actual payoffs in the game are not equal to the payoffs presented in the payoff matrix. Such a player receives an additional payoff when cooperating and because of this it might happen that cooperation is no longer a dominated strategy. In such a way it is possible that cooperation yields higher payoffs than defection.

A different approach aimed at explaining cooperation is based on RL models. In contrast to the theories that rely on social relations and concepts of cooperation, RL models try to demonstrate that even in the absence of social interactions cooperation can occur as it leads to higher payoffs. Some of the most influential models are presented in Section 7.3.

## 7.3   Reinforcement learning models of IPDG playing

RL theory accounts for human game playing in terms of rewards and punishments that people receive or expect to receive (Sutton and Barto, 1998). RL models try to demonstrate that cooperation can emerge based on payoff maximization. The positive outcomes increase and the negative outcomes decrease the probability of repetition of the associated strategy. In general, RL models consist of a probabilistic decision rule and a learning algorithm for updating the probabilities in the decision rule. The received game payoffs are evaluated as positive or negative reinforcements and the move choice probabilities are updated accordingly. Typically, in IPDG modeling, the search for solutions is "backward-looking" (e.g., see Macy, 1990) and the influence of CI is not taken into account in typical RL-based models. Most models use a single payoff matrix and decide about the probability to cooperate only based on past experience in the games assuming full knowledge of all the payoffs during decision making.

In the following subsections, some of the most influential models will be presented and discussed.

### 7.3.1   The Bush–Mosteller (BM) model

One of the simplest RL models for IPDG is the Bush–Mosteller (BM) model (Macy, 1991). The model implements two main mechanisms of learning: approach and avoidance. This approach implies a self-reinforcing equilibrium, also called "satisficing" behavior (March and Simon, 1958). The equilibrium is reached when a pair of strategies yields payoffs that are mutually rewarding, including the situation when both players receive less than their optimal payoff (such as the payoff R for mutual cooperation in PD games), as long as the payoffs exceed the so-called "aspiration level." The aspiration level determines when a received payoff is evaluated as a reward or punishment. Payoffs above the aspiration level will result in satisficing. Avoidance implies an aversive self-correcting equilibrium characterized by "dissatisficing" behavior. Dissatisficing means that both players will try to avoid an outcome that is better than their worst possible payoff (such as payoff S in PD games), as long as this payoff is below their aspiration level (Macy, 1991). The BM model was used in social dilemmas such as Stag Hunt, Chicken, and IPDG. In this model, regimes of stable cooperation can be obtained.

### 7.3.2    The model of Erev and Roth (ER)

In Erev and Roth (1998) and Erev *et al.* (1999), a model (hereafter called the ER model) alternative to the BM model has been proposed. The ER model is based on the so-called "matching law," which states that the players will choose between two possible moves in a ratio that matches the ratio of the rewards received after each of the moves. Applied to social dilemmas, the matching law predicts that players will learn to cooperate to the extent that the payoff for cooperation exceeds that for defection, which is only possible if both players happen to cooperate and defect at the same time (given R > P).

Like the BM, the ER model is stochastic, but it uses the concept of "propensity." Propensities and probabilities represent different entities – propensities are a function of the cumulative satisfaction and dissatisfaction with the associated moves, and probabilities are a function of the ratio of propensities. A probabilistic choice rule is needed to translate propensities into actual moves in the IPDG.

The general framework of learning in games is that players continually adjust their strategies in response to observations and outcomes from past experience. But when we speak of learning from non-experienced (but "imagined") situations, we call that fictitious play (see Section 7.3.3).

### 7.3.3    Fictitious play and the Experience-Weighted Attraction (EWA) model

The Experience-Weighted Attraction (EWA) model (Camerer *et al.*, 1999) combines elements of two seemingly different approaches. The first one is the belief learning approach consisting of the inclusion of weighted "fictitious play" payoffs. The latter were introduced in Brown (1951) as a possible explanation for Nash equilibrium play. It was proposed in Brown (1951) that players would simulate playing the game in their minds and update their future play based on this simulation (see also Fudenberg and Levine, 1995; Berger, 2005). In other words, in EWA, the payoff related to a move that was not chosen, is also used for the calculation of positive or negative reinforcements.

The second approach in the EWA model is the standard RL approach. It assumes that strategies are reinforced by previous payoffs, and the propensity to choose a strategy depends on this. Players who learn by reinforcement do not generally account for the payoffs they have missed. Also, they care only about the payoffs their strategies yielded in the past, not about the history of the play that created those payoffs.

Each approach uses a different kind of information. Belief-based models do not especially reflect past success (reinforcements) of chosen strategies. Reinforcement models do not reflect the history of how others played. The EWA model approach includes both by incorporating these two kinds of information (Camerer *et al.*, 1999).

In the EWA model, each strategy has an "attraction" associated to it. Attractions are weights given to alternative strategies. The attraction levels are updated according to the payoffs provided by the chosen strategies and to the payoffs which could have been provided by the non chosen strategies. These attractions decay after each period, and are normalized by a factor, which captures the amount of experience players have accumulated. The attractions to strategies are then mapped into the probabilities of choosing those strategies (Ho *et al.*, 2008).

One of the main points of EWA model is that belief learning and RL are not basically different. The only important difference between belief and reinforcement models is the extent to which they assume that players include fictitious payoffs in evaluating the possible game strategies.

The EWA model was later extended to the so-called "sophisticated" EWA model which includes predictions about the strategy of its opponent to maximize its payoff (described in Section 7.4.1). Such a decision-making mechanism is anticipatory in nature (see Pezzulo *et al.*, 2008). Anticipatory models of game playing are discussed in the next section.

## 7.4    Anticipatory models for social dilemmas

Anticipatory agents are agents making decisions based on predictions, expectations, or beliefs about the future. It is widely accepted that anticipation is an essential component of complex natural cognitive systems (e.g., see Pezzulo *et al.*, 2008). Rosen (1985) defined an anticipatory system as follows: "a system containing a predictive model of itself and/or its environment, which allows it to change state at an instant in accord with the model's predictions pertaining to a later instant."

Sutton and Barto (1981) discuss the anticipatory decision-making process in terms of actions taken in an internal model of the world. The internal model (Sutton and Barto, 1981) allows for the prediction of the world behavior as a function of actions available to the agent. The internal model is used to select a behavior interactively, similar to the interaction of the agent with the environment. Trials and errors achieving a best result in the external environment are replaced by internal (or fictitious) trials and errors due to the presence of the internal model. The action chosen corresponds to the best anticipated result from the internal model. For the internal model to be useful, it must provide safety or speed unachievable in the external world.

In the following subsections two anticipatory models used in iterated games will be presented.

### 7.4.1    The sophisticated EWA model

The sophisticated EWA model (Camerer *et al.*, 2002), extends the adaptive EWA model (see Section 7.3.3) to account for sophisticated learning and strategic teaching. The "sophisticated" players use the adaptive EWA model, presented in Section 7.3.3, to forecast what the moves of the other players will be and then choose the strategies giving the highest expected payoffs. When the opponents are sophisticated EWA models, their interaction can be very complex as shown in Camerer *et al.* (2002).

When playing iterated games with the same adaptive EWA opponents, sophisticated EWA players have the incentive to "teach" them. This strategic teaching gives rise to repeated-game equilibriums and reputation formation behavior through the interaction between the players (Camerer *et al.*, 2002).

### 7.4.2    The Best-Response with Signaling (BRS) model

The Best-Response with Signaling (BRS) model was put forward by Isaac *et al.* (1994). It incorporates rational forward-looking decisions and signaling of cooperative intentions (Janssen and Ahn, 2003). The BRS model is another model with a decision-making mechanism, based on anticipation.

To explain the dynamics of behavior in the iterated public goods game, Isaac *et al.* (1994) propose a hypothesis that individuals are involved in a forward-looking decision problem. A player may benefit from signaling of their cooperative intentions to the other players. Signaling is in the form of a slight increase in the cooperation rate of a player, and it serves as a "proposal" for the opponents to cooperate more in the games that follow – a sort of social interactions among players.

In the BRS model, an aspiration level is calculated for each round and for each player. When a player's aspiration level is reached, the player is satisfied. When the experienced utility obtained in a round is higher than or equal to the aspiration level, the individual is satisfied and repeats the decision they made in the previous round. If the aspiration level has not been reached, the model makes a probabilistic choice over all the possible actions. The move choice probability decision rule is the same as in the EWA models the only difference is that attractions are now substituted for the calculated values of the move alternatives.

The BRS model resembles EWA model and the RL models for IPDG (BM and ER models, Sections 7.3.1 and 7.3.2, respectively) in that they all have aspiration levels representing a payoff evaluation threshold. BRS also shares important common features with the sophisticated EWA such as the forward-looking mechanisms for game playing, implying "awareness" of the presence of opponents whose strategy is also accounted for by payoff-maximization. However, BRS and sophisticated EWA models have not been used in IPDG. A model which has been applied in IPDG is presented in the next subsection.

### 7.4.3    The model of Taiji and Ikegami

Another prediction-based model, especially designed to investigate cooperation in IPDG, is the model of Taiji and Ikegami (1999). It is able to recognize the dynamics of the opponent's strategy, and make predictions about the opponent's future moves with the help of a dynamic recognizer (Pollack, 1990). Therefore, it has an internal model of the way its opponent plays and can optimize its moves according to this internal model to maximize its payoff.

The model of Taiji and Ikegami exists in two versions which are called "Pure Reductionist Bob" and "Clever Alice." "Pure Reductionist Bob" anticipates the opponent's strategy using the dynamic recognizer. He believes that his opponents behave by following simple algorithms, like finite automata, and is trying to infer these algorithms from their behavior.

"Clever Alice" in turn assumes that the opponent behaves like "Pure Reductionist Bob." She knows that the opponent is making a model of her and she decides on what her next action will be taking into account that model.

To decide the player's next action, a prediction of the opponent's future action based on the dynamic recognizer apparatus is used in both versions of the model. "Pure Reductionist Bob" chooses his forward actions in several fictitious future games. He also predicts the opponent's actions and the expected score is evaluated. The process is repeated for all possible sequences of actions of a certain length and Bob chooses the action with the highest score as the best action.

"Clever Alice" chooses her forward actions and she predicts the opponent's actions assuming that he behaves like "Pure Reductionist Bob." Again the process is repeated for all possible variations of future sequences of C and D moves and she chooses the one with the highest score.

In the simulations, when the model played against the same model (not against simple computer strategies), the IPDG always converged to mutual defection after some time. However, the model managed to discover the mutual cooperation strategy as the most profitable against a TFT computer opponent.

All the payoffs during the game sessions belonged to one and the same PD matrix and in such setting no dependence of cooperation on CI could arise. Another issue is that for evaluation of fictitious payoffs the model checks all possible sequences of future moves C and D with the corresponding predicted opponent's responses and own payoffs. This mechanism seems quite unlikely to occur in real-life game playing due to the very high cognitive load. Thus, the authors explore effects of the play of their model in idealized settings.

## 7.4.4 Summary

The models for IPDG players and other iterated games, presented in this section, give insights in the various ways standard backward-looking RL can be applied to maximize payoffs based on positive or negative past rewards. On the other hand, it was shown that the payoff-maximising behavior can be based on past experience together with prediction of the opponent's moves and evaluation of possible (but not received) payoffs by fictitious play. In all anticipatory models, the addition of predictive decision-making mechanisms has improved essentially their performance.

But what all these models lack so far, is the sensitivity to the current game payoff matrix, which is necessary to account for when exploring the sensitivity of the cooperative behavior to the variations in the PD game matrix (i.e., to CI; see Section 7.2.2). The model of Hristova and Grinberg (2005a) was especially designed to account for such effects and is presented in the next section.

## 7.5    Context-Sensitive Reinforcement Learning (CSRL) model

### 7.5.1    *The model*

The experimental data (e.g., Rapoport and Chammah, 1965; Hristova and Grinberg, 2004, 2005a) demonstrate that cooperation rate is influenced by the CI of the games. On the other hand, there is evidence that the cooperation rate also depends on the context created by the CIs of the other games played in a particular game session (see Section 7.5.2.4). The experimental results showed that the cooperation rate depends not only on the current game payoffs and CI but also on the CIs of the other games in the game set. If a PD game with a given CI is presented in the context of games with higher CI values, people tended to cooperate significantly more than in the other experimental conditions. The obtained results gave good evidence for context effects during PD game playing (Hristova and Grinberg, 2004, 2005a).

To account for these findings, we proposed a model (Hristova and Grinberg, 2005a) that incorporates mechanisms accounting for the current game's CI and payoffs and in the same time is sensitive to the other games in the game set (and can account for the context effects observed).

The model is inspired by subjective expected utility theory which is one of the most important variations of the expected utility theory (for a discussion, see Schoemaker, 1982). The main distinctive feature of the subjective utility theory is that it allows for subjective probabilities of the outcomes (probabilities in the classical expected utility theory are considered to be objective probabilities). The treatment of probabilities as subjective rather than objective is very important in cases when the objective probabilities cannot be determined in advance. In this view the probabilities are degrees of belief about an outcome.

Another major extension of the classical expected utility theory is the stochastic model of choice developed by Luce (1959). In this model the preferences are treated as probabilistic rather than fixed deterministic choices – instead of identifying one of the alternatives as the chosen option, each alternative has a probability to be selected.

Building upon these theories, in the model presented here the preferences are treated as probabilistic choices. The expected values of the possible moves (C and D) are computed and then used to determine the probability of move C. The subjective expected value of a given move is computed as a sum of the subjective expected values of the two possible game outcomes associated with this move (CC and CD game outcomes for move C; DC and DD game outcomes for move D).

The subjective value of a game outcome (CC, CD, DC, and DD) is calculated by using:

- the current game payoff for that outcome – R for CC, S for CD, T for DC, and P for DD;
- the weight associated with that game outcome, computed on the basis of previous payoffs received in the same outcome;
- the estimated probability that the opponent will make the move specific for that outcome.

Thus, the value of a move depends not only on the possible payoffs but also on the predictions (expectations) about the opponent's move. A similar model has been used in a different context (one-shot PD games) in a work by Antonides (1994). The model presented here can be viewed as based on the general framework of the subjective utility theory but with a dynamic determination (learning) of the utilities and of the expectations about the opponent's move. The model can be regarded as an application of this theory in the domain of game playing.

The model is expressed as follows:

$$V(C) = w_{CC}\, Poff(CC)\, P_{op}(C) + w_{CD}\, Poff(CD)\, (1 - P_{op}(C)) \tag{7.1}$$

$$V(D) = w_{DC}\, Poff(DC)\, P_{op}(C) + w_{DD}\, Poff(DD)\, (1 - P_{op}(C)) \tag{7.2}$$

where $V(C)$ and $V(D)$ stand for the subjective values of moves C and D; *Poff* (CC), *Poff* (CD), *Poff* (DC), and *Poff* (DD) are the current payoffs R, S, T, and P, corresponding to the four possible game outcomes; $P_{op}(C)$ is the probability for cooperation predicted for the opponent; $w_{CC}$, $w_{CD}$, $w_{DC}$, and $w_{DD}$ represent weights (importance) of the game outcomes. These weights are computed as running averages over the payoffs assigned to those outcomes and thus depend on previous payoffs.

The weights are updated after each game as follows:

$$[w_{XY}]_{new} = (1 - \alpha)\, [w_{XY}]_{old} + \alpha\, Poff(XY) \tag{7.3}$$

where X and Y stand for the possible moves (C or D); *Poff*(XY) the received payoff in outcome XY; $\alpha$ a parameter satisfying the condition $0 < \alpha < 1$. Indexes *new* and *old* refer to the current and past moments, respectively.

The predicted probability for the opponent to cooperate $P_{op}(C)$ is also obtained as a running average over the past opponent's moves according to the equation:

$$[P_{op}(C)]_{new} = (1 - \beta)\, [P_{op}(C)]_{old} + \beta\, M_{op} \tag{7.4}$$

where $\beta$ is a parameter satisfying the condition $0 < \beta < 1$; $M_{op}$ the opponent's move with D coded as 0 and C as 1 in order to get values for the probability $P_{op}(C)$ between 0 and 1.

The decision for a move is calculated with a formula similar to the one of the ER model (Section 7.3.2) as the propensities are replaced by the calculated subjective values $V(C)$ and $V(D)$:

$$P(C) = \frac{V(C)}{V(C) + V(D)} \qquad (7.5)$$

where $P(C)$ is the probability for move C.

## 7.5.2 Simulations and experiments

A study was designed to test for context effects in IPDG playing for human participants and the model. The goal was to investigate if and how the contexts determined by different CI distributions are influencing actual game playing. We used a design that allows us to directly evaluate change in playing for games covering the full CI-scale, when the context determined by different CI distributions, is manipulated.

### 7.5.2.1    Design and PD games

The payoff matrices were randomly generated with payoffs held within certain limits of magnitude. The CI is a quantity invariant with respect to the possible linear transformations of the payoffs (see Section 7.2.2). However, Oskamp and Perlman (1965) claimed that the average payoff $((T + R + P + S)/4)$ is a very important factor with significant effect on the level of cooperation. Another reason to keep the payoffs in certain limits is that participants could pay more attention to games with higher payoffs than to games with the same CI but with much smaller payoffs.

Taking all these considerations into account, we generated the games so that T was between 22 and 78 points (mean 53), R was between 15 and 77 points (mean 45), P was between 4 and 47 points (mean 17). For simplicity we set $S = 0$. In the same time, in each payoff matrix the defining inequalities for the PD games are preserved (i.e., $T > R > P > S$ and $2R > T + S$; see Section 7.2 for details).

Twenty five of the PD matrices (5 games with each of the CIs – 0.1, 0.3, 0.5, 0.7, and 0.9) were used as "probe" games.

The study consists of three experimental conditions that differed in the CI distributions of the games played (see Figure 7.3):

- Full-CI-range condition – games from the Full CI-range (CIs equal to 0.1, 0.3, 0.5, 0.7, and 0.9), intermixed with the probe games;
- High-CI-range condition – games from the High CI-range (CIs equal to 0.7 and 0.9), intermixed with the probe games;

- Low-CI-range condition – games from the Low CI-range (CIs equal to 0.1 and 0.3), intermixed with the probe games.

In each experimental condition there was the same number of PD games (270 games). The probe games were the same in all three context conditions and were placed on the same places in the game sequences – each 10th game was a probe game.

The aim of using such a design was to measure cooperation for the full CI scale during play while changing the context as slightly as possible (especially for the High-CI and the Low-CI ranges).

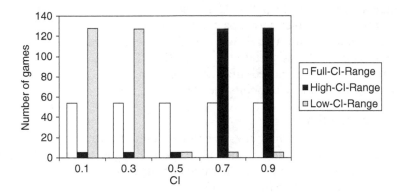

*Figure 7.3   Distribution of PD games played in different experimental conditions: Full-CI-range (white bars), High-CI-range (black bars), and Low-CI-range (gray bars)*

### 7.5.2.2   Experimental procedure

Each subject was tested individually and played 270 PD games against the computer. The computer used a probabilistic version of Tit-For-2-Tats (TF2T) strategy that takes into account the two previous moves of the player and plays the same as the participant with probability 0.8. For instance, in the case when the subject makes one and the same move during the last two games, the computer plays the same move with probability 0.8. In the case of two different moves by the participant in the last two games, the probability for cooperation by the computer is 0.5. This was done to allow the subject to choose her own strategy (followed by the computer) without easily become aware of the computer's strategy.

The game was presented to the participants in a formal and a neutral formulation to avoid other factors and contexts as much as possible. The terms "cooperation" or "defection" were not mentioned in the instructions to further avoid influences other than the payoff matrix. On the interface, the moves were labeled in a neutral manner as "1" and "2." Subjects were not informed about the existence of CI.

The payoffs were presented as points, which were transformed into real money and paid at the end of the experiment. After each game the subjects got feedback about their and the computer's choice and could monitor permanently the total number of points they have won and its money equivalent. The subjects received information about the computer's payoff only for the current game and had no information about the computer's total score. This was made to prevent a shift of subjects' goal – from trying to maximize the number of points to trying to outperform the computer. In this way, the subjects were stimulated to pay more attention to the payoffs and their relative magnitude and thus indirectly to CI.

There were in total 72 participants (34 males and 38 females). In the Full-CI-range condition participated 24 subjects, in High-CI-range condition – 26 subjects, and 22 subjects took part in the Low-CI-range condition. All were university students with an average age of 23 (ranging from 18 to 35). Each subject was randomly assigned to one of the three experimental conditions and after being instructed played five training games. All subjects were paid according to the number of points they have obtained.

### 7.5.2.3    Simulations

The model was run 30 times in each context condition (Full-CI-range, High-CI-range, and Low-CI-range) with the respective game set from the experiment. The model played against the same computer player as the participants in the experiment (probabilistic TF2T).

In the simulations, the averaging parameters of the model ($\alpha$ and $\beta$) were both fixed to 0.5. The initial cooperation probability $P(C)$ was set to 0.5 and the initial perceived probability of playing C for the opponent ($P_{op}(C)$) was also set to 0.5. These values seem psychologically plausible as in the beginning of the game players probably do not posses clear preferences between choices, or expectations about the play of the opponent (especially in the neutral presentation of the PD games used). As in the experiment, the model played five training games. In the training games $P(C)$ was kept equal to 0.5. On the basis of the training games the initial values for the w's and $P_{op}(C)$ are calculated. Then for the rest of the simulation the w's and $P_{op}(C)$ are calculated using (7.4) and (7.5).

It should be stressed that the same set of parameters was used in all three context conditions without any additional fitting. Thus the differences between the three conditions are due to the dynamic properties of the model as defined by (7.1)–(7.3).

### 7.5.2.4    Experimental results

Here we present the main findings from the experimental data. More detailed description could be found in Hristova and Grinberg (2005a).

Participants cooperated in 44% of the games in the High-CI-range condition, in 34% of the games in the Full-CI-range condition, and in 23% of the games in the Low-CI-range condition. The cooperation rates in different experimental conditions were significantly different ($F(2,69) = 7.7$, $p = 0.001$). The observed difference in the cooperation rates was expected, as participants in different groups played PD games with different CIs and this result confirms previous findings. A non-trivial hypothesis was that if playing is context sensitive, one would expect a difference between the cooperation rates for games with the same CI presented in different experimental conditions. The results obtained support the presence of context influences. Players cooperate more in PD games with CI = 0.7 and 0.9 when they play only highly cooperative games (in the High-CI-range condition) than when they play the same games mixed with other less cooperative games (the PD games with CI = 0.7 and 0.9 in the Full-CI-range condition). On the other hand, players cooperate less in PD games with CI = 0.1 and 0.3 when they play only low-cooperative games (in the Low-CI-range condition) than when they play the same games mixed with other, more cooperative games (the PD games with CI = 0.1 and 0.3 in the Full-CI-range condition).

When playing probe games covering the full CI-scale, all participants cooperated more with increasing CI. However, there were differences in the mean level of cooperation, influenced by the context distributions of the games played. Participants in the High-CI-range conditions cooperated more in the probe games, than participants in the Low-CI-range and in the Full-CI-range conditions.

All these results give good evidence for context effects during IPDG. On the one hand, as well known, the cooperation rate is influenced by the CI of a given game; but on the other hand, it also depends on the context created by the CIs of the other games played in a particular game session.

### 7.5.3 Comparison between model and experimental data

The predictions of the model were analyzed for the presence of context effects and the same pattern of context influences as in the experimental data was found. Then the model predictions were compared with the data from the experiment with human participants. Several characteristics were selected for the comparison to estimate how well the model describes the experimental data: overall cooperation, the cooperation with respect to CI, and the cooperation in the probe games. We also used the distribution of game outcomes (CC, CD, DC, and DD) obtained during the game sessions as another criterion of goodness of fit.

#### 7.5.3.1 Cooperation rates

Model cooperation rates are presented in Figure 7.4. They are very similar to the corresponding data from the experiment – the cooperation rates are highest in the High-CI-range condition and lowest in the Low-CI-range condition.

The same context effect as the one observed in the experimental data was found: the cooperation rates for games with the same CI are significantly different between the experimental conditions (all $p < 0.001$). The CSRL model cooperates more in PD games with CI = 0.7 and 0.9 in the High-CI-range condition than when in the Full-CI-range condition. On the other hand, the model cooperates less in PD games with CI = 0.1 and 0.3 in the Low-CI-range condition than in the Full-CI-range condition.

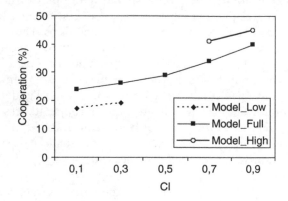

*Figure 7.4    CSRL model results for the mean cooperation in PD games with different CI in each context condition: Full-CI-range (solid line), High-CI-range (bold line), and Low-CI-range (dotted line)*

The games in which the subjects or the model cooperated is compared with a repeated measures analysis of variance with CI as a within-subjects factor and the treatment condition (experiment vs. model) as a between-subjects factor. The analysis is performed for each context condition separately. For all of them, the only statistically significant factor is CI ($p < 0.001$) (see Figure 7.5). The influence of the treatment condition (experiment vs. model) is not statistically significant, nor is the interaction between CI and the treatment condition. In summary, the model predictions and the experimental data are statistically undistinguishable. This is an evidence of the good matching of model and experiment.

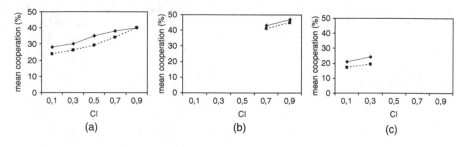

*Figure 7.5    Comparison for the cooperation rates between the model predictions (dotted lines) and the experimental data (solid lines) in each context condition: (a) Full-CI-range; (b) High-CI-range; and (c) Low-CI-range*

### 7.5.3.2  Cooperation in the probe games

The model cooperation rates for the probe games, which cover the full CI scale, are very similar to the experimental ones. As in the previous subsection, the number of probe games in which the subjects or the model cooperated is compared with a repeated measures analysis of variance with CI as a within-subjects factor and the treatment condition (experiment vs. model) as a between-subjects factor. In all context conditions, the only statistically significant factor is again CI ($p < 0.001$) (see Figure 7.6) and the influence of the treatment condition (experiment vs. model), and the interaction between CI and the treatment condition, are not statistically significant. The experimental and theoretical results are again statistically undistinguishable, which indicate a good description of the experimental data by the model.

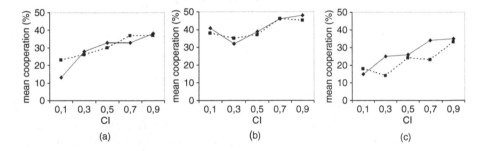

*Figure 7.6*  *Comparison between the model predictions (dotted lines) and the experimental data (solid lines) for the mean cooperation in the probe games for each context condition: (a) Full-CI-range; (b) High-CI-range; and (c) Low-CI-range*

### 7.5.3.3  Number of different game types

As discussed earlier, there are four possible game outcomes – CC, CD, DC, and DD. The same total number of C and D moves in two pairs of players can give two different distributions of game outcomes. Thus the number of different game outcomes is a relatively independent characteristic of the game dynamics and must be accounted as a separate characteristic by a computational model.

The means for the number of different game outcomes, in the simulation and in the experiment, were compared for each context condition (Full-CI-range, High-CI-range, and Low-CI-range). A $\chi^2$ test was used to evaluate the fit of the model to experimental data (see Figure 7.7). For the experimental and the model data, the distributions of different game outcomes in the Full-CI-range condition ($\chi^2 = 3.86$, $p = 0.277$) and in the Low-CI-range condition ($\chi^2 = 5.54$, $p = 0.1.36$), are undistinguishable statistically. In the High-CI-range condition the $\chi^2$ test is significant ($\chi^2 = 15.9$, $p = 0.001$), still it is seen in Figure 7.7b that the model results approximate the experimental data well.

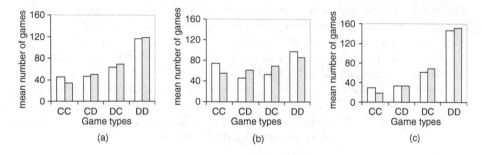

(a)                              (b)                              (c)

*Figure 7.7    Comparison between the model predictions (gray bars) and
the experimental data (white bars) for the number of different
game types in each context condition: (a) Full-CI-range;
(b) High-CI-range; (c) Low-CI-range*

### 7.5.3.4    Summary

Context effects related to different CI distributions of PD games were investigated. To compare the cooperation rate in different context conditions, a set of probe games was incorporated in the game sequence of each experimental condition. Thus, it became possible to determine the influence of the context for the whole CI scale. The experimental results showed that the cooperation rate depends not only on the current game payoffs and CI but also on the CIs of the other games in the game set. If a PD game with a given CI is presented in the context of games with higher CI values, people tended to cooperate significantly more than in the other experimental conditions. The results obtained in the experiment give good evidence for context effects during PD game playing.

To explore the observed context sensitivity, a computational CSRL model was used (Hristova and Grinberg, 2005a). The model takes into account the previous moves and corresponding payoffs to determine dynamically the utility of the game outcomes, the prediction about the probability for cooperation of the opponent, and the current game payoffs. This form of the model turned out to be very sensitive both to CI of the given game and to the context created by games with different CIs. The agreement between the theoretical results and the experiment is very good despite the fact that only three parameters define the model and they were kept fixed and the same for all three conditions. The model reproduced quite accurately the dependence of cooperation on CI and context and the cumulative number of different game outcomes. This result suggests that the quantities included in the model are essential for the explanation of the phenomena observed and must be taken into account in more realistic models.

## 7.6 Connectionist model for IPDG (Model-A)

We took into account the experience with existing models and designed a connectionist model for IPDG, called Model-A (Lalev and Grinberg, 2007). The model was intended to represent and predict the game dynamics (see Pollack, 1990 for a similar approach) including predictions about the moves of the opponent and expected payoff. A representation of the game payoff matrix structure is also present in the model. Another key feature is an anticipatory decision-making module that uses the predictions and representations to maximize the future payoff of the model.

The model presented in this section can be considered as a step from a pure computational modeling of human decision making in IPDG, toward a more cognitive modeling approach.

### 7.6.1 The model

The core of the model architecture is the Elman Simple Recurrent Network (SRN) module (Elman, 1990). SRN is used to generate predictions about the game outcome (moves and payoff), given the payoff matrix of the game and the moves of the player and opponent, and the received payoff. The SRN network is modified to incorporate an autoassociator, concerning the game payoff matrix, thus enforcing the analysis and influence of the payoff structure in the predictions of the model. The output of the SRN-based module is used by the move choice module which evaluates the best move based on fictitious plays generated by the SRN network. The detailed description of the model and its use is described in the following text.

#### 7.6.1.1 Inputs and outputs

All the inputs of the network were rescaled to fit in the interval [0, 1]. As can be seen in Figure 7.8, the following information is presented at the input nodes at each cycle: the possible payoffs for the current game matrix (excluding the payoff S which was always 0), the payoff received in the previous game, the player's and opponent's moves in the previous game.

The previous moves are coded as (01) – for C, and (10) – for D. The current game payoffs T, R, and P have to be reproduced by the model at the output (implementing an in-built autoassociator). There are two reasons to include this component in the network architecture. The first reason is to force the network to form representations of the games in the context of specific outcomes and thus account for the payoff structure in the decision-making process. The second one is to implement a fictitious play mechanism, which can make predictions about future fictitious games by using the output of the model as fictitious input in the fictitious play generation (see Lalev and Grinberg, 2007, for details).

The outputs of the player's and the opponent's move nodes ("Sm" and "Cm," respectively), are interpreted as probabilities for cooperation (the

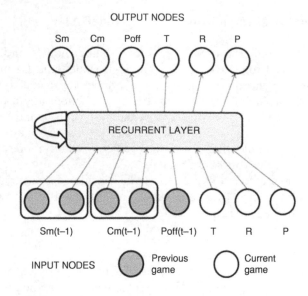

*Figure 7.8    Schematic view of the recurrent neural network and its inputs
and outputs/targets. Notation: $S_m(t-1)$, $C_m(t-1)$, $Poff(t-1)$
and the player's and opponent's moves, and the player's payoff
from the previous game. The quantities without explicit time
dependence refer to the current game*

output activations are given by a sigmoid activation function giving values in
the interval [0, 1]). The payoff ("*Poff*") node represents the expected payoff
player's payoff from the current game.

The model is expected to outperform the predicting capabilities of the
CSRL model (see Section 7.5) based on the potential of the SRN and
the autoassociator neural networks. A neural network of this type can extract
the interdependencies among the inputs, and between the values of these inputs
in time.

For instance, a specific game outcome (CC, CD, DC, or DD) goes with a
specific payoff (correspondingly, R, S, T, or P) and with specific moves of the
players. SRN is a predictor neural network and any pattern in the moves can
be intercepted and learned. The same holds for any persistent correlation
between the moves and/or the payoffs (or the kind of payoff).

### 7.6.1.2    Move choice

The move choice is made by an anticipatory mechanism which evaluates fictitious plays outcomes, generated by the SRN based module. This module
makes use of the predictive power of the SRN and tries to guess how the game
will proceed in the case of the two possible initial moves C or D. So, two
sequences of five fictitious games are generated before making a move and the
payoffs for each of them are summed with a discount. The first sequence begins

with a C move, and the second one begins with a D move (whatever the activation of node "$S_m$," see Figure 7.8). Only the first move is fixed in any sequence. The SRN has as first inputs the current game payoffs T, R, and P, and the moves and the player's payoff from the previous game, as explained in the previous subsection. The anticipatory mechanism used in our model is simpler than the one used by Taiji and Ikegami (1999), where all the possible sequences of C and D moves are taken into account.

As the model's move is known at the beginning of the first fictitious games (C and D), the opponent's move is generated with the probability predicted by the network (see node "$C_m$," in Figure 7.8). Once the moves of both players are known the payoff for the player is obtained from the game payoff matrix.

In the second fictitious game the input consists of the player's move (node "$S_m(t-1)$") (C or D, depending on the sequence), the generated opponent's move (node "$C_m(t-1)$") and the respective payoff (node "$Poff(t-1)$"), and generated by the SRN payoffs T, R, and P (output nodes "T," "R," and "P," see Figure 7.8). The idea behind the use of the network game output is related to the fact that the games in the experiment are randomized, so the prediction of the next game would be inefficient. So, the network uses the current game to generate the next fictitious games. This is repeated three more times with the only difference that the player's move is generated using the cooperation probability given by the activation of node "$S_m(t-1)$." The discounted sums of payoffs from the two sequences (*Poff_C* for initial move C and *Poff_D* for initial move D) are calculated as follows:

$$Poff_{C,D} = \sum_{t-1}^{5} Poff_{C,D}(t)\beta^{t-1} \tag{7.6}$$

where $Poff_{C,D}(t)$ is the value of the payoff at moment $t$, for initial move C or D and $\beta$ the discount parameter ($0 \le \beta \le 1$) which controls the importance of remote fictitious game payoffs. If $\beta$ is 0, only the first fictitious payoff would matter, and if $\beta$ is 1, all the payoffs will be equally important. In the present IPDG simulations $\beta$ was set to 0.7.

The probability for cooperation is calculated using the soft-max function:

$$P(C) = \frac{e^{Poff_C/K}}{e^{Poff_C/K} + e^{Poff_D/K}} \tag{7.7}$$

where $P(C)$ is the player's probability for cooperation. The parameter $k$ determines the sensitivity of the soft-max function with respect to the differences between $Poff_C$ and $Poff_D$. In all the simulations, presented below, $k$ has been chosen to be 0.1.

## 7.6.2    Comparison of simulation with experiment data

To evaluate the capabilities of the model, a series of simulations were performed using the settings from Section 7.5.2.2, from experiments with human participants. The network was trained using back-propagation and overlapping inputs, consisting of five consecutive games and their outcome – the current game and four previous games – forming a micro-epoch. For all outputs of the SRN module of Model-A, the training signal was supplied by the game payoffs of the current game, the opponent's move, and the payoff received by the player. In the present version of the model, the target for the player's move is given by the move choice module, described in Section 7.6.1.2. In some cases (not presented here), the target signal can come from a real play, by using the actual moves of the human participant as training signal for the model.

Data from the first part of the experiment in Hristova and Grinberg (2004) was used (see Section 7.5.2.2, for details). For the comparison between experiment and model, the settings of the experiment were kept the same in the simulations. In this experiment (see Hristova and Grinberg, 2004), 30 participants played 50 PD games against the TF2T computer opponent. The participants were stimulated to pay more attention to the payoffs and their relative magnitude and thus indirectly to CI. Games of different CI, ranging from CI = 0.1 to CI = 0.9, were presented in random order.

### 7.6.2.1    Comparison of model and experiment cooperation and payoffs

The mean cooperation between Model-A and human participants was different only in the Low-CI condition ($F = 4.13$, $p = 0.047$) (see Figure 7.9a). In the other two conditions there were no significant differences in cooperation. No significant differences were found for the mean payoffs in the three conditions (see Figure 7.9b).

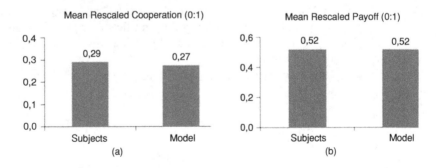

*Figure 7.9    Comparison of mean cooperation (a) and mean payoff (b) in IPDG for Model-A (Lalev and Grinberg, 2007) and the experiment (Hristova and Grinberg, 2004) for the three experimental conditions*

### 7.6.2.2   Dependence of cooperation on CI

In the simulations, a main effect of CI on cooperation rates was observed ($F = 16.908$, $p < 0.01$). Figure 7.10 shows a comparison of the cooperation rate dependence on CI, between the model and the experiment. There were no statistical differences between the mean cooperation of subjects and the model at all CI levels, and there was no main effect of the type of player (the model or human) on cooperation ($F = 0.386$, $p = 0.856$).

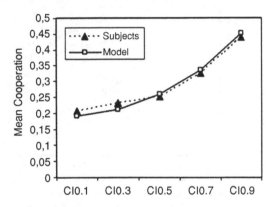

*Figure 7.10    Mean cooperation at different levels of CI for human subjects and the model*

We consider the ability to reproduce such details in the experiment very important for the model validity assessment. To understand the presence or lack of CI dependence, we analyzed the hidden layer activations during IPDG playing. As discussed earlier in this paper, we included the autoassociator part in the SRN module, to enforce the representation of the payoff structure at the hidden layer thus hoping to help the network to account for it (and hopefully for CI). As expected, such representations of the games' CI really emerged in the hidden layer nodes of the network. But the analysis showed that these representations were only partially responsible for the CI dependence. The largest part is due to the move choice module involving the structure and payoffs of the current game (see Section 7.6.1.2).

As shown in this section, Model-A seems to be adequate to account for the experimental findings and its properties are further explored in the next section.

### 7.6.3   The role of anticipation on cooperation

In Figure 7.11, the dependence of cooperation on CI is shown in the case of two different predictions of the model about the opponent's behavior – cooperate and defect. As expected, the model cooperated more when it had predicted that the opponent will cooperate, compared to when it had predicted that it will defect ($F = 23.12$, $p < 0.01$) (see Figure 7.11).

In Figure 7.12, the dependence of the mean cooperation on the game CI is given. It is seen that even in this case of predicted opponent's move D, the model displays a non-zero probability for cooperation, increasing with CI (see also Figure 7.11). There are several factors working together for the model to behave in such a way: The first one is that the PD games' normalization ($T = 1$, $S = 0$, and R and P distributed in this interval) entailed a strong

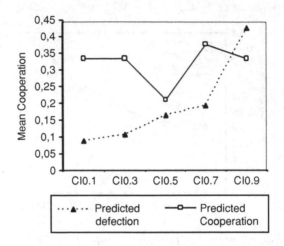

*Figure 7.11    Cooperation of the model at different levels of CI depending on the prediction for the opponent's move (rounded prediction for the opponent's move (C or D) for the current game). Solid line: opponent's move is predicted to be C; dashed line: opponent's move is predicted to be D*

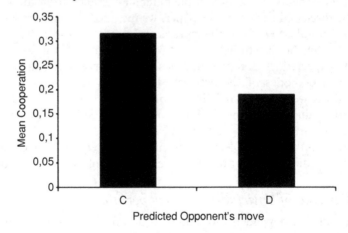

*Figure 7.12    Mean cooperation of the model as a function of the predictions for the opponent's move (the predictions for the opponent's move were either D or C if 0.5 is taken as a threshold for a D or a C move)*

negative correlation of P with CI ($r = -0.92, p < 0.01$). For example, in the case when the CI was 0.9, P was equal to 0.03, and when CI was 0.1, P was 0.83. Thus, the probability of cooperation, calculated in the move choice module, increase with decreasing P (as a payoff received from a DD outcome).

Another factor is related to the fact that the soft-max functions (see (7.7)) which gives probabilities close to 0.5 when $Poff_C$ and $Poff_D$ have close values. This is the case for instance for CI = 0.9, when S = 0 and P = 0.03, for normalized games.

Moreover, the cooperation of the model increased with the expected opponent's probability for cooperation for the relatively large value of $\beta$ (0.7), which imply larger anticipation (see Section 7.6.1.2). To analyze this trend, the predictions of the model, for the move probabilities of the opponent, were divided into five ranges and the mean cooperation averaged for each of them (see Figure 7.13). It is seen form Figure 7.13 that in average, the model cooperated more when it predicted a larger probability of cooperation for the opponent.

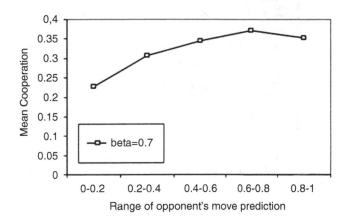

*Figure 7.13 Cooperation of the model at five successive predictions of the cooperation probability for the opponent*

## 7.6.4 Summary

A connectionist model was presented, incorporating the principles and the main features of the CSRL model. The detailed comparisons of experiments with human participants show a good agreement between model and experiment. It was shown that anticipation leads to larger cooperation and the interplay between predictions about the opponent's behavior influence the behavior of the model player, leading to higher cooperation for higher probability for cooperation by the opponent.

In Table 7.1, a comparison between the models discussed in the preceding sections is given. The comparison is based on the main features whose

*Table 7.1  Comparison of models for iterated social dilemmas*

| Model | Authors | Designation (games modeled) | Back-ward-looking | Aspiration level | Interaction with the opponent | Fictitious play | Forward-looking | Game matrix (present game played) |
|---|---|---|---|---|---|---|---|---|
| BM | Macy and Flache | IPDG, Chicken, Stag hunt | Yes | Yes | No | No | No | No |
| ER | Roth and Erev | IPDG | Yes | Yes | No | No | No | No |
| EWA | Camerer and Ho | Public goods, etc. | Yes | Yes | No | Yes | No | No |
| Sophisticated EWA | Camerer, Ho, and Chong | Public goods, etc. | Yes | Yes | Yes | Yes | Yes | No |
| BRS | Isaac, Walker, and Williams | Public goods | Yes | Yes | Yes | No | Yes | No |
| "Alice," "Bob" | Taiji and Ikegami | IPDG | Yes | No | Yes | Yes | Yes | No |
| CSRL | Hristova and Grinberg | IPDG | Yes | Payoff prediction | No | No | Yes | Yes |
| Model-A | Lalev and Grinberg | IPDG | Yes | Payoff prediction | No | Yes | Yes | Yes |

importance was discussed in the previous sections. All models use backward-looking mechanisms. The aspiration level in various forms seems to be also a widely shared feature. In CSRL and Model-A the aspiration level is not explicit but is related to the model expectations about next payoff in the game. Only the model CSRL and Model-A can account for the payoff matrix of the game with variable CI.

## 7.7 Multi-agent simulations

As seen from the comparison of experiments with human participants, Model-A presented in the previous section gives a good account for human playing in IPDG against a TF2T player. In this section, we present the results from simulations of the interactions in a society of artificial Model-A players. The aim of the simulations was to investigate what is the role of anticipation in a society of payoff-maximizing agents on cooperation and coordination among them (Grinberg and Lalev, 2009).

### 7.7.1 Simulated agent societies

Groups of ten agents played IPDG in a simulated social environment in randomly assigned pairs. The length of the IPDG interaction sessions was 100 games for a pair of players. The PD game payoff matrices used in the simulations were identical to the ones used in the previous sections, i.e., with CI ranging from 0.1 to 0.9.

In a society, only one pair of agents at a time played a whole game session. The pairs were chosen randomly with replacement so it was possible that one or both players from the previous IPDG session also played in the current one. There were 50 sessions in a simulation. After the end of a session, the agents kept the SRN weights and they were used as initial weights for the IPDG series with the next opponent. The sequences (four previous games) of the last inputs and targets were also kept for each particular agent as a memory from the previous session. These served as initial inputs and targets in the next IPDG sessions for the agents. When a new session began in the sequence of inputs the values of the new PD game's payoffs were used in the input vector along with the values for the last payoff, last own and opponent's moves.

The overall performance of all players in the society determined its specific states and processes. When starting a new IPDG session, each player was influenced by its experiences in previous sessions with other opponents from the same society. In these simulations no mixing of agents from different societies has been done. This simulation scheme was chosen to have some common basis for comparisons with the simulations with Model-A alone and with the experimental results reported earlier. To investigate the role of anticipation, several parameters of the agents were varied like the number of the recurrent network's hidden units, the training method, and the importance and number of fictitious games used for move evaluation (the parameter $\beta$ in (7.6)).

We considered five societies of agents by varying their capabilities to predict future opponent's moves and received payoffs:

- Low-anticipation society – agents without anticipation of payoffs and opponent's move beyond the present PD game, i.e., $\beta = 0$ in (7.6);
- Model-A-30 society – agents implementing exactly Model-A (30 hidden units) from Lalev and Grinberg (2007);
- Model-A-50 society – agents with a larger number of hidden units (50 hidden units), with expected increased predictive power;
- Pseudo-rehearsal society – agents with 50 hidden units and the pseudo rehearsal training method used (see Ans *et al.*, 2002 for details) with even larger predictive power (the method circumvents the neural networks' catastrophic interference problem and improves the learning and therefore the predictive capability of the model by a rehearsal procedure using pseudo training vectors);
- High-anticipation society – agents with 50 hidden units and strengthened anticipation predispositions: the number of fictitious games was set to 10 (twice as more as in the usual training procedure of Model-A) as well as the importance of remote games was increased by setting the discount parameter to 0.9 (the usual value is 0.7).

### 7.7.2   Simulation results and discussion

To compare the five societies of agents, with various anticipation capabilities, we have concentrated on the following characteristics: cooperation rate, payoffs, distribution of game outcome types, and coordination in cooperation (sequences of games in which both agents cooperated simultaneously).

#### 7.7.2.1   Cooperation rates

In a simulated society, agents played on average ten IPDG sessions, being randomly assigned to pairs of players. Agents retained their learned strategies between interaction sessions as explained in Section 7.7.1. Thus, with each next session, their experience accumulated. The influence of experience on the mean cooperation in a session is shown in Figure 7.14.

Figure 7.14 also shows the importance of the anticipatory capabilities of the agents for sustaining reasonable levels of cooperation over time. The most constant level of cooperation (although not the highest) was exhibited in the Pseudo-rehearsal society, which has the maximal predictive power. The lowest cooperation (almost 0) was observed in the Low-anticipation society. In all other cases, the cooperation rate showed a tendency of gradual decrease with time or low cooperation rate for all sessions.

In Figure 7.15, the mean cooperation in the agent societies is presented. It is seen that cooperation increases with anticipation capabilities and reaches about 0.3 for the High-anticipation and Pseudo-rehearsal societies, while in the Low-anticipation society it is below 0.05. There was no significant difference between the mean cooperation of the High-anticipation

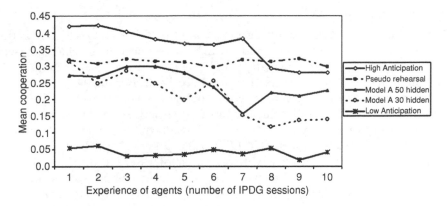

*Figure 7.14   Mean cooperation rates of (Low- to High-anticipatory agents) as a function of experience measured in terms of the number of IPDG sessions in a society*

and Pseudo-rehearsal simulations (F = 1.45, p = 0.231) (see Figure 7.15). These two societies had the highest cooperation rates among the societies as there was a significant difference between the mean cooperation of the Pseudo-rehearsal society and the Model-A-50 society ($F = 18.72$, $p < 0.01$). There was also no difference in the cooperation rates between the agents from the Model-A-30 and Model-A-50 societies ($F = 1.93$, $p = 0.168$). Their mean cooperation rates were higher than the mean cooperation in the Low-anticipation agent society as in the comparison between Model-A-30 and the Low-anticipation societies, $F = 69.95$ and $p < 0.01$ (see Figure 7.15).

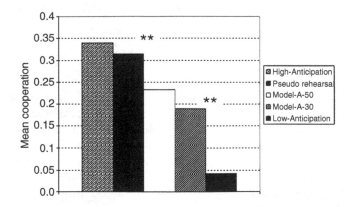

*Figure 7.15   Mean level of cooperation in simulations of different societies (** indicates significance p < 0.01)*

### 7.7.2.2   Payoffs

The mean payoff received by agents is another interesting characteristic because the agents use payoff-maximizing move selection mechanism (see Figure 7.16). The High-anticipation and the Pseudo-rehearsal societies did not differ in the mean payoffs that were received ($F = 0.004$, $p = 0.953$) which seem to be the largest among the societies. They received statistically significantly higher payoffs than Model-A-50 society ($F = 7.82$, $p < 0.01$). The payoffs of the society with Model-A-30 agents did not significantly differ from those of society Model-A-50 ($F = 2.36$, $p = 0.128$). The Low-antici-pation society got the lowest payoffs as its payoffs were lower than Model-A-30 society's ($F = 62.21$, $p < 0.01$). In general, the comparison of the two analyses of cooperation and payoffs (see Figures 7.15 and 7.16) show that higher anticipation leads to higher cooperation and respectively higher payoffs.

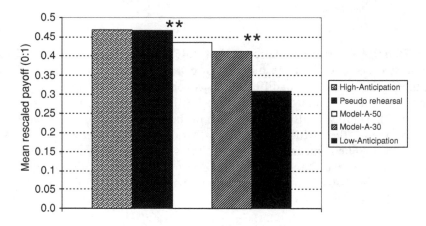

*Figure 7.16    Mean rescaled payoffs in simulations of different societies*
*(\*\* indicates significance p < 0.01)*

### 7.7.2.3   Game outcomes

As expected from the results about cooperation of the previous subsection, the High-anticipation and the Pseudo-rehearsal societies showed the largest number of CC games (although a little bit more than 10%) and the smallest number of DD games (see Figure 7.17). The number of CC games was not significantly different for these two simulated societies ($F = 0.74$, $p = 0.39$). On the other hand, the DD game outcomes were more for the Pseudo-rehearsal society than in the High-anticipation society ($F = 4.99$, $p < 0.05$). For the other societies (Model-A-50, Model-A-30, and Low-anticipation societies), the number of CC games was significantly lower with the Low-anticipation society having the smallest number (see Figure 7.17). Concerning mutual defection (DD) the situation is the inverse. In the High-anticipation society the smallest number of DD games was observed. The largest mean number of DD games per IPDG session (more than 90% of the games) was reached in the Low-anticipation society simulation. For the DD game outcomes there was no significant difference between Model-A-50 and Model-A-30 societies ($F = 1.8$, $p = 0.183$).

A tendency of increase of the mean number of CC games per simulation is observed with increasing anticipatory predisposition of the agents in different societies (see the definition of the societies in the beginning of Section 7.7.1). The opposite is valid for the mean number of DD games per simulation, regarding the anticipatory propensities of agents in the societies (Figure 7.17).

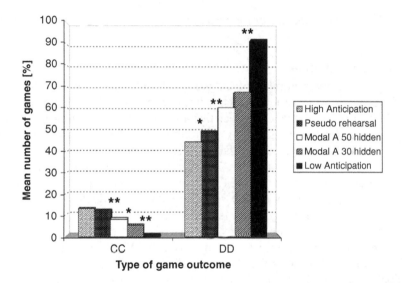

*Figure 7.17    Mean number of CC and DD game outcomes calculated for the agent societies (\* indicates significance p < 0.05; \*\* indicates significance p < 0.01)*

For each agent society, we calculated the mean cooperation rates of agents for games with a specific CI (see Figure 7.18). It was interesting to see if the dependence on CI will be preserved in the societies of agents in which they played among themselves and not against a TF2T opponent as it was the case in the experiment replication (see Section 7.3). In all societies the monotonously increasing dependence of cooperation on the CI is clearly observed except for the Low-anticipation society. This confirms again the role of anticipation in getting this dependence as in the experimental results with human subjects (Rapoport and Chammah, 1965).

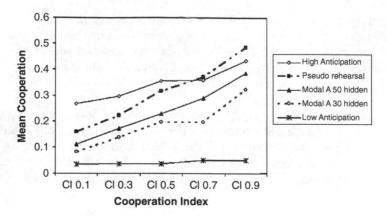

*Figure 7.18    CI dependence of the mean cooperation rate in the agent societies*

### 7.7.2.4    Coordination

We adopted the mean number of CC games played in a row per IPDG session as a first measure of the level of coordination between the agents. In Figure 7.19, the statistics for the agent societies are presented. The longest CC coordination lasted for five games and was found only in the High-anticipation and Pseudo-rehearsal societies. Four-games-long sequences were observed also in latter two and also in the Model-A-30 society. In the Low-anticipation society no sequences longer than two were found. Although the sequences are not very long (especially compared to DD sequences some of which were 100 games long) the influence of anticipation is clearly seen.

This conclusion is confirmed by the analysis of the number of agents in a society that participated in a CC game sequence of given length (see Figure 7.19). It is seen, for example, that only 70% of the agents from the Low-anticipation society ever played a CC game, whereas for all other societies this percentage equals 100 (see Figure 7.20). Moreover, a considerable number of agents with sequences of CC games longer than two are observed only in societies with anticipation.

### 7.7.3    Evolution of simulated societies

The agent societies from the previous subsection did not evolve. The question considered in this section is whether anticipatory properties can emerge in evolving agent societies and under what conditions. Therefore, we ran and explored several simulations of agents with evolving anticipatory parameters (Grinberg and Lalev, 2009). The aim of the evolution (and the fitness function) was either to achieve a highly cooperative society, or a "rich" society, with the highest possible group payoffs. An additional attempt was made to compare the behavior within a 10-agents' society, and of only 2 agents interacting with each other, with the same overall number of games per agent in both cases. Then we investigated the evolved parameters paying attentions to the anticipatory capabilities of the agents.

We maintained the general settings and rules of societies as described in Section 7.7.1. All agents in a society were identical with respect to the model

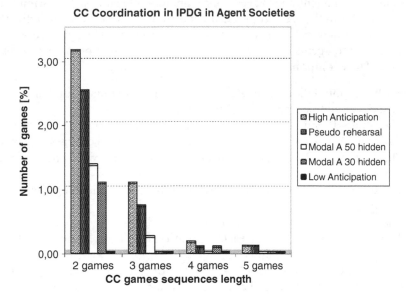

*Figure 7.19*   *Agents' coordination in terms of the mean length of the series of mutual cooperation (CC games) per IPDG session averaged over 50 IPDG sessions in each of the agent societies*

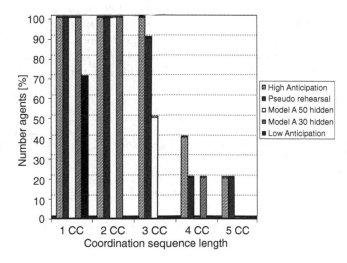

*Figure 7.20*   *Relative number of agents that played a series of CC games of a given length for each agent society*

used. This included the number of considered fictitious games, discount factor $\beta$, number of hidden units, etc. Thus each society was characterized by a single model and a single set of parameters for this model and represented an individual. The agents of each society interacted by playing in pairs sequences of IPDG until each agent had played approximately 500 games (10 sessions of 50 games each with the same opponent). There were several such individuals (societies with identical agents and parameters) that were run in a generation. During each run agents could learn but had their anticipation parameters kept constant. The evolutionary algorithm was the following:

- Evolving populations consisted of only ten individuals (or ten separate societies) due to computational limitations;
- The fitness function evaluated how far the individual was from the ideal case, e.g., 100% cooperation or which societies had the largest payoffs;
- The best five individuals (societies) of a population were chosen to give procreation;
- There were 100 generations for each simulation (maximizing cooperation or payoff);
- The initial values of the parameters responsible for anticipation were deliberately chosen low for each simulation: fictitious games – 1; $\beta$ – 0.1; and number of hidden units – 5;
- The ranges of parameters allowed were: for fictitious games – [1, 10]; for $\beta$ – [0, 1]; and for the number of network hidden units – [1, 50].

The Matlab genetic algorithms tool was used to carry out the simulations.

### 7.7.3.1 Evolving cooperative simulated societies

In the first simulation the goal was to explore the evolution of anticipatory capabilities when cooperation among the agents is stimulated by the environment, i.e., when higher cooperation means higher fitness. The general observation is all three evolved parameters increased their values leading to higher anticipatory capabilities. The number of fictitious games typically reached 9 at

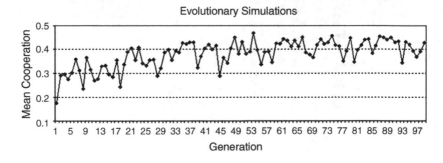

*Figure 7.21    Mean cooperation of the agent societies with highest cooperation in 100 consecutive generations*

the end of the simulation, the discount factor $\beta$ reached 1, and the number of hidden units was about 35–40. Thus under the conditions of the simulation, the anticipatory properties of the agents evolved (grew) to achieve the higher cooperation required by the environment. As seen in Figure 7.21, the cooperation rates reached a plateau of mean cooperation of about 40%.

It turned out that there was no influence of the number of agents in a society on the fitness value reached. In both cases, when the society consisted of ten or only two agents, they reached about 40% of cooperation for the finally evolved societies.

### 7.7.3.2   Evolving payoff-maximizing simulated societies

Using the same setting, we changed the fitness function to a payoff maximizing one. The expected result after evolution was that payoff maximizing, mediated by cooperation, would lead to increase in the anticipation of the agents as in the previous simulation.

We observed the same phenomenon of increasing anticipatory capabilities in the course of evolution. Compared to the finally evolved parameters in the cooperation-maximizing condition some differences have been noticed. Apparently, anticipation developed less in the payoff-maximizing condition, as the number of fictitious games reached was typically 6, as compared to 9 in the cooperation-maximizing condition. Similarly, $\beta$ reached a value of about 0.6 in the payoff-maximizing condition against a value close to 1 in the cooperation-maximizing condition.

As for the influence of the number of agents in a society (2 or 10), it turned out that the number of hidden units was higher in the 2-agents' case, than in the 10-agents' case (45 and 22, respectively), in this fitness condition. This was the only parameter difference between the two types of evolved payoff-maximizing societies. But it obviously had a big impact on the overall performance regarding the societies' fitness. In the 2-agents case, the mean fitness measured in normalized payoff (1 being the maximum) was 0.38, whereas in the 10-agents case, the mean fitness was much higher – 0.6.

### 7.7.4   Discussion

As visible from these simulations, anticipation is a solution to the problem of sustainable high cooperation and payoffs. The evolving societies of agents reached anticipatory properties very close to those we initially predefined in the simulations from Section 7.7.1. Concerning the number of agents in a society, the results were similar in the cooperation-maximizing condition. But anticipatory capabilities, as evidenced by the number of hidden units, the number of forward-looking games, and the value of the discount factor, developed more in the cooperation maximizing than in the payoff-maximizing simulation, despite the connection between them in IPDG.

There was an interesting observation when evolving payoff-maximizing societies. Judging from the results, the mean payoff for a society could reach higher values when the society is composed of more agents (10 in the simulation), than when it has less members (2 in the simulation). The understanding of this phenomena needs further exploration.

## 7.8   Summary and general discussion

This chapter presents the main progress, made so far, in the exploration of how people play games from an essentially cognitive science perspective. One of the goals of this research program is to be able to build reliable models of game players which account for the behavior of human participants in experiments. A more distant goal is to use such models in multi-agent artificial societies based entirely on games (including the PD game). In our opinion (see also Sun, 2006), the cognitive models behind the agents should be tested and validated empirically.

The results presented here, show to what extent such an approach is promising. Several aspects of PD game playing have been considered both theoretically and empirically and have been shown to be important for explaining the decision-making process and cooperation in particular. Such aspects are context sensitivity (e.g., the payoff structure of the game and priming), selective use of information, existence of more than one type of players in the same group of players, and most importantly – the effects of anticipation on cooperation and coordination. The features of the two models explored (CSRL and Model-A) seem to be necessary for a fair account of the experimental findings and will be preserved in future models.

The theoretical results seem to indicate that the higher the anticipatory ability is, the higher the cooperation rate and the coordination in cooperation between agents are. In the same time, anticipatory agents opposed to each other get involved into sophisticated behavior which needs further investigation. The empirical exploration of the role of anticipation on cooperation, based on the prediction of our simulation is in progress and the results will be reported shortly. Another interesting question, which must be explored further, is the evolution of anticipatory mechanisms when cooperation and coordination are part of the fitness function. The results reported here, show that when the fitness function of an environment requires cooperation or payoff maximization, the anticipatory capabilities of the agents grow (especially in the former case).

The implementation of a multi-agent platform which will enable the deeper investigation of artificial societies based on the models presented here is being currently developed and behavioral experiments with human participants are in preparation.

## Acknowledgements

The work presented in this chapter has been supported by FP6 EC MindRACES project and the euCognition network.

# References

Ans, B., Rousset, S., French, R. M. and Musca S. (2002). 'Preventing catastrophic interference in multiple-sequence learning using coupled reverberating Elman networks'. *Proceedings of the 24th Annual Conference of the Cognitive Science Society*. New Jersey: LEA.

Antonides, G. (1994). 'Mental accounting in a Sequential Prisoner's Dilemma game'. *Journal of Economic Psychology*, **15**, 351–374.

Axelrod, R. (1984). *The Evolution of Cooperation*. New York: Basic Books.

Berger, U. (2005). 'Fictitious play in $2 \times N$ games'. *Journal of Economic Theory*, **120**, 139–154.

Brown, G. W. (1951). 'Iterative solution of games by fictitious play'. In: Koopmans, T. C. (Ed.), *Activity Analysis of Production and Allocation* (pp. 374–376). New York: Wiley.

Camerer, C. F. and Ho, T.-H. (1999). 'Experience-weighted attraction learning in games: Estimates from weak-link games'. In: Budescu, D., Erev, I., and Zwick, R. (Eds.), *Games and Human Behavior: Essays in Honor of Amnon Rapoport* (pp. 31–52). New Jersey: LEA.

Camerer, C. F., Ho, T.-H., and Chong, J.-K. (2002). 'Sophisticated Experience-Weighted Attraction learning and strategic teaching in repeated games'. *Journal of Economic Theory*, **104**, 137–188.

Chater, N., Vlaev, I., and Grinberg, M. (2008). 'A new consequence of Simpson's paradox: Stable cooperation in one-shot prisoner's dilemma from populations of individualistic learning agents'. *Journal of Experimental Psychology: General*, **137**, 403–421.

Colman, A. (1995). *Game Theory and its Applications in the Social and Biological Sciences*. Oxford: Butterworth-Heinemann Ltd.

Colman, A. (2003). 'Cooperation, psychological game theory, and limitations of rationality in social interaction'. *Behavioral and Brain Sciences*, **26**, 139–153.

Cooper, R., De Jong, D., Forsythe, R., and Ross, T. (1996). 'Cooperation without reputation: Experimental evidence from Prisoner's Dilemma games'. *Games and Economic Behavior*, **12**, 187–218.

Dawes, R. (1980). 'Social dilemmas'. *Annual Review of Psychology*, **31**, 169–193.

Elman, J. L. (1990). 'Finding structure in time'. *Cognitive Science*, **14**, 179–211.

Erev, I., Bereby-Meyer, Y., and Roth, A. E. (1999). 'The effect of adding a constant to all payoffs: Experimental investigation and implications for reinforcement learning models'. *Journal of Economic Behavior and Organization*, **39**, 111–128.

Erev, I. and Roth, A. (1998). 'Predicting how people play games: Reinforcement learning in experimental games with unique, mixed strategy equilibria'. *American Economic Review*, **88**, 848–879.

Erev, I. and Roth, A. (2001). 'Simple reinforcement learning models and reciprocation in the Prisoner's Dilemma game'. In: Gigerenzer, G. and Selten, R. (Eds.) *Bounded Rationality: The Adaptive Toolbox*. Cambridge: MIT Press.

Fudenberg, D. and Levine, D. K. (1995). 'Consistency and cautious fictitious play'. *Journal of Economic Dynamics and Control*, **19**, 1065–1090.

Furnham, A. and Quilley, R. (1989). 'The Protestant work ethic and the prisoner's dilemma game'. *British Journal of Social Psychology*, **28**, 79–87.

Gärdenfors, P. (2006). *How Homo Became Sapiens: On the Evolution of Thinking*. Oxford: University Press.

Grinberg, M. and Hristova, E. (2009). 'SARL: A computational reinforcement learning model with selective attention'. In: Taatgen, N. A. and van Rijn, H. (Eds.), *Proceedings of the 31st Annual Conference of the Cognitive Science Society*, pp. 821–826. Austin, TX: Cognitive Science Society.

Grinberg, M., Hristova, E., and Popova, M. (2006). 'Applicability of eye-tracking information acquisition methods for studying the strategy dynamics in the iterated Prisoner's Dilemma game'. Position paper in the workshop: What have eye movements told us so far, and what is next? *28th Annual Conference of the Cognitive Science Society*.

Grinberg, M. and Lalev, E. (2008). 'Anticipation in coordination'. In: Pezzulo, G., Butz, M. V., Castelfranchi, C., and Falcone, R. (Eds.), *The Challenge of Anticipation: A Unifying Framework For the Analysis and Design of Artificial Cognitive Systems. LNAI*, vol. 5225, pp. 219–239. Berlin: Springer.

Grinberg, M. and Lalev, E. (2009). 'The role of anticipation on cooperation and coordination in simulated Prisoner's Dilemma game playing'. In: Pezzulo, G., *et al.* (Eds.), *ABiALS 2008, LNAI*, vol. 5499, pp. 209–228.

Ho, T-H, Wang, X., and Camerer, C. F. (2008). 'Individual differences in the EWA learning with partial payoff information'. *The Economic Journal*, **118**, 37–59.

Hofstadter, D. R. (1985). *Metamagical Themas*. New York: Basic Books.

Hristova, E. and Grinberg M. (2004). 'Context effects on judgment scales in the Prisoner's Dilemma game'. *Proceedings of the 1st European Conference on Cognitive Economics. ECCE1*, Gif-sur-Yvette, France. Available from http://ceco.polytechnique.fr/COLLOQUES/ECCE1/index2.htm#acces

Hristova, E. and Grinberg, M. (2005a). 'Investigation of context effects in Iterated Prisoner's Dilemma game'. In: Dey, A., Kokinov, B., Leake, D., and Turner, R. (Eds.), *Modeling and Using Context, LNAI*, vol. 3554, pp. 183–196. Berlin: Springer.

Hristova, E. and Grinberg, M. (2005b). 'Information acquisition in the Iterated Prisoner's Dilemma game: An eye-tracking Study'. In: Bara, B., Barsalou, L., and Bucciarelli, M. (Eds.), *Proceedings of the 27th Annual Conference of the Cognitive Science Society*, pp. 983–988. New Jersey: LEA.

Isaac, R. M., Walker, J. M., and Williams, A. W. (1994). 'Group size and the voluntary provision of public goods: Experimental evidence utilizing large groups'. *Journal of Public Economics*, **54**, 1–36.

Janssen, M., and Ahn, T. K. (2003). 'Adaptation vs. Anticipation in Public-Good Games.' Paper presented at the Annual Meeting of the American Political Science Association, Philadelphia, PA: Philadelphia Marriott Hotel Online.

Knoepfle, D. T., Tao-yi Wang, J., and Camerer, C. F. (2009). 'Studying learning in games using eye-tracking'. *JEEA*, **7**, 388–398.

Komorita, S., Hilty, J., and Parks, C. (1991). 'Reciprocity and cooperation in social dilemmas'. *Journal of Conflict Resolution*, **35**, 494–518.

Kreps, D., Milgrom, P., Roberts, J., and Wilson, R. (1982). 'Rational cooperation in the finitely repeated Prisoner's Dilemma'. *Journal of Economic Theory*, **27**, 245–252.

Lalev, E. and Grinberg, M. (2007). 'Backward vs. forward-oriented decision making in the Iterated Prisoner's Dilemma: A comparison between two connectionist models'. In: Butz, M. V., Sigaud, O., Pezzulo, G., and, Baldassarre, G. (Eds.), *Anticipatory Behavior in Adaptive Learning Systems. LNAI*, vol. 4520, pp. 345–363. Berlin: Springer.

Luce, R. D. (1959). *Individual Choice Behavior: A Theoretical Analysis.* New York: Wiley.

Macy, M. W. (1990). 'Learning theory and the logic of critical mass'. *American Sociological Review*, **55**, 809–826.

Macy, M. W. (1991). 'Learning to cooperate: Stochastic and tacit collusion in social exchange'. *American Journal of Sociology*, **97**, 808–843.

Macy, M. W. and Flache, A. (2002). 'Learning dynamics in social dilemmas'. *PNAS*, **99**, 7229–7236.

March, J. G. and Simon, H. A. (1958). *Organizations.* New York: Wiley.

Neyman A. (1999). 'Cooperation in repeated games when the number of stages is not commonly known'. *Econometrica*, **67**, 45–64.

Oskamp, S. and Perlman, D. (1965). 'Factors affecting cooperation in a Prisoner's Dilemma game'. *Journal of Conflict Resolution*, **9**, 359–374.

Pezzulo, G., Butz, M. V., Castelfranchi, C., and Falcone, R. (Eds.) (2008). *The Challenge of Anticipation. LNAI*, vol. 5225. Berlin: Springer.

Pollack, J. B. (1990). 'The induction of dynamical recognizers'. *Machine Learning*, **7**, 227–252.

Rapoport, A. and Chammah, A. (1965). *Prisoner's Dilemma: A Study in Conflict and Cooperation.* Ann Arbor: University of Michigan Press.

Rosen, R. (1985). *Anticipatory Systems.* New York: Pergamon Press.

Sally, D. (1995). 'Conversation and cooperation in social dilemmas. A meta-analysis of experiments from 1958 to 1992'. *Rationality and Society*, **7**, 58–92.

Schoemaker, P. (1982). 'The expected utility model: Its variants, purposes, evidence and limitations'. *Journal of Economic Literature*, **20**, 529–563.

Sun, R. (Ed.) (2006). *Cognition and Multi-agent Interaction: From Cognitive Modeling to Social Simulation.* New York: Cambridge University Press.

Sutton, R. S. and Barto, A. G. (1981). 'An adaptive network that constructs and uses an internal model of its world'. *Cognition and Brain Theory*, **4**, 217–246.

Sutton, R. S. and Barto, A. G. (1998). *Reinforcement Learning: An Introduction.* Cambridge, MA: MIT Press.

Taiji, M. and Ikegami, T. (1999). 'Dynamics of internal models in game players'. *Physica D*, **134**, 253–266.

Trivers, R. (1972). 'The evolution of reciprocal altruism'. *Quarterly Review of Biology*, **46**, 35–37.

Vlaev, I. and Chater, N. (2006). 'Game relativity: How context influences strategic decision making'. *Journal of Experimental Psychology: Learning, Memory, and Cognition*, **32**, 131–149.

*Chapter 8*
# Virtual humans made simple

*Jakub Gemrot, Cyril Brom, Rudolf Kadlec, Michal Bída,*
*Ondřej Burkert, Michal Zemčák, Radek Píbil*
*and Tomáš Plch*

## Abstract

Virtual humans, or intelligent virtual agents, born from the mould of academic, industrial, and science-fiction communities, have been surrounding us for more than a decade. Theoretical knowledge allowing many practitioners and enthusiasts to build their own agents is now available. However, the initial learning curve imposed by this specialized knowledge is quite steep. Even more fundamentally, there is a lack of freely available software assisting in the actual building of these agents. In this chapter, we introduce the Pogamut 3 toolkit. Pogamut 3 is a freeware platform for building the behaviour of virtual humans embodied in a 3D virtual world. It is designed for two purposes: first, to help newcomers to build their first virtual agents; second, to support them as they advance in their understanding of the topic. On the one hand, Pogamut 3 allows for rapid building of simple reactive agents and on the other hand, it facilitates development of challenging agents exploiting advanced artificial intelligence techniques. Pogamut 3 is primarily tailored to the environment of the Unreal Tournament 2004 video game, but it can be connected to other virtual world as well. It is suitable both for research and educational projects.

## 8.1 Introduction

People are fascinated by the idea of artificially created man. Who would not know Shelley's Frankenstein (1818), Jewish golems (e.g. Meyrink, 1915) or Czech Robots (1921)? The past century allowed for materialisation of some of these ideas. Undoubtedly, the two most notable kinds of artefacts that can be connected with those fantasies are software programs with some, or mimicking some, human-level cognitive abilities, such as chess-playing programs and robots. Recently, yet another kind of artefacts has emerged: *intelligent virtual agents* (IVAs) inhabiting *virtual reality* or, in other words, *virtual humans*.

Virtual humans are a kind of software agents (Wooldridge, 2002) inhabiting virtual worlds. Conceptually, they are typically conceived as consisting of two parts that are intervened with each other – a *virtual body* and a *virtual mind*. The virtual body is subject to the laws of the virtual world (e.g. gravity) and has predefined senso-motoric capabilities. These capabilities are used by the virtual mind to control the body: the mind is embedded in full sensory-motor loop.

A virtual world is a relatively ecologically plausible model of the real world, it provides a semi-continuous stream of rich sensory data, e.g. at 15 Hz, as well as a complex environment for acting. At the first glance, this is similar to the embodiment of the robots. However, at a second look, several distinctions become apparent: going from the robotic world to the virtual one brings one closer to reality in one sense at the cost of losing some detail in another – this is a trade-off. In a robotic platform, one works with low-level, many would say relatively plausible, environmental input; however, the whole environment is typically small and artificial (e.g. a simple arena with a couple of obstacles or a sloping pavement). Even though several attempts to use robots in real environments have emerged recently, consider, e.g. the Darpa challenge (Darpa, 2009), robots in these experiments still act in relatively limited domains. In virtual worlds, the environmental structure – the walls, objects and their positions – is typically represented in an explicit manner, hence its extraction for the purpose of a virtual human is much simpler than that in robotics. This is implausible, but consequently, the virtual worlds can be much larger and more complicated than environments of robots. Still, in a typical 3D world, it is possible to use sub-symbolic inputs, such as ray casting, which is a mechanism through which a virtual human can see the underlying geometry of the virtual world similar to the way rats use their whiskers.

When it comes to the developmental cost, the virtual humans are relatively cheap comparing to robots. However, the development of a virtual world and its simulator is quite expensive contrary to most robotic environments.

We are not saying that 'virtual humans' are not similar to 'robots' but are maturing the way the robots did by becoming used in many industrial applications (but in different ways than robots are!). Besides, researchers from various fields ranging from neurobiology to enactive artificial intelligence use robots as test beds for investigating and refining their ideas concerning various cognitive or motor skills, thereby producing detailed computational models. Virtual humans are not widely used for this purpose yet, except for artificial intelligence for video games; however, this technology is now mature enough to be exploited in this way (Brom & Lukavský, 2008; Jilk *et al.*, 2008). Arguably, it may be beneficial to investigate an abstract, system-level model, e.g. of localisation and navigation, episodic and spatial memory and decision-making or communication, using a virtual human rather than a robot. Some have also argued that virtual humans in video games are a befitting test bed for tackling the 'grand goal of AI' to build general human-level AI system (Laird & van Lent, 2001).

However, what a newcomer to the field, be it a researcher, a teacher, a student, or just an enthusiast, should do? Where to start? The problem is that the initial learning curve is steep and, presently, there is a lack of systematic support of education of newcomers in the IVAs community (Brom *et al.*, 2008), probably because the discipline is younger than robotics. For example, one cannot buy a virtual human like an e-puck robot.

This chapter concerns itself with our work that addresses the educational issue of bringing the virtual humans to a larger audience. The topic of this chapter is the Pogamut 3 toolkit we have been developing for the past three years, a freeware software platform for building behaviour of intelligent virtual agents embodied in a 3D virtual world. Pogamut 3 was intentionally created to facilitate beginning with IVAs, both for research and educational purposes. Additionally, Pogamut 3 can also support its users when they advance in their understanding of the topic. On the one hand, it allows for a rapid building of simple reactive virtual agents, and on the other hand, it facilitates development of challenging agents exploiting advanced techniques such as neural networks, evolutionary programming, fuzzy logic or planning.

Pogamut 3 is primarily tailored to the 3D virtual environment (VE) of the video game Unreal Tournament 2004, but it can be connected to other virtual worlds as well. Capitalising on its flexible architecture, Pogamut 3 integrates: (1) an integrated development environment (IDE) with a support for debugging; (2) a library of senso-motoric primitives, path-finding algorithms and reactive planning techniques for controlling virtual humans; and (3) tools for defining and running experiments. The platform documentations include several tutorials, videos and example projects. Support from an online forum is also provided.

Presently, the platform is most suitable for investigating control mechanisms of virtual humans for video games; however, it can be used by anyone interested in setting up experiments utilising up to 10 virtual humans inhabiting a middle-sized virtual world (e.g. 500 m$^2$). Pogamut 3 can also be used as a tool, in which university and high school students can practice their skills of building virtual agents (Brom *et al.*, 2008).

Presently, the platform is not suitable for simulations of large crowds. Also the graphical part and natural language processing have been out of our scope (e.g. there is only limited support for facial expressions and conversations).

We are now extending the platform with features that will facilitate research and education in the domains of virtual storytelling and computational cognitive psychology. The most notable extensions we are currently working on are:

1. seamless integration of the general cognitive architecture ACT-R (Anderson, 2007)
2. a tool allowing for the control of gestures
3. a generic story manager
4. a generic module enhancing characters with emotions
5. a generic episodic memory module
6. a connection of the Pogamut 3 to another 3D VE.

The next part of the chapter will briefly overview intelligent virtual agents: their application domains, most notable issues already addressed by the community and a few tools complementing Pogamut that are available. Pogamut 3 has two faces: it can be viewed both from the perspective of those who want to build control mechanisms of virtual humans (for whatever purpose) and from the standpoint of programmers willing to modify parts of the platform itself or extend it. After a brief introduction to Pogamut 3, this chapter will continue describing how to use the platform from the perspective of a virtual human developer. Then, the platform will be described from the point of view of a programmer. Finally, extensions of Pogamut 3, our current work in progress, will be reviewed. The most up-to-date information concerning this work can be found at the project's web page (Pogamut, 2009), where Pogamut 3 can also be downloaded. Teachers considering utilisation of the platform in their classes are recommended to consult (Brom *et al.*, 2008).

## 8.2   Intelligent virtual agents

The notion of IVAs began to emerge about two decades ago. Today, most IVAs are capable of real-time perception and autonomous acting in a *graphical virtual environment* (VE). These environments are predominantly 3D, even though 2D or 2½D worlds[1] are sometimes used as well, predominantly for prototyping purposes. IVAs imitate a range of human-like or, less frequently, animal-like qualities. They often respond emotionally and possess various high-level cognitive abilities, such as communication skills, social awareness, learning or episodic memory (Brom & Lukavský, 2008; IVA, 2009). Frequently, VEs inhabited by IVAs are dynamic and unpredictable similar to the real world. The VEs allow for some degree of interaction among agents and agents and humans.

During the past years, IVAs expanded to many domains, including commercial and serious video games (e.g. Aylett *et al.*, 2005a), virtual tutoring systems, therapeutic applications, virtual storytelling, cultural heritage applications, the movie industry, the cognitive science research, the military, the medical training, virtual companions applications and industrial applications. Building of IVAs is very hard. Besides graphical issues, which are a world of their own, there is the cognitive and behavioural side of the problem. Consider IVAs from FearNot! (eCircus, 2009; Figure 8.1), which is one of the largest non-military serious game featuring IVAs delivered by the academic community to date. FearNot! is an anti-bullying game targeted at primary school pupils and its IVAs represent children of the same age. These IVAs possess many abilities. At the first place, they are able to perceive their surrounding and they can execute actions that change the environment. They are able to navigate in the environment and plan their actions with respect to goals they have (Aylett *et al.*, 2006). They can interact with each other and they can communicate with a child using a

---

[1] 2½ worlds feature two full dimensions and one reduced dimension. For instance, depth planes of many arcade games represent such a reduced dimension.

text-based natural language interface. They have personalities, express emotions (Aylett *et al.*, 2005b) and feature autobiographic memory (Dias *et al.*, 2007).

*Figure 8.1    Screenshot from FearNot! – in the course of a bullying episode
[ adapted from eCircus ( 2009 )]*

The development of these agents took many man-years. However, the project (and many similar projects) accumulated a large amount of knowledge of how to build agents and with appropriate authoring tools, the development time can be substantially reduced. This would allow more people to start their own projects. Even though such tools are not readily available yet, at least not for public use, it seems that the research community started to reflect their necessity. For example, in virtual storytelling, a discipline neighbouring to the field of IVAs and overlapping with it partially, Pizzi and Cavazza (2008) put it: 'Recent progress in Interactive Storytelling has been mostly based on the development of proof-of-concept prototypes, whilst the actual production process for interactive narratives largely remains to be invented. Central to this effort is the concept of authoring ...'. These authors also review activities focused on development of authoring tools for that domain. Authoring tools have been built also for development of video game agents (reviewed in Kadlec *et al.*, 2007). The discipline of cognitive science modelling is represented by the CAGE toolkit (CAGE, 2007). There is also Netlogo (Wilensky, 1999), an excellent entry-level tool for building simple agents and running social simulations, and tools for learning programming by means of virtual agents, such as Alice (1995–2009). Also several toolkits for developing Machinimas are available, e.g. Movie Sand BOX (Kirschner, 2009).[2]

---

[2] Machinimas are movies generated using real-time 3D, typically video game, engines.

Neither any of these applications nor Pogamut 3 presents a tool allowing for building agents of the FearNot! level of complexity in a reasonable time. Nevertheless, many of them, including Pogamut 3, have made important steps towards this ideal.

While Pogamut 3 directly capitalises on the knowledge gained by many researchers during development of previous video game tools and it outperforms these tools in many respects (Kadlec *et al.*, 2007), it is rather complementary to other tools than being a rival. We depart from the other tools in several ways such as:

1. We explicitly focus on the support of newcomers, a feature only Alice and Netlogo and perhaps some Machinima tools share with Pogamut 3. However, these projects tend to support development of agents only by scripting; that is the agents' behaviour needs to be programmed step-by-step. Instead, in Pogamut 3, one can develop agents using programming paradigms befitting for controlling agents in dynamic environments better than scripting, such as reactive planning. Pogamut 3 even allows one to equip agents with cognitively plausible 'minds'. On the other hand, computer graphic issues are deeply addressed in many Machinima tools as opposed to Pogamut 3; Netlogo can run tens of simple agents, hence it can be exploited in agent-based ethological modelling (Bryson *et al.*, 2007), and Alice offers better support than Pogamut 3 for general teaching of programming.

2. We explicitly focus on simplifying the process of development of IVAs high-level behaviour. In this context, high-level behaviour is typically conceived as human-like behaviour requiring high-level cognitive abilities and lasting for minutes or tens of seconds, as opposed to low-level behaviour being more akin to motor behaviour with a shorter time span. For example, high-level behaviour is finding an object in a house, which includes recalling where the object is and going for it while possibly avoiding obstacles. Low-level behaviour is picking the object up, which includes many animation issues such as computing the trajectory of the hand reaching for the object. Conversational characters are also out of the scope of Pogamut 3. Our focus being high-level behaviour as opposed to low-level behaviour and conversational behaviour, we complement many projects stemming from the fields of computer graphics, including Alice and Machinima tools, and virtual storytelling, such as Scenjo (Spierling *et al.*, 2006).

3. Time spans of simulations that can be run in Pogamut 3 are not longer than hours or possibly days of simulation time. Essentially, Pogamut 3 employs abstraction of the human world. These two features – the time span and the ground-level abstraction – are shared by many other toolkits. In fact, most of them enable only running even shorter simulations. However, due to its design philosophy, simulations in Netlogo can employ different levels of abstractions and consequently simulate much longer intervals. For example, one can run a population dynamic model in Netlogo.

4. Pogamut 3 is not a side product of a research prototype as is the case for many other authoring tools. We directly focus on the development of our toolkit, hence it is well build and tested. Actually, some published authoring tools have never been released for public.

## 8.3 Pogamut 3 introduction

The Pogamut platform has already undergone a few major upgrades that completely changed the way a developer uses it and interacts with it. These updates reflected our major goals outlined earlier in the text. Still, all versions of Pogamut employ the same underlying architecture. This architecture features four components: (1) a simulator of the virtual world, (2) a graphical user interface (GUI) for IVAs' development, (3) a library of base classes for coding of IVAs' behaviour, including IVAs' senso-motoric primitives, (4) a middleware handling communication between (1) and (3). Roughly, the design process of developers building their IVAs is as follows: First, a developer will create an IVA, either using an external editor or in the GUI directly. In either case, he or she will employ the code from the Pogamut's library. Second, the developer will use the GUI as a launcher for running the IVA in the virtual world and for its debugging and testing.

The main scope of this chapter is Pogamut 3, the third major release of Pogamut. This section briefly reviews its predecessors, Pogamut 1 and 2, and then overviews the architecture of Pogamut 3.

### 8.3.1 History of the Pogamut platform

The first version of the Pogamut platform, Pogamut 1, worked with the Unreal Tournament 1999 video game (Epic, 1999) and it was designed as a Python library with a simple agent launcher GUI, which had only limited debugging capabilities. Its successor, Pogamut 2 (Kadlec et al., 2007), is based on Java and its GUI is a full-fledged IDE developed as a NetBeans plugin. Pogamut 2 also moved to more advanced Unreal Tournament 2004 (UT2004) (Epic, 2004). As opposed to Pogamut 1, the new version has been extensively evaluated both by our team (Kadlec, 2008; Brom et al., 2008) as well as by others (e.g. Small, 2008; Tence & Buche, 2008). It was also picked for the 2K BotPrize challenge at Australia[3] where the task assigned to the participants was to develop an agent (using Pogamut 2 platform) that can gool the panel of judges that it (the agent) is human.

It was found that the main drawbacks of Pogamut 2 were: (1) It was tightly bound to the world of the UT2004, which is a violent action game, thus limiting the field of the research to the gaming AI. Indeed, the users often brought up the topic whether the Pogamut platform could support more VEs (Brom et al., 2008). (2) The lack of documentation of the UnrealEngine, the engine of

---

[3] 2K BotPrize competition , URL: http://botprize.org [27.1.2009].

the UT2004, together with the multi-threaded design of the core of Pogamut 2 library brought an unnecessary burden for the developer. These two points led us to a decision to completely remake Pogamut 2 to simplify its core library and to make it a general platform for the development of virtual cognitive agents instead of just UT2004 agents.

## 8.3.2    Terminology

For historical reasons, we will use the terms *agent* and *bot* interchangeably for IVAs developed in Pogamut 3, as the bot is a term frequently used in the context of video games. Importantly, IVAs community predominantly employs the mind–body dualism as a design metaphor. That is IVAs are conceived as having two components: the mind, which is a set of data structures and algorithms responsible for controlling the IVA and the body, which is a set of data structures and algorithms for graphical portrayal of the IVA's body. For the former, we will use the term a *bot's logic* or a *bot's mind*. For the latter, we will employ a frequently used term *avatar*. We will also use the following two terms interchangeably to refer to environments where IVAs reside: a *virtual world* and a *virtual environment*.

## 8.3.3    Pogamut 3 overview

The Pogamut 3 new heart lies in the *GaviaLib* (General Autonomous Virtual Intelligent Agents Library). The GaviaLib (see Figure 8.2) is designed as a middleware that mediates communication between the VE (left side of the figure) and the bot's logic (right side of the figure). The Pogamut 3 platform still features UT2004 integration that makes the reference implementation of the GaviaLib bindings to a specific virtual world (see Figure 8.2). However, Pogamut's integration with different environments is now possible, as described in Sections 8.5 and 8.6 in more detail.

The communication pipeline from the virtual world to a bot's logic (transmitting virtual world events) and vice versa (transmitting bot's commands) can be easily reconfigured by a programmer willing to connect GaviaLib to another virtual world or to extend the communication protocol with UT2004. At the same time, this pipeline is hidden by a simple high-level application programming interface (API) from the eyes of regular IVA developers. On the one hand, this limits the developers to using only given VEs and on the other hand, the API allows them to access bot's senso-motoric primitives easily, so the developers only write what and when the bot should do and need not care about how the low-level communication really works. The GaviaLib interface also allows for controlling the bot's avatar remotely across the network. The GaviaLib may also be utilised by various auxiliary modules such as the reactive planner POSH (Bryson, 2001), which helps with the development of a bot's logic, or an emotion module.

The next section will introduce Pogamut 3 from the perspective of a regular IVA developer, that is refraining from low-level implementation details. After

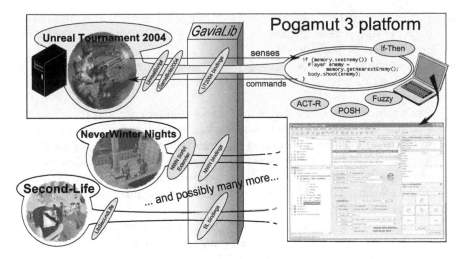

*Figure 8.2    Pogamut 3 platform overview picturing the generic GaviaLib library that acts as a mediator between the virtual world and the agent. The figure consists of screenshots from games Unreal Tournament 2004, NeverWinter Nights and Second Life copyrighted by Epic Games, Inc., BioWare, Inc. and Linden Lab respectively*

reading this part, the reader should have a basic idea about building of agents for UT2004 in Pogamut 3. Then, technical aspects of Pogamut 3 will be described for programmers willing to dive below the high-level GaviaLib API to exploit the full functionality of GaviaLib.

## 8.4    Using Pogamut 3 – Pogamut NetBeans plugin

Present-day software developers expect not only a runtime support in the form of libraries of base classes with a code they can utilise to obtain the required functionality, they also expect a design-time support by means of an IDE. This applies for a web building, database development, etc. and there is no reason why development of IVAs should be any different. Pogamut 3 offers this functionality in a form of a plugin for NetBeans, a generic freeware Java IDE. The plugin provides the following (see Figure 8.3):

- Empty bot template projects – make it easy for new users to start coding their first bot.
- Example bots – introduce various bot programming paradigms; one can derive own bots from these examples easily.
- List of UT04 servers – shows all servers registered in the IDE and their current status (e.g. RUNNING/PAUSED/DOWN, etc.) (Figure 8.3, Window 4).

- List of running bots – helps to keep a track of all running bots, shows their current status (e.g. RUNNING/PAUSED/EXCEPTION/FINISHED, etc.) (Figure 8.3, Window 4).
- Introspection of bot's variables – assists a developer in observing and changing parameters and variables of the agent during runtime (Figure 8.7).
- Bot's properties – give a quick access to variables common for all agents, e.g. position, velocity, orientation and health (Figure 8.3, Window 3).
- Log viewers – show incoming and outgoing messages together with important decisions made by the bot's logic (Figure 8.3, Window 5).
- Bot remote control – allows a developer to move the bot manually to any desired location and to send it arbitrary commands using a text field (Figure 8.3, Window 2).
- Server control – allows a developer to pause/resume the server, change the map remotely, save replays, change game speed, etc. (Figure 8.3, Window 6).

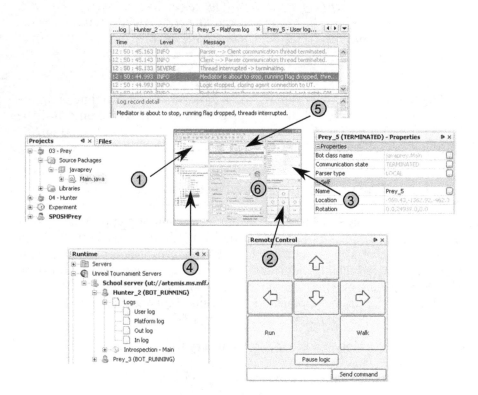

*Figure 8.3    Pogamut 3 NetBeans plugin (6) and its parts (1-5). The parts are described in the text*

These features were purposely designed to provide support during the agent development cycle, which can be conceived as having five stages:

1. inventing the agent,
2. implementing the agent,
3. debugging the implemented agent,
4. tuning the parameters of the agent and
5. validating the agent through series of experiments (see Figure 8.4).

The rest of this section illustrates how exactly Pogamut 3 supports the latter four of these stages.

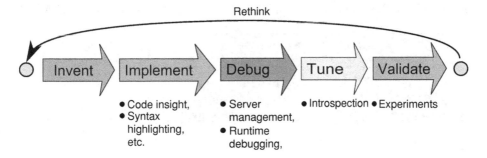

*Figure 8.4   The bot development cycle*

## 8.4.1   Implementing the model

Before a developer starts to implement an agent, he or she has to grasp two notions: the notions of *action selection mechanisms* (ASMs) and of the *communication* between GaviaLib and virtual world. The former is typically conceived as a formalism in which all what the agent can possibly do is represented *plus* algorithms deciding what the agent should actually do based on these representations and the state of the agent at a particular instant. In a sense, an ASM is the top of the agent's mind, albeit the mind possesses other capabilities as well, such as perception and memory. Note that sometimes, in the context of virtual agents, the representation of possible behaviour is referred to as *procedural memory*, which is, in a simplified way, analogical to the notion of procedural memory in neuropsychology.

The second thing a developer should understand is the logic behind the connection between a bot's mind and the virtual world. Since we are in the realm of autonomous agents, everything what the agent knows about its surrounding has to come through its senses or be given *a priori*. Obviously, the agent has to have a kind of memory to retain this information. Similarly, all that the bot's mind can do to move it's body is to issue a motor command (recall the mind–body design metaphor). How exactly this communication works? Let us start with incoming messages. For explanatory reasons, we will employ the psychological distinction between sensations and perceptions. By analogy, the incoming messages from the virtual world can be conceived as sensations. The GaviaLib features a low-level API for handling these sensations. Most of the GaviaLib library can then be perceived as a periphery, which makes high-level percepts from these low-level sensations. These high-level percepts, or sensory primitives, are accessible via the high-level API by the ASM situated at the very top of GaviaLib (see Figure 8.5). This high-level API hides complexity of the periphery, and it is this API a typical designer has to work with. Similar mechanisms are employed in the opposite way. There is a high-level interaction API providing high-level motoric primitives that are translated by GaviaLib to low-level outgoing messages. Importantly, the periphery also holds a kind of working memory available via the high-level API.[4]

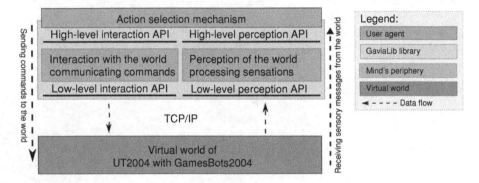

*Figure 8.5    Abstraction of the GaviaLib*

This section describes development of IVAs from the perspective of the high-level API. The 'periphery' is detailed in Section 8.5.

Now, we have the notions of ASMs and communication and we can return to our main issue: how to implement a bot using Pogamut 3 easily. At the first place, the designer has to choose an ASM that will control their agent. The platform currently supports the development of the UT2004 bot's behaviour in

---

[4] The psychological distinction between sensation, a low level process of sampling the surrounding of an organism by its receptors, and perception, a high-level process interpreting sensations, is mirrored here only partially. An agent's ASM typically works with 'percepts' (i.e. the high-level sensory primitives). Here, we may see the psychological analogy. However, virtual environments typically neither radiate physical energies to be detected by the agent's receptors nor model complex chemical processes to assist in modelling chemical senses, but mediate information more akin directly to outputs of receptors or some cells upstream in the processing pathways (e.g. a presence of an entity at a particular distance: think of robotic proximity sensors or a rat's whiskers; see also Footnote 6). In fact, virtual worlds also often send a high-level information akin to percepts to the agent, such as 'there is a ball number 23 with features $<X>$ in the direction a'. The reasons are technical: to simplify the design and to speed up the computation. Still, not all percepts send by the virtual environment arrive to the agent's ASM, the core of its mind. Some may not pass through an 'attention filter' for example. This brings us to he notion of a *possible* percept.

Java, Python, behaviour-oriented reactive ASM POSH ('Parallel-rooted, Ordered Slip-stack Hierarchical') (Bryson, 2001), and fuzzy rules. We have also been adding the cognitive architecture ACT-R (Anderson, 2007), as detailed in Section 8.6.3.

Each type of ASM has its own advantages. Using the plain Java or Python code gives the user the complete power of the underlying GaviaLib library. POSH language is similar to Halo 2 behavioural trees (Isla, 2005) (or more correctly Halo 2 behavioural trees are similar to the POSH reactive plans) and gives the user a simple access to the high-level API and enables him or her to employ decision trees as the algorithm for selection of actions to execute. The ACT-R may be used to create psychologically plausible agents. Fuzzy rules are more suitable for research projects on ASMs.

The Pogamut 3 plugin together with the Netbeans IDE provides out of the box syntax highlighting and code completion for Java and recently the same have been provided to Python.[5] The Pogamut 3 plugin also offers analogous support for the POSH, namely syntax highlighting and alternative graphical editor (Figure 8.6) for editing the POSH plans. Provision of the alternative behaviour editor makes the platform accessible not only for undergraduates but for the high school students as well.

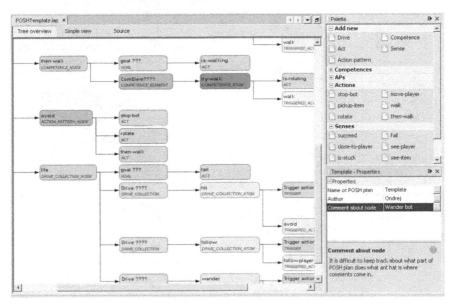

*Figure 8.6   The editor of POSH plans*

Regardless of the choice of the ASM, the designer will use some features of *UT2004 integration*, which is a set of classes tailoring common GaviaLib

---

[5] As of NetBeans version 6.5.

classes to the virtual world of UT2004, i.e. the so called *bot modules*. The bot modules are the classes that provide the high-level senso-motoric API. There are three kinds of bot modules: (1) command modules, (2) event modules and (3) advanced modules.

### 8.4.1.1    Command modules

The command modules are merely wrappers for low-level messages, the so-called command objects (Figure 8.7) that are sent to the UT2004. The command modules spare the user of the necessity to instantiate a command object every time he or she wants the bot to do something in the world. They also aggregate similar commands into one command module, thereby providing an API for moving and shooting. The provided command modules are:

- *Simple commands* – basic locomotion and shooting commands suitable for beginners
- *Locomotion* – all possible movement commands including jumping and strafing
- *Shooting* – advanced shooting commands including shooting at a target or using alternative weapon modes.

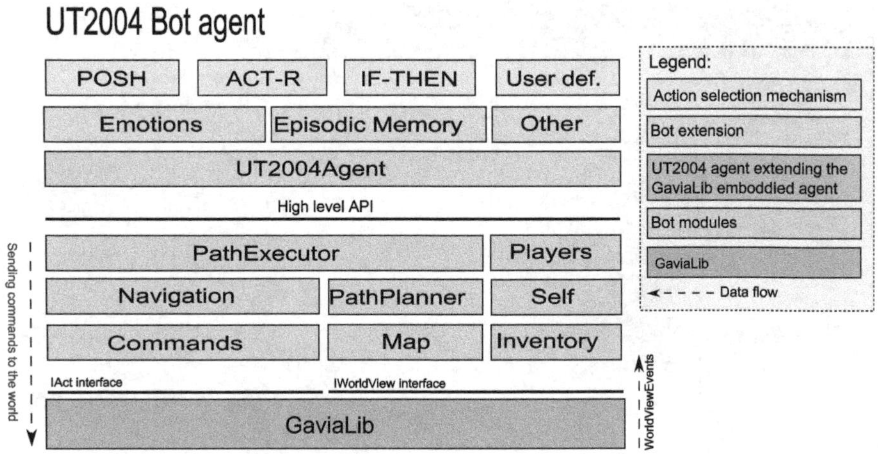

*Figure 8.7    Bot modules and agent extensions built on top of the GaviaLib. Compare this figure with Figure 8.5*

### 8.4.1.2    Event modules

Event modules represent the sensory primitives. Each module takes care about a different category of world events such as information about the bot, the map or the bot's inventory (what items the bot has). The provided event modules are:

- *Self* – provides internal sensors (level of health, armour, location in the world, etc.)

- *Map* – provides a list of navigation points and items and their positions
- *Players* – provides information about other avatars in the world; whether our agent can see them or not, their location, a weapon they are holding, etc.

### 8.4.1.3   Advanced modules

Advanced modules are usually built upon other command or event modules providing a specific functionality. At this moment, there exist two such modules – *Path planner* and *Path executor*. The *Path planner* module works over the *Map* module allowing the user to easily find the path in the world to a desired location by means of A* or Floyd–Warshall algorithm. The path is then followed by the *Path executor*, which is periodically issuing necessary commands so the bot may reach a target location.

These senso-motoric primitives will be employed by a developer when developing an ASM. The ASM can also utilise various auxiliary advanced modules, such as episodic memory or an emotion module (see Figure 8.7).

Description of how to develop the core of the agent's mind is out of scope of this chapter. The reader is recommended to consult our tutorials, which are available online (Pogamut, 2009), or an introductory paper.

## 8.4.2   Debugging

After an agent is created, it should be tested – a process that includes debugging. Actually, the processes of creation and testing are typically intertwined to some extent. Pogamut 3 supports this by enabling the user to use standard NetBeans Java debugger (step over, trace into, etc.). Pogamut 3 may pause the simulation inside UT2004 at any time, allowing the user to debug one iteration of the agent's mind. Besides the classic debugging, the Pogamut 3 toolkit introduces other tools such as log management, which is useful for inspecting the events the agent has received and the commands it has issued recently; variable introspection, which assists in inspecting the internal state of the agent; and remote UT04 server management. The IDE also offers an option to change the speed of the simulation or to stop it. This feature helps the designer to consult logs on line.

Typically, during debugging, a user will spend most time observing behaviour of his or her bot directly in the UT2004 environment. The UT2004 game provides a mode for observing the world from the spectator point of view, but this is not always convenient for debugging as many underlying information such as the navigation graph or the bot's intended path are not visible. Additionally, the spectator mode forbids the user to fly through walls. Thus we have extended this native mode with several visualisation tools (see Figure 8.8).

- Visualisation of navigation points on the map together with the visualisation of the reachability grid on the map – gives a quick overview of a bot's navigation capabilities; places without navigation points are usually unreachable (Point 1).

- A list of all avatars in the world – serves as a quick overview of avatars present in the world, both computer-driven as well as human-driven (Point 4).
- Visualisation of individual avatars on the map – allows a user to see avatars through walls, which assists in finding a desired bot in the environment quickly (Point 2).
- Visualisation of a bot's trace lines – assists in development of steering algorithms and algorithms of low-level sensation;[6] green colour indicates that the ray does not collide with anything, red indicates a collision (Point 3).
- Visualisation of the bot's field of view – helps to identify objects the bot can see; the yellow lines enclose the field of view while the white line indicates the bot's heading (Point 5).
- Visualisation of the bot's intended path – helps with debugging of the bot's navigation (Point 6, blue lines).
- Visualisation of the bot's health – helps with debugging of bots for the death-match game mode of UT2004.

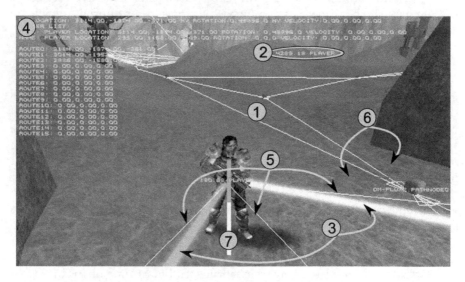

*Figure 8.8    Debugging features for the UT2004 environment. The individual features are described in the text (Copyright Epic Games, Inc.)*

## 8.4.3    Tuning the parameters

Behavioural models are typically quite sensitive to the settings of various parameters. It is convenient to have a support for changing these parameters in runtime, avoiding the need to recompile the agent's code each time a change is

---

[6] The trace lines, or ray casting lines (Fig. 8.8), serve to give a bot information about distances towards surrounding objects and walls; think of a rat's whiskers or proximity sensors of robots.

made. The IDE allows a developer to set those parameters through the Intro-spection window without the need for restarting the agent (see Figure 8.9).

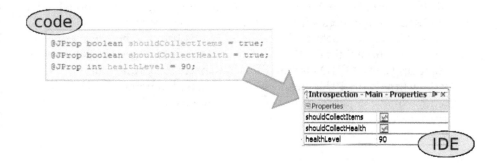

*Figure 8.9    An example of a bot's parameters introspection*

Presently, the introspection is available for agents with a Java-based, Python-based or POSH-based ASM. Java-based agents must have their para-meters annotated with `@JProp` annotation inside the code, while Python-based agents have all their class fields exported automatically. Their values are periodically updated at runtime and can be altered as well. The user may simply pause the simulation, set new values and resume the simulation.

### 8.4.4    Experiments

A developer can define various experiments in the IDE to evaluate a finished bot. Each experiment can model a particular situation, in which the bot can be run multiple times. The main idea is to allow the developer to separate the bot's logic and various test cases, a reminiscence of the so-called unit testing in Extreme Programming (XPprogramming.com).

The module for experiments enables the developer to define custom actions for various events that are based on the bot's internal variables and on the state of the environment. The API for experiments provides methods for the speci-fication of an experimental scenario including a map of the VE, a number of bots, their starting locations and their equipment. The IDE also allows users to configure breakpoints in the course of an experiment.

### 8.4.5    Beyond simple experiments – the Pogamut GRID

A solid statistical validation of a bot's behaviour requires many repetitions of an experiment. Running thousands of experiments on one computer can take days or even weeks, thus it is convenient to run the experiments in parallel on multiple computers. We have developed the Pogamut GRID for this purpose. The Pogamut GRID is based on the JPPF framework (JPPF, 2008). It sup-ports management of many UT2004 servers running in pseudo real time.

The grid architecture follows the architecture of any grid application built on top of the JPPF (Figure 8.10). The *Client* computer sends the definition of experiments to the *Driver*, which is the gateway to the GRID. The *Driver* resends the experiment to *Nodes*. The *Nodes* represent the computational units of the GRID – computers with UT2004. Each experiment is executed on the *Node* where it has been delivered. The result is sent back to the *Client* after the experiment had been finished. Powerful nodes can run multiple UT2004 servers simultaneously. The GRID also monitors system resources and pauses some of the nodes when their CPU load is close to the maximum point of sustaining the correctness of the experiment (otherwise the course of the experiment could differ from a run on a single computer).

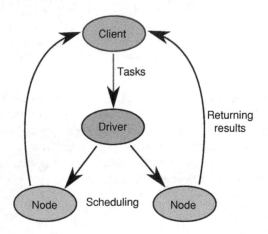

*Figure 8.10    The GRID architecture. The Client sends the definition of experiments to the Nodes through the Driver. Results of the experiments are returned directly to the Client*

### 8.4.6    Example bots

The Pogamut 3 NetBeans plugin also features a set of example bots that support first steps with the platform. These examples are tailored to be used during the class courses utilising Pogamut. Reading example bot's code is perhaps the most convenient way to begin with the Pogamut platform. The bots range in complexity from simple ones, e.g. a pathfollower, to the complex one, the Hunter bot that features a full set of behaviours that are needed for the deathmatch game mode (collecting items, combat behaviour, etc.). Examples build upon each other – every example bots' code introduces the user to a few (one to three) GaviaLib features. These features include:

- motoric library of the bot (running, jumping, shooting, etc.)
- bot's sensors and the event handling model (seeing the enemy, querying the game state, etc.)

- path-retrieving and path-following, obstacle avoidance with ray casting
- bot's customisations (its shooting skill level, appearance in the game, etc.)
- the POSH reactive planner.

### 8.4.7 Summary

Pogamut 3 brings together all basic features necessary for newcomers, to start their first projects with IVAs, as well as for advanced developers seeking for a platform that would assist them in their development and research: a virtual world simulator, the GaviaLib library containing the code that solves many tedious technical issues and provides clear high-level API, and NetBeans–IDE together with the set of example bots. Beginners can find a workflow example in Appendix A, online tutorials and the web forum useful (Pogamut, 2009).

From the educational perspective, it is important that the VE presently integrated with Pogamut 3, the UT2004 game, appeals to many (typically male) students who like to play computer games. Pogamut 3 introduces them to IVAs from an unknown perspective; from inside, so to speak. It allows them to *build* agents that they could only interact with by that time, thus giving them a gentle knock out of the door of the gaming world into the world of behavioural modelling. Importantly, Pogamut 3 can be used as a tool for high school computer science courses (Brom *et al.*, 2008).

It is no less important that from the educational perspective, UT2004 is a violent action game. This clearly limits possible applications of Pogamut 3, as also showed by the users' feedback (Brom *et al.*, 2008), and questions whether it is applicable for girls as well. The issue of connecting Pogamut 3 to another environment is detailed in Section 8.6. The section also reviews our current work in progress aiming to make Pogamut 3 more suitable for the domains of virtual storytelling, serious games and cognitive science research.

Now, we turn our attention towards technicalities of Pogamut 3, that is we introduce it from the perspective of a programmer willing to know more about how the GaviaLib library and the communication with UT2004 work.

## 8.5 Developing Pogamut 3

Developing Pogamut 3 section concerns itself with the things under the roof of the high-level API of the GaviaLib library. It describes individual components of GaviaLib and GameBots2004. After reading this section, a programmer should have a basic idea how to modify either of these two parts of Pogamut 3.

### 8.5.1 GameBots2004

The UT2004 video game, developed by Epic Games and Digital Extremes, is one of the few video games with the so-called partly opened code. The core of the game comprises UnrealEngine, providing basic functionality for running the virtual world, such as rendering 3D graphical models. To execute various functions of this core, the authors of the game built UnrealScript, a native scripting language of the game. The game mechanics, including bots' logics, is programmed in UnrealScript. While the code of UnrealEngine is not opened, the code written in UnrealScript is. This fact and additional support[7] allowing for creation of new maps and other models easily enable users to create new extensions and blend them with the original game content. This feature makes UT2004 very appealing for the community of advanced players with programming or computer graphics skills, and, not surprisingly, also for researchers. The game extensions are called *mods*.

GameBots2004 (GB2004) is a special mod for UT2004. Its main purpose is to make the rich virtual world of UT2004 available for virtual agents developed *outside* the UnrealScript. GB2004 achieves this by providing a network TCP/IP text protocol for controlling in-game avatars by external programs.

The original GameBots project (Adobbati *et al.*, 2001) was started by Andrew N. Marshal and Gal Kaminka at the Information Sciences Institute of University of Southern California. Their GameBots used older version of the video game – Unreal Tournament 1999. The goal of the project was to enable the use of the environment of the Unreal Tournament game for research in artificial intelligence.

The GameBots project was continued by Joe Manojlovich, Tim Garwood and Jessica Bayliss from RIT University (RIT, 2005). They ported original GameBots to Unreal Tournament 2004.

In the meantime, first Pogamut GameBots branch emerged due to the debugging of the old Marshal and Kaminka GameBots version. This branch was used in the Pogamut 1. Later, this branch went through major code refactorisation, implementation of new features, correcting bugs and extending the original communication protocol. Finally, it was ported by our team to UT2004 for Pogamut 2 and it is also used for Pogamut 3 (with minor extensions). This branch has been named GameBots2004. As far as we know, GameBots2004 is the only branch still developed at the time of writing this chapter.

The GB2004 mod runs a server, which provides a means for controlling avatars inside UT2004. Typically, every agent features one instantiation of the GaviaLib library and connects to this server as a client. Whenever a connection to GB2004 server is made, GB2004 spawns a new avatar into the UT2004 world and the server becomes the eyes and the ears of the agent, periodically sending sensory information to the agent while accepting its commands (Figure 8.11).

The GB2004 communication protocol is detailed in the online supplementary material (Pogamut, 2009).

---

[7] Namely UnrealEd, which is the UT2004 virtual world editor.

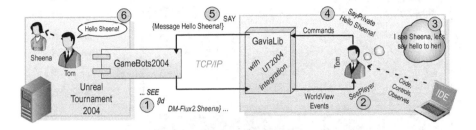

*Figure 8.11    The high-level architecture of the Pogamut 3 platform with GB2004 and UT2004 integration picturing one iteration of sense-reason-act cycle. On the left side, there is the UT2004 server with GB2004 mod and two bots (Tom and Sheena) inhabiting the virtual world. The GB2004 notices that Tom can see Sheena, therefore it exports this information with SEE message (1). SEE message is then translated into a Java object inside GaviaLib. Thanks to the UT2004 integration and presented to the Tom's ASM via the high-level API in the form of so-called WorldView event SeePlayer (2). The Tom's ASM decides that he should greet Sheena (3) and issues the SayPrivate command (4). The command is translated into the low-level SAY message (5) that is understood by the GB2004, which orders Tom's avatar to do it (6)*

## 8.5.2    GaviaLib

The real core of the Pogamut 3 framework lies in the GaviaLib, which acts as a middleware between a virtual world and the core of agents' minds. The important point is that GaviaLib is defined as a generic library for different virtual worlds and agents utilising different ASMs.

However, Pogamut 3 also features a set of classes extending GaviaLib to enable development of UT2004 agents. This extension is called UT2004 integration.

GaviaLib itself is composed of the following three parts (see Figure 8.12):

1. Perception of the world – this part serves to process information messages from the virtual world that Pogamut 3 is connected to.
2. Interaction with the world – this part sends commands to the virtual world.
3. Agent abstractions – this part comprises stubs of agents, defining high-level interface for agent's minds.

UT2004 integration customises each of these parts for the purposes of UT2004 agents.

1. Perception of the world – this part tailors the handling of incoming communication to the protocol employed by GB2004.

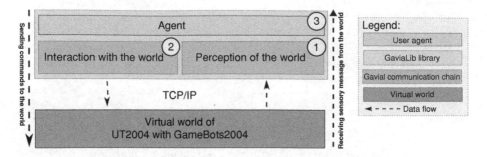

*Figure 8.12    GaviaLib architecture concept – the library consists of three parts described in the text. Compare this figure with Figure 8.5*

2. Interaction with the world – this part tailors the outgoing commands of GaviaLib to the communication protocol of GB2004.
3. Agent abstractions – it provides abstract classes for three types of UT2004 agents together with the high-level API for bot developers; essentially, this API presents the bot modules described in the previous section.

Note that agents from diverse worlds can differ in classes not only for perception and interaction but also for agent abstractions. The reason is that high-level information about what an agent knows about its virtual world is represented in these classes and these representations can be different for different virtual worlds. As an example, consider high-level terrain representation. While UT2004 uses navigation points, the world of Quake III (Id software, 1999), another action game, which is sometimes utilised as a VE, describes the terrain by means of convex areas.

A developer can either use the high-level API (3) to communicate with a virtual world, or dive bellow this API and use perception (1) and interaction (2) classes to control a bot's avatar more directly. While the former option is suitable for beginners, the latter is advisable only for advanced developers. A programmer willing to connect Pogamut 3 to a new VE will need to customise (1–3) for the purposes of this environment.

Generally, GaviaLib utilises two Java technologies:

1. Java Management Extension (JMX, 2008) for remote agent instrumentation and
2. Guice [8] (2008) for dependency injection.

The rest of this section overviews the three parts of GaviaLib and then details the whole communication pipeline from an agent's mind to a virtual world and other way round.

---

[8] Guice is the Google lightweight dependency injection framework based on the Java annotations.

### 8.5.2.1 Perception of the world

This part of the library presents the abstraction of an agent's world view – its sensory periphery. The important assumption we made is that the virtual world is event-driven, which means that its state changes can be described by instances from a finite set of possible events. GaviaLib distinguishes between two types of events: object events and general events. The former are events describing the state of an object that resides in the world as perceived by the agent ('the agent sees a green ball'), or the change of this state ('the green ball has changed its position'). The latter are other events that the agent can possibly be aware of, e.g. 'agent hears a clap of hands' and 'agent hits a wall'.

The events are processed by the main library component called the *EventDrivenWorldView* (*EDWV*). This component also represents the agent's current known state of the world, offering the following functionality:

● maintaining a list of known world objects and updating their states via object events,
● allowing the agent to set up various listeners for incoming both general and object events,
● a locking mechanism, which enables the agent to freeze its current state of the *EDWV* for reasoning purposes.

### 8.5.2.2 Interacting with the world

The second part of the library is quite simple; it merely translates high-level motor commands of an agent's ASM to text messages that GB2004 understands.

### 8.5.2.3 Agent abstraction

This part of GaviaLib presents stubs for defining bots' logics. These stubs provide basic start/pause/resume/stop methods for controlling the bot's avatar and describe a high-level state of the bot (e.g. running, terminated, etc.). The library distinguishes between three types of agents:

1. an observer agent,
2. a ghost agent,
3. an embodied agent.

*Observer agent*
The observer agent is the abstraction of a bodiless agent that cannot perform any physical action in the virtual world – it can only receive world events. This kind of an agent can be utilised as an observer storing statistics about the world and agent behaviours, for instance, counting the number of conversations between agents.

*Ghost agent*
The ghost agent is the abstraction of a bodiless agent that can not only perceive the world but also act. It may be utilised to create a world controller agent. For

instance, this type of agent can be used in storytelling as a story director (Magerko, 2006) that is not present in the world but may observe actors and guide them (e.g. alternating their goals).

### Embodied agent

The embodied agent is an abstraction of classical embodied agent. Technically, it differs from the ghost agent in only one feature: it has the body and hence an avatar in the virtual world.

#### 8.5.2.4    GaviaLib in detail

GaviaLib is described in detail in this section. This part is quite technical and the reader willing to be spared these subtleties is suggested to skip to the next section. For explanatory reasons, the library is described together with the UT2004 integration. Figure 8.13 provides detailed overview of the library architecture. At the top of the figure, there is an agent that uses high-level GaviaLib API. The UT2004 world together with GameBots2004 is pictured at the bottom.

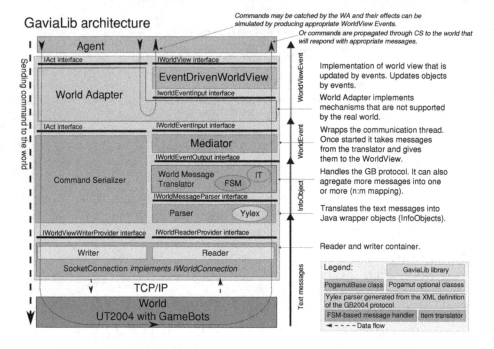

*Figure 8.13    The architecture of the GaviaLib together with the UT04 integration. This figure presents a detailed view on the architecture depicted in Figure 8.5*

We will start the description in the point when a message is generated by GameBots2004 and sent via TCP/IP to the agent client. This message utilises the GameBots2004 communication protocol, which has a text format. The generated massage is, first, accepted by the class *SocketConnection* and, second, passed over to the *Parser* that transforms the text content of the message into a Java object called *InfoObject* (which is a wrapper for parsed textual message). *InfoObject* is forwarded to the *WorldMessageTranslator*, whose main function is to translate *InfoObects* into *WorldEvent* objects and categorise them either as object or global events (see 8.5.2.1). While the semantic of *InfoObject* is based on the needs of UT2004, the semantic of *WorldEvent* objects is tailored to the needs of agents' ASMs. This translation may include aggregation of more *InfoObjects* into one *WorldEvent* object, or generation of more *WorldEvent* objects based on one *InfoObject*, for instance more PathNode *Info-Messages* may form one Path *WorldEvent*.

The whole process described in the preceding paragraph is controlled by one thread. This thread is wrapped in the *Mediator* object. Actually, the *Mediator* reads world events periodically and sends them up to the *WorldAdapter*.

The *WorldAdapter* is an optional component of the diagram. It serves for three purposes. First, it can provide a place for altering information sent by the world to the agent, if needed. For example, it can implement an attention filter. Second, a virtual world simulator may actually lack some functionality and the *WorldAdapter* can provide it. For instance, UT2004 does not compute many drives that may a developer need for his or her agents, such as stress or hunger. These drives can be implemented by the *WorldAdapter*. Finally, the *WorldAdapter* passes the *WorldEvent* objects up to the *EDWV*.

The *EDWV* is the most complex object of the GaviaLib, which has already been introduced in Section 8.1. As stated, it implements three mechanisms, which function together as a kind of working memory: (1) It maintains a list of known world objects and updates their states via object events; (2) It allows the agent to set up various listeners for incoming events; (3) It implements a locking mechanism, thus the agent may freeze the state of the *EDWV* for reasoning purposes – one usually wants to reason over the fixed world state (but this is not a mandatory behaviour).

The output of the *EDWV* is used by bot modules introduced in Section 8.1, which provides the high-level API for ASMs.

The communication from the agent towards the UT2004 is much simpler. The agent's ASM can control the agent's avatar by means of motor primitives. There is a fixed set of them as defined by GB2004. Technically, the ASM needs to send a so-called *CommandObject* wrapping a desired motor primitive to the *WorldAdapter*. This functionality is provided by command modules (Section 8.1).

Every *CommandObject* can have one of the two fates. First, it may be handled solely inside the *WorldAdapter*, which may consequently react by generating a new *WorldEvent* object back for the *EDWV*. This possibility applies for situations in which the *EDWV* implements functionality missing in the virtual world, such as simulating the missing hunger drive. Second, the *CommandObject* may be passed through the *CommandSerializer* to the

*SocketConnection* that will send it down to the UT2004 where the command is handled by GB2004.

### Object technologies

All objects instantiated from classes of the GaviaLib are strictly interacting via interfaces and all objects are created using dependency injection.[9] Every interface defines a contract [10] for a class implementing it. As the GaviaLib is utilising Guice technology[11] for instantiation of objects, it is a matter of one line of code to swap an object of a default class implementing a chosen interface with a user-created object of a class implementing this interface. For instance, if a developer wishes to implement new 'eat apple' *CommandObject* that is not supported by the UT2004 world, he or she may create own implementation of the *WorldAdapter* that will catch every 'eat apple' *CommandObject* and generate appropriate *WorldEvent* objects for *EDWV* and low-level commands for UT2004 avatar removing the apple from the agent's inventory and decreasing the agent's hunger drive.

### Flexibility of GaviaLib

The employed object technologies and the design of GaviaLib make the architecture of the library quite flexible. For instance, GaviaLib does not require the virtual world to communicate via TCP/IP. GaviaLib works with standard Java objects – *Reader* and *Writer* – which are provided by the GaviaLib's interfaces *IWorldReaderProvider* and *IWorldWriterProvider*. We may also totally omit the *Parser* and the *WorldMessageTranslator* classes in cases the virtual world is run inside the same Java Virtual Machine running the GaviaLib and connect the virtual world directly to the *Mediator* via *IWorldEventOutput* interface.

### Remote control of GaviaLib agents

The GaviaLib utilises the JMX framework that may be used for the remote management of the agents. Thus the agents may be controlled from different Java Virtual Machines even across the network. This feature is used by the IDE but it also allows for building resource-consuming agents, for instance agents with planners, and running them on multiple machines while managing all of them from one machine.

## 8.6 Limitations and prospect

The integration of Pogamut 3 with UT2004 and the support provided by GaviaLib classes make Pogamut 3 most suitable for gaming AI projects. This suits some but is unsuitable for others. Apart from other things, it is disputable

---

[9] Dependency injection means that objects are constructed with all needed objects specified within the constructor call.

[10] Contract is an expected behaviour of the object implementing a specific interface. The semantic of a contract is typically specified in the interface documentation.

[11] See Footnote 8.

whether a gaming world is the right environment for exploration of the ways how IVAs' behaviours can be built.

We are now extending Pogamut 3 with features that simplify the platform's usage also in the fields of virtual storytelling and computational cognitive psychology. This section reviews the most notable extensions.

## 8.6.1   Support for artificial emotions and gestures

When a virtual agent expresses emotions, the user can typically better understand its internal state, and thereby, whole scenario in which the agent acts. Not all present-day applications, for example, action games such as Unreal Tournament, actually need emotions to help elucidate an agent's internal state, but many do. For instance, emotions are important for serious games such as FearNot! (eCircus, 2009). Present version of Pogamut 3 supports neither modelling nor expressing emotion. However, we have been working on this issue along three lines. First, we have been integrating Pogamut 3 with a generic emotion model. Second, we have been developing new graphical models for agents' avatars expressing emotions. Third, to offer higher flexibility of avatar animation, we have decided to give users a way of influencing avatars' body animations. The first and the third points will now be detailed.

### 8.6.1.1   Emotion modelling

We have been working on integration of the ALMA model with the Pogamut 3 platform. ALMA (Gebhard, 2005) is an emotion model based on OCC (Ortony *et al.*, 1988) cognitive theory of emotions, a theory on which many virtual agents' projects capitalise. ALMA models a character's effect with OCC emotion categories and three-dimensional mood; the dimensions are pleasure, arousal and dominance. For defining an agent's unique emotion reactions, ALMA exploits 'Big five' personality model (McCrae & John, 1992). This personality model affects both the emotions as well as the mood of the agent. The ALMA model also features a powerful GUI where the user can inspect an agent's current emotions and mood and specify the agent's parameters and personality.

### 8.6.1.2   Body animation

Besides equipping agents with emotions, the challenge is to allow users to control UT2004 avatars directly from the IDE of Pogamut 3 at the level of atomic animations, parametric animations and their sequences; this is not possible presently. Examples of atomic animations include 'arm flexing', 'arm twisting' or 'arm abduction'. A parametric animation is 'pointing with a finger towards a target'. We have been extending Pogamut 3 with this feature. To allow the maximum flexibility to the user, the graphical libraries and the parser will be situated on the side of the IDE, not the virtual world.

Technically, the graphical libraries will exploit Behaviour Markup Language (BML), which was developed by (Mindmakers, 2005-8) and is intended to become 'a framework and an XML description language for controlling the communicative human behaviour of embodied conversational agents (Mind-makers, 2008). It provides a convenient way to create and parse sequences of behaviours, including gestures, facial animations and speech as well.

Communication with the virtual world will be carried in slightly modified FBA bitstream format specified in MPEG4. MPEG4 describes a protocol that allows definition and transportation of parametric animations using very little bandwidth. A user will be able to fabricate own frames, that is to control animations directly. FBA bitstream is quite modest itself, but to achieve even less traffic, we decided to include a variable frame rate. Most common human movements such as waving or clapping hands can be easily interpolated without considerable impact on believability.

## 8.6.2    Interactive storytelling

Interactive storytelling is a modern discipline that joins researchers from various fields attempting to challenge the paradigm of conventional linear stories by creating dynamically changing narratives in which users can interact. These attempts have many facets. For example, one goal is to develop a text-based system for automatic narrative generation, in which a user can make important decisions about the course of the story online. Another goal is to build an application unfolding stories in VEs inhabited by IVAs. The latter is also our scope. The starting point is a tool that would assist in rapid development of short scenarios in which several virtual characters and possibly human avatars interact. An example of such a scenario is buying a cinema ticket or asking another agent for information. Not many people realise that specifying such scenarios and their unfolding typically requires different techniques than those that are utilised by ASMs of typical video game agents. While there already exists a couple of tools for building some storytelling applications (reviewed in Pizzi and Cavazza, 2008), we are not aware of any free tool allowing for building such scenarios.

### 8.6.2.1    Technical details

One useful way of conceiving control architectures of virtual storytelling applications is the two-layered 'individual characters – story manager' conceptual framework (Magerko, 2006). Basically, besides embodied agents, there is a bodiless agent in the VE that coordinates behaviour of the embodied agents to unfold the desired scenario correctly. The embodied agents can be autonomous to some extent, but the story manager is typically able to alter their goals and needs.

The million dollar issue here is to unfold a correct story, or to generate one automatically, despite a user's interactions, that is the so-called 'narrative-interactive' tension. We do not address this question directly. Instead, our goal is to build an infrastructure for developing story directors in Pogamut 3, including a language, called StorySpeak, for the description of how to unfold particular scenarios (Gemrot, 2009). Using this infrastructure, a developer will be able to challenge the issue of 'narrative-interactive' tension.

The language alone extends AgentSpeak(L) (Rao, 1996) and gets inspiration from its Java implementation Jason (Bordini *et al.*, 2007). Basically, StorySpeak will allow for specifying joint agents' plans. The language is built without any assumptions about the underlying virtual world, meaning its core will not contain any sensory-motor primitives directly. However, we will provide also UT2004 implementation, which will define the sensory-motor primitives for this world and which will also feature a set of predefined small joint plans such as greeting.

### 8.6.3   ACT-R extension (PojACT-R)

Recently, it has been proposed that computational cognitive science modelling can benefit from implementing models using IVAs (Brom & Lukavský, 2008; see also CAGE, 2007; Jilk *et al.*, 2008; see also Laird & van Lent, 2001 for similar arguments concerning building human-level AI systems). That is a neuro-/psychologically plausible model can be integrated within an IVA's 'mind' and then tested in scenarios in virtual worlds. Advantages of this approach have been discussed in Brom and Lukavský (2008). One potential way how to do this is to adopt a general cognitive architecture, such as Soar (Newell, 1990), ACT-R (Anderson, 2007) or LIDA (Franklin *et al.*, 2005), as the main architecture of an IVA's mind. These architectures are somewhat plausible and tend to be modular, hence one can embed his or her model in one of the modules without losing much of its original plausibility, as Jilk *et al.* (2008) did for a model of visual cortex.

To allow the cognitive science modelling community to use Pogamut 3 in such a way, we have started to work on integration of ACT-R with Pogamut 3.

#### 8.6.3.1   Technical details

ACT-R is originally designed for LISP but few years ago Anthony Harrison from Naval Research Laboratory started the jACT-R project (Harrison, 2008), a Java port of ACT-R theory. jACT-R supports running of models described by the old-style LISP syntax as well as a new XML syntax. We have been integrating jACT-R with GaviaLib, developing a generic template for PojACT-R agents (i.e. Pogamut – jACT-R). This template features classes translating world events into PojACT-R messages, which can be interpreted by jACT-R as sensations of agents. In a similar fashion, the template supports translation of jACT-R motor commands to GaviaLib motor primitives allowing jACT-R to control the agent's avatar.

### 8.6.4   Episodic memory

From the psychological point of view, episodic memory (Tulving & Donaldson, 1972) represents personal history of an entity. Episodic memories are related to particular places and moments, and connected to subjective feelings and current goals.[12] It was argued that episodic memory is an important component for many kinds of virtual characters, ranging from role playing games' agents to

---

[12] In the context of virtual agents, some prefer to refer to autobiographic memory instead of episodic memory. From the psychological point of view, some distinctions between these memories can be drawn; however, we will disregard these subtleties here for the terminology is not settled among psychologists anyway.

pedagogical agents and virtual companions (Castellano *et al.*, 2008). All of these agents are supposed to 'live' for a long period of time; hence it is important that they remember what has happened. Moreover, experimenting with IVAs enhanced by episodic memory can provide us with an interesting feedback on neuro-/psychological theories underpinning our understanding of episodic memory. Importantly, this understanding advanced immensely in the past 30 years, but is still limited.

We have been working for a while on a generic model of episodic memory for virtual characters (Brom & Lukavský, 2008). Originally, the project was independent on the Pogamut platform. Now, we are integrating it with Gavia-Lib to allow Pogamut's users to develop agents with episodic memory easily. The model possesses five parts: (1) a memory for events, (2) a component reconstructing the time when an event happened, (3) a topographical memory, (4) an allocentric and (5) an egocentric representations of locations of objects. Its main functional features include: detailed representation of complex events (e.g. cooking dinner) over long intervals (days) in large environments (house), forgetting and development of search strategies for objects in the environment. The model is partly grounded in behavioural data, increasing the agent's plausibility.

### 8.6.4.1   *Dating of episodic memories*

We will briefly illustrate the intricacies of episodic memory modelling on the example of memory for time of events (the component (2)). Every agent with episodic memory should have the notion of time; otherwise it is not able to tell when an event happened. The notion of time has many facets, for instance the agent should be able to remember dating of events and to compare recency of two events. Importantly, a believable agent cannot just remember absolute time for it is known that humans are relatively poor in dating (Friedman, 1993). The notion also includes the agent's ability to adapt to a new way of life, e.g. noticing a change of a time zone. The agent should also be able to speak using relative time concepts such as 'after a lunch' or 'morning'. However, in terms of absolute time units, 'morning' is context-dependent – Monday morning is typically sooner than Sunday morning. All of these issues, and many others, should be addressed, and preferably validated on real users (i.e. users should be asked whether it seems to them that an IVA in question behaves believably). We have been addressing these issues exploiting a set of special neural networks (Burkert, 2009).

### 8.6.5   *Genetic behaviour optimisation*

Fine-tuning of parameters of a bot can be a tedious task with a slim chance of success. We have made several experiments applying genetic algorithms to this problem to prove that it can be, at least to some extent, automatised using the Pogamut platform. We have conducted two kinds of experiments: evolving high-level behavioural structures for DeathMatch and Capture the Flag Unreal Tournament game modes using genetic programming (Koza, 1992), and evolving low-level movement tasks using the NEAT algorithm (Stanley & Miikkulainen, 2002). Detailed discussion of these experiments can be found in Kadlec (2008). What is important for present purposes is that this work demonstrated that at least some evolution computation problems could be addressed in Pogamut 3. This work employed the Pogamut GRID.

### 8.6.6 Other virtual environments

As already said, the environment of Unreal Tournament 2004 limits possible use of Pogamut 3 to gaming AI. To overcome this limitation, we are adding a new environment, specifically the Virtual Battle Space (BI, 2008). Even though this project originated as a gaming VE, recently, it has been also used for military purposes, and because it features large civil environments and many models of avatars, it can be used for the purposes of serious games and storytelling applications. We are also considering other VEs, e.g. Second Life or Defcon.[13]

## 8.7 Conclusion

We end where we started, with the realisation of how little has been done so far for newcomers willing to enter the field of virtual humans. Our contribution lies in the development of the Pogamut 3 toolkit addressing this issue directly. Pogamut 3 is not a universal authoring platform, but it allows rapid development of high-level behaviour of virtual characters for beginners as well as those who are already advanced. Presently, the toolkit is most suitable for gaming AI projects, but several ongoing subprojects make it suitable also for general development of human-level AI systems, for computational modelling for cognitive neuro-/psychology and for virtual storytelling.

## Appendix – A workflow example

This appendix shows how to create the first bot for UT2004 using Pogamut 3 in several steps. This bot features a Java-based ASM. Individual steps are demonstrated in the online tutorials (Pogamut, 2009).

1. Run Pogamut 3 (consult the online video tutorial)
2. Run UT2004
3. Add a new UT server instance to the IDE. Later, you will find your running bot in the corresponding *treeview*.
4. Create a new empty bot project. The IDE contains empty project templates with all the necessary files properly configured. After creating the empty project, you can instantly run the bot without any ASM and see it standing in the virtual world.

---

[13] Defcon is a real-time strategy game where player controls one of the nuclear powers and scores points by destroying opponents strategic targets and cities.

The newly created bot project will contain a new agent class that extends the base *UT2004Agent* class of the Pogamut 3 platform. This class contains several methods that needs to be implemented, most importantly:

- createInitializeCommand() – this method is used to set the agent's name, skin, skills, etc.
- botInitialized(ConfigChange config, InitedMessage init) – this method is called after the agent connects to the game
- doLogic() – this is the main method called periodically by GaviaLib; the bot's ASM is implemented here
- botKilled(BotKilled event) – this method is called each time when the bot is killed; the method is usually used for resetting internal variables.

One of the simplest possible example bots available in Pogamut 3 is the *Follow bot*. This bot is programmed to follow any player it sees. The doLogic() method of the *Follow bot* is:

```
@JProp

boolean shouldTurn = true;


protected void doLogic() throws AgentException {
    // find visible friendly player
        Player  player  =  players.getVisibleFriends().values().
        iterator().next();
    // if the player exists
        if (player != null) {
            // then follow him
    locomotion.moveTo(player);
        } else {
            // if none player is visible then turn around to
            find any
                if (shouldTurn) locomotion.turnHorizontal(10);
            } // of else
    } // of doLogic
```

The players variable references the sensory module, which manages information about players in the game. The locomotion variable references motor module with methods for basic movement. These two modules are a part of the

UT2004 integration, which extends the general GaviaLib library. Now, you can:

1. Execute the bot, check the bot's logs and observe its behaviour directly in the game.
2. Return to the IDE and adjust values of parameters that have been previously marked with the @JProp annotation. Observe how this affects the bot's behaviour.
3. If need, pause the execution of the bot and use the arrows to get it to a desired location.
4. Decrease the game speed in server control window and resume the bot. The game will run slower. Observe details of the bot's decisions in the logs.

## Acknowledgements

This work was partially supported by the project CZ.2.17/3.1.00/31162 that is financed by the European Social Fund, the Budget of the Czech Republic and the Municipality of Prague. It was also partially supported by the Program 'Information Society' under project 1ET100300517, and by the research project MSM0021620838 of the Ministry of Education of the Czech Republic.

## References

Adobbati, R., Marshall, A.N., Scholer, A., Tejada, S.: 'Gamebots: A 3d virtual world test-bed for multi-agent research'. In: *Proceedings of the 2nd International Workshop on Infrastructure for Agents, MAS, and Scalable MAS*, Montreal, Canada. (2001) Available from URL: http://gamebots.sourceforge. net/ [26 February 2010].

Anderson, J.R.: *How Can the Human Mind Occur in the Physical Universe?* NY, Oxford University Press. (2007).

Aylett, R.S., Louchart, S., Dias, J., Paiva, A., Vala, M.: 'Fearnot! – An experiment in emergent narrative'. In: *Proceedings of IVA 2005*, Berlin, Springer Verlag *Lecture Notes in Artificial Intelligence*, vol. 3661. (2005a) pp. 305–316.

Aylett, R.S., Paiva, A., Woods, S., Hall, L., Zoll, C.: 'Expressive characters in anti-bullying education'. *Animating Expressive Characters for Social Interaction,* In: L. Canamero and R. Aylett, (eds.): Amsterdam/Philadelphia, John Benjamins. (2005b).

Aylett, R.S., Dias, J., Paiva, A.: 'An affectively-driven planner for synthetic characters'. In: *Proceedings of ICAPS 2006,* Menlo Park, CA, AAAI Press. (2006).

Bohemia Interactive Australia: Virtual Battle Space 2. (2008) Available from URL: http://virtualbattlespace.vbs2.com [15 January 2010].

Bordini, R.H., Hübner, J.F., Wooldridge, M.: *Programming Multi-Agent Systems in AgentSpeak Using Jason.* Chichester, John Wiley & Sons, Ltd. (2007).

Brom, C., Lukavský, J.: 'Episodic memory for human-like agents and human-like agents for episodic memory'. Technical Reports FS-08-04, *Papers from the AAAI Fall Symposium Biologically Inspired Cognitive Architectures.* Westin Arlington, VA, AAAI Press. (2008) pp. 42–47.

Brom, C., Gemrot, J., Burkert, O., Kadlec, R., Bída, M.: '3D immersion in virtual agents education'. In: *Proceedings of First Joint International Conference on Interactive Digital Storytelling ICIDS 2008, Lecture Notes in Computer Science 5334,* Erfurt, Berlin, Springer-Verlag. (2008) pp. 59–70.

Bryson, J.J.: *Intelligence by Design: Principles of Modularity and Coordination for Engineering Complex Adaptive Agent.* PhD thesis, MIT, Department of EECS, Cambridge, MA. (2001).

Bryson, J.J., Ando, Y., Lehmann, H.: 'Agent-based modelling as scientific method: a case study analysing primate social behaviour'. *Philosophical Transactions of the Royal Society.* London B 362 (1485). (2007) pp. 1685–1698.

Burkert, O.: *Connectionist Model of Episodic Memory for Virtual Humans.* Master thesis. Department of Software & Computer Science Education. Charles University in Prague, Czech Republic. (2009).

*CAGE: Cognitive Agent Generative Environment.* (2007) Available from URL: http://arlington.setassociates.com/cage/ [15 January 2010].

Castellano, G., Aylett, R., Dautenhahn, K., Paiva, A., McOwan, P.W., Ho, S.: 'Long-term affect sensitive and socially interactive companions'. *Fourth International Workshop on Human-Computer Conversation,* Bellagio, Italy, 6–7 October 2008. (2008).

Dias, J., Ho, W.C., Vogt, T., Beeckman N., Paiva, A., Andre, E.: 'I know what I did last summer: autobiographic memory in synthetic characters'. In: *Proceedings of Affective Computing and Intelligent Interaction 2007,* Berlin/ Heidelberg, Springer. (2007) pp. 606–617.

Defense Advanced Research Projects Agency (DARPA 2009). Available from URL: http://www.darpa.mil/grandchallenge/index.asp.

eCircus: FearNot! an anti-bullying application. (2009) Available from URL: http://www.macs.hw.ac.uk/EcircusWeb/ [15 January 2010].

Epic Games: Unreal Tournament 1999. (1999) Available from URL: http://www.unreal.com [15 January 2010].

Epic Games: Unreal Tournament 2004. (2004) Available from URL: http://www.unreal.com [15 January 2010].

Franklin, S., Baars, B.J., Ramamurthy, U., Ventura, M.: 'The role of consciousness in memory'. *Brains, Minds, and Media,* vol. 1. (2005) pp. 1–38.

Friedman, W.J.: 'Memory for the time of past events'. *Psychological Bulletin,* 113(1). (1993) pp. 44–66.

Gebhard, P.: 'ALMA – a layered model of affect'. In: *Proceedings of the Fourth International Joint Conference on Autonomous Agents and Multiagent Systems (AAMAS'05)*, Utrecht. (2005) pp. 29–36.

Gemrot, J.: *Behaviour Coordination of the Virtual Characters*. Master thesis. Department of Software & Computer Science Education. Charles University in Prague, Czech Republic. (2009).

Guice. (2008) Available from URL: http://code.google.com/p/google-guice/ [15 January 2010].

Harrison, A.: jACT-R project. (2008) Available from URL: http://www.jactr. org [15 January 2010].

Id software: Quake III arena. (1999) Available from URL: http://www. quake3arena.com/ [15 January 2010].

Isla, D.: 'Handling complexity in Halo 2'. *Gamasutra Online,* November 3, 2005. (2005).

IVA: 9th International Conference on Intelligent Virtual Agents. (2009) Available from URL: http://iva09.dfki.de/ [15.1.2010].

Jilk, D.J., Lebiere, C., O'Reilly, R.C., Anderson, J.R.: 'SAL: An explicitly pluralistic cognitive architecture'. *Journal of Experimental and Theoretical Artificial Intelligence,* vol. 20. (2008) pp. 197–218.

JMX: Java Management Extensions Technology. (2008) Available from URL: http://java.sun.com/javase/technologies/core/mntr-mgmt/javamanage-ment/ [15 January 2010].

JPPF framework. (2008) Available from URL: http://www.jppf.org [15 January 2010].

Kadlec, R.: *Evolution of Intelligent Agent Behaviour in Computer Games*. Master thesis. Charles University in Prague, Czech Republic. (2008) Available from URL: http://artemis.ms.mff.cuni.cz/main/papers/GeneticBots_MSc_Kadlec. pdf [15 January 2010].

Kadlec, R., Gemrot, J., Burkert, O., Bída, M., Havlíek, J., Brom, C.: 'Pogamut 2 – a platform for fast development of virtual agents' behaviour'. In: *Proceedings of CGAMES 07*. La Rochelle, France. (2007).

Kirschner, F.: 'MovieSandBox'. (2009) Available from URL: http://www. moviesandbox.net [15 January 2010].

Koza, J.R.: *Genetic Programming: On the Programming of Computers by Means of Natural Selection*. Cambridge, MA, MIT Press. (1992).

Laird, J.E., van Lent, M.: 'Human-level AI's killer application: Interactive computer games'. *AI Magazine*, 22(2). (2001) pp. 15–26.

LIREC: living with robots and interactive companions. *The European Community's Seventh Framework Programme*. (2008) Available from URL: http://www.lirec.org [15 January 2010].

Magerko, B.: 'Intelligent story direction in the interactive drama architecture'. *AI Game Programming Wisdom III,* Charles River Media. Hingham, MA. (2006) pp. 583–596.

McCrae, R.R., and John, O.P.: 'An introduction to the Five-Factor Model and its applications'. *Journal of Personality,* vol. 60. (1992) pp. 175–215.

Mindmakers forum. (2005–2008) Available from URL: http://www.mindmakers. org/index.jsp [15 January 2010].

Mindmakers: BML draft 1.0. (2008) Available from URL: http://wiki.mindmakers. org/projects:bml:draft1.0 [15 January 2010].

Newell, A.: *Unified Theories of Cognition.* Cambridge, MA, Harvard University Press. (1990).

Ortony, A., Clore, G.L., Collins, A.: *The Cognitive Structure of Emotions.* Cambridge, MA, Cambridge University Press. (1988).

Pausch, R. (head), Burnette, T., Capeheart, A.C., Conway, M., Cosgrove, D., DeLine, R., Durbin, J., Gossweiler, R., Koga, S., White, J.: 'Alice: Rapid prototyping system for virtual reality' *IEEE Computer Graphics and Applications.* (1995–2009) Available from URL: http:// www.alice.org/ [15 January 2010].

Pizzi, D., Cavazza, M.: 'From debugging to authoring: adapting productivity tools to narrative content description'. *First Joint Conference on Interactive Digital Storytelling* (ICIDS), Erfurt, Germany, November 2008. pp. 285–296.

AMIS: Pogamut. (2009) Available from URL: http://artemis.ms.mff.cuni.cz/ pogamut [15 January 2010].

Rao, A.S.: *AgentSpeak(L): BDI Agents Speak Out in a Logical Computable Language.* Berlin/Heidelberg, Springer. (1996).

RIT: GameBots Branch. (2005) Available from URL: http://www.cs.rit.edu/ ~jdb/gamebots/ [15 January 2010].

Shelley, M.: *Frankenstein; or, The Modern Prometheus.* London, Harding, Mavor & Jones. (1818).

Small, R.K.: 'Agent Smith: A real-time game-playing agent for interactive dynamic games'. In: *Proceedings of the 2008 GECCO Conference Companion on Genetic and Evolutionary Computation,* Atlanta, GA. (2008) pp. 1839–1842.

Spierling, U., Weiß, S., Müller, W.: 'Towards accessible authoring tools for interactive storytelling'. In: Göbel, S., Malkewitz, R., Iurgel, I. (eds.) *TIDSE 2006. Lecture Notes in Computer Science,* vol. 4326, Heidelberg, Springer. (2006) pp. 169–180.

Stanley, K.O., Miikkulainen, R.: 'Evolving neural networks through augmenting topologies', vol. 10, Cambridge, MA, MIT Press. (2002) pp. 99–127.

Tence, F., Buche, C.: 'Automatable evaluation method oriented toward behaviour believability for video games'. *International Conference on Intelligent Games and Simulation (GAME-ON'08).* (2008) pp. 39–43.

Tulving, A., Donaldson, W.: *Organization of Memory.* New York: Academic Press. (1972).

Wilensky, U.: NetLogo. Center for Connected Learning and Computer-Based Modeling, Northwestern University. (1999) Available from URL: http://ccl. northwestern.edu/netlogo/ [15 January 2010].

Wooldridge,   M.: *An Introduction to Multiagent Systems,* Chichester, John Wiley & Sons. (2002).

XPprogramming.com: Extreme Programming. Available from URL: http://www.xprogramming.com/xpmag/whatisxp.htm [15 January 2010].

*Chapter 9*

# More from the body: embodied anticipation for swift readaptation in neurocomputational cognitive architectures for robotic agents

*Alberto Montebelli, Robert Lowe and Tom Ziemke*

## Abstract

The coupling between a body (in an extended sense that encompasses both neural and non-neural dynamics) and its environment is here conceived as a critical substrate for cognition. We propose and discuss the plan for a neurocomputational cognitive architecture for robotic agents, so far implemented in its minimal form for supporting the behavior of a simple simulated robotic agent. A non-neural internal bodily mechanism (crucially characterized by a timescale much slower than the normal sensory-motor interactions of the robot with its environment) extends the cognitive potential of a system composed of purely reactive parts with a dynamic action selection mechanism and the capacity to integrate information over time. The same non-neural mechanism is the foundation for a novel, minimalist anticipatory architecture, implementing our *bodily anticipation hypothesis* and capable of swift readaptation to related yet novel tasks.

## 9.1 Introduction: an age of scientific synthesis

In the history of humanity we observe ages that prove extremely prolific in the production of novel ideas. Analogous to most natural phenomena, the social process of cultural evolution seems to proceed through cycles. Phases of high intellectual achievements are followed by periods of sedimentation, when new understanding, in the form of brand new theories, seeds the social community, becoming accepted and exploited in terms of innovative applications and technologies. It would be insensible to deny that the 20th century, so politically questionable, was characterized by a terrific intellectual enrichment in the scientific field. The science of the mind participated in this cultural process with obvious implications due to its striking immaturity. Nevertheless, a large body of theoretical work has been proposed and now urges a systematic effort of synthesis to introduce actual progress.

A common thread links several different sources of inspiration that strike us as researchers in cognitive systems. It can be formulated as: *the whole is more than (and qualitatively different from) the sum of its parts*. Far from esoteric or vague, this assumption received sound mathematical formalization by the science of non-linear dynamic systems (e.g., [1–3]) and pragmatic validation by physics in plenty of important findings, currently resulting in popular applications. It constitutes one of the core theoretical milestones of contemporary science and induced a more systemic attitude in the scientific study of complex systems, as opposed to the traditional reductionist approach.[1]

The very same idea, in the form of intuition, permeates the general spirit of *gestalt psychology* [4,5] and *ecological psychology* [6], and the work of influential psychologists and philosophers like Piaget, Vygotsky, and Merleau-Ponty. It also appears, with a more structured approach, in the seminal work of early cybernetics [7–9]. In its modern form, it originated a whole new scientific paradigm, namely the Dynamic Systems approach to the study of human and animal cognition (e.g., [10–13]).

Although perhaps stating the obvious, it is worth making explicit the relation between cognitive architectures and the phenomenon of biological cognition. Why, in the first place, should researchers interested in cognitive systems be interested in biological cognition? The quick answer is that, to date, nature offers by far the most remarkable examples of cognitive systems, in terms of performance, autonomy, generality, and so on. Admittedly, this might sound naive. Yet, scratching beneath the surface of such an answer, we start realizing that biological cognition is deeply entangled in a specific form of organization of the living [14–16]. The exploration of this specificity might offer new perspectives and new understanding on cognition as a general phenomenon, as well as offering new ways to approach its practical implementation.

The remainder of our chapter is organized as follows. Section 9.2 introduces the role of the body and its environment in the enactment of a cognitive process. The following two sections are linked to recent experimental results obtained in our Lab. Section 9.3 demonstrates the modulatory effect of non-neural bodily dynamics on the production of adaptive behavior. In Section 9.4, after introducing the theoretical context for anticipatory behavior in natural and artificial agents, we argue that non-neural internal dynamics can be powerfully involved in the anticipatory process. In the discussion, Section 9.5, we

---

[1] Our argument is not aimed at the trivialization of the scientific importance and achievements of the reductionist method. Indeed, the *divide et impera* approach proved as a powerful tool for scientific analysis. Nevertheless, it was culturally determined and fully justified by the computational limitations of the past centuries. The method is based on the decomposition of the system in elementary parts that are separately analyzed and finally recomposed. However, not all explanations can be reductionistic, and new phenomena can emerge at the global level of organization that cannot be foreseen at the level of isolated components. Nowadays, the large availability of computers endowed with appropriate computational power allows us to perform a non-linear analysis that overcomes some of the intrinsic limitations of the linear methods, thus producing results of extended validity.

will critically analyze the ideas presented. The final Section 9.6 recapitulates the main contributions of our chapter.

## 9.2 On coupling bodies, environments and finding a (scientific) place for emotions

Direct consequence of a more systemic view of cognition is the critical revision of the roles of body and environment in the cognitive process [17]. From mere input–output devices in a disembodied theory of the mind, their role has received a crucial upgrade. The systemic view conceives body and environment of the cognitive agent as constitutive of a largely distributed cognitive process, backing the brain in its operation by constantly offering cognitive support and tools [18]. Thus, the cognitive process is the result of the activity of the brain–body–environment triad, whose components, coupled in a global dynamic, are equally necessary to the creation of the mental process [11, 19, 20]. The body can be interpreted as an enduring pre/post-processor of neural information [19], and its interaction with the environment stores a wealth of knowledge about the "how to" of a cognitive activity [21]. Research in embodied and situated cognition investigates in theoretical and experimental terms the role of the body and of the environment in the cognitive process [22–25]. Therefore, these approaches tend to conceive and analyze cognition as a broadly distributed process, emphasizing attunement in cognitive aggregates, rather than in localized and proprietary processes. In this light cognitive robotics seems like the perfect candidate for the dirty work of generating an experimentally grounded synthesis, as it forces researchers to take very seriously the interplay among coupled bodies, control systems, and environments [16, 26, 27].

Alongside the indisputable role of the body, projected toward its environment, there is a less obvious, less visible, and consequently often neglected internal dynamic component of the body. We are referring to the plethora of background bio-regulatory mechanisms, aimed at the maintenance of a viable metabolic balance necessary for the organism's survival. An increasing number of researchers investigate the potential cognitive role of this hidden dynamic. Early philosophical insights by, for example, Baruch Spinoza and the seminal work by William James [28], gave inspiration to contemporary neurologist Antonio Damasio for a view of cognition that is deeply rooted in a hierarchy of bodily processes and consistent with state-of-the-art neurological findings [29–31]. According to Damasio, emotions emerge from the complex hierarchy constituted by levels of *automated homeostatic regulation* – the basic evolutionary organization for the maintenance of the living organism (ref. Figure 9.1). Metabolic regulation (e.g., endocrine/hormonal secretion, muscle contraction facilitating digestion), basic reflexes (e.g., basic tropism or taxes), and the immune system constitute the lower level of the machine. At a higher hierarchical level come behaviors related to pleasure/reward or pain/punishment (e.g., feeling pain triggers a specific pattern of protective behaviors), drives and

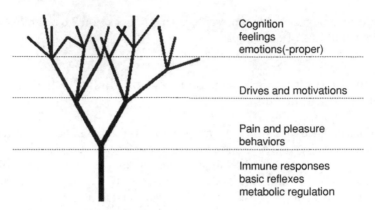

Cognition
feelings
emotions(-proper)

Drives and motivations

Pain and pleasure
behaviors

Immune responses
basic reflexes
metabolic regulation

*Figure 9.1     Damasio's representation of the levels of automated homeostatic regulation [after Reference 31]*

motivations (e.g., hunger, thirst, curiosity, play, and sex). One step further in the hierarchy we find *emotion proper* (e.g., joy, sorrow, fear) as a subset of the homeostatic reactions that is triggered by *emotionally competent stimuli* (ECS), either actual or imagined. ECS are such in virtue of the evolutionary history or of the ontogenesis of the organism. Finally, at the top of the hierarchy, from the current body state mapped in cortical body maps emerge (either conscious or unconscious) *feelings*. Feelings are perceptions of a certain state of the body, together with the perception of a certain mode of thinking and of attuned thoughts with certain themes. Similar approaches constitute the core motivations of *somatic theories of emotions* [32–34].

Conceiving emotions in physical (rather than mental) terms constitutes the entry point for their appealing robotic rendition. Domenico Parisi points to the necessity of a deep investigation of the relation between the control system and what happens inside the body [27]. The emphasis on bodily parameters affecting bodily processes can be traced back further to the cyberneticist W. Ross Ashby, who focused on the behavioral consequences of a set of *essential variables*, critical to the organism's survival (e.g., sugar concentration in the blood and body temperature). The organism's implicit necessity to restrict their range within viable limits determines the random creation of new adaptive behaviors [8] (for an introductory discussion, see Reference 26). To implement robots endowed with genuine autonomy, agency, intentionality, and grounded in a meaningful relation to their environment, focusing on the cognitive implications of bio-regulatory processes might be a promising direction for scientific explorations [16,26,35–37]. Indeed, internal robotics in the *here and now* is not sufficient for modeling emotions. It requires the presence of ECS that derive from the coupling of body and environment in an *adaptive history of interactions*. Consequently, a somatosensory theory of emotions only makes sense in a broad systemic perspective, as the result of the complex interactions between body and environment.

As a matter of fact, all the above is in apparent contrast to the traditional perspective on AI and cognitive science, i.e., the presumption that the description of the world in terms of related symbol structures and logical processing on such structures is the necessary and sufficient condition for general intelligent action by appropriate instances of physical systems [38]. A concept mapped in cognitive robotics onto the linear *sense-plan-execute* scheme [39], and conceptually akin to the functional approach of traditional computational neuroscience, focused on specific and decontextualized subdomains. In an intuitive image by Chiel, the change in perspective is as significant as the passage from the central control exerted by a musical director on a classic orchestra, to the self-governed coordination of several musicians in a jazz ensemble [19].

## 9.3 Bodily neuromodulation: a minimalist experiment

To start assessing the potential cognitive relevance of non-neural internal dynamics in a simple robotic agent, we recently carried out a series of computer simulations, reported in detail in References 40,41. A simulated standard Khepera robot was free to move in an empty square arena. It received its sensory input from an array of light and infrared sensors. Two identical light sources, centrally located in the arena, provided the environment with a stationary light gradient. The robot could also sense its simulated energy level (e.g., the level of a battery charge), subject to linear decay, from a maximum value down to zero. This sensor has a biological analog in glucose neurons involved in glucose homeostasis [42], whose firing rate is correlated to changes in the local concentration of glucose. The entering of an invisible area centered under one of the two light sources, randomly selected for each replication, would provide an instantaneous full energy recharge. A simple feedforward artificial neural network (ANN) with no hidden layers would generate the motor output driving the left and right activation of the two wheels. Its parameters (weights and biases) were evolved using a standard evolutionary algorithm. The fitness function rewarded at each time step the maintenance of positive levels of energy, but this was only accredited when the robot moved outside the recharging area, to penalize static behaviors. At the end of each generation, the best individuals were selected for reproduction to create the new population for the evolutionary algorithm.

Indeed, the evolved agents performed well on this elementary task. The evolved agent would approach one light. In case it did not receive its reward it would move to the next light and as its energy level jumped to a maximum it would remain dynamically engaged in a to and fro movement across the boundary of the recharging area. The interesting part of the analysis came when, setting aside the evolutionary task, we used the energy level as control parameter of the system [11].[2] We clamped the energy level to a fixed value for

---

[2] We use the term system (here and in what follows) with reference to the triad composed by the coupled robot's body (including its energy mechanism), its environment, and its sensory-motor map (i.e., its ANN).

the whole duration of each replication (therefore, during this analysis the energy mechanism was frozen and the recharging area absent in the replication), and systematically explored the robot's behaviors for values ranging from empty to full in the different replications. Consequently, we were able to map the behavioral repertoire of the evolved agent as a function of its energy level. We observed three main classes of behavioral attractors (ref. Figure 9.2, left panels):

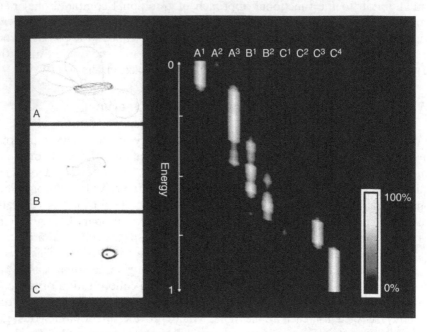

*Figure 9.2*    *Left: The three panels exemplify prototypical trajectories, after transient exhaustion, of the agent with clamped energy level (see text) freely moving inside its arena. Panel "A," exploratory behaviors: the agent engages in large loops from one light source to the other (the positions of the light sources in the environment are represented by red stars) and at times exploring the peripheral space. Panel "C," local behaviors: the agent's loops are closely bound to a single light source. Panel "B," hybrid behaviors: the behavioral attractors consist of both exploratory and local loops. Right: The intensity of the pixels for each column (corresponding to attractors belonging to classes A–C, as specified by their labels on the top row) represents the relative frequency of the behavioral attractor as a function of the energy level. Observe the clear dominance of attractors in class A for energy levels in the interval [0.0, 0.4]. Similarly, attractors in class C dominate in the energy interval [0.7, 1.0]. The hybrid forms in class B characterize intermediate levels of energy. Data from 500 replications (10 for each energy level) [after Reference 40]*

*exploratory behaviors* (i.e., the agent engages in large loops from one light source to the other – attractor class "A"), *local behaviors* (the agent's loops are closely bound to a single light source – class "C"), and *hybrid behaviors* (combining the characteristics of both exploratory and local attractors – class "B"). These three behavioral attractors proving to be neatly distributed as a function of the energy level. Exploratory behaviors dominated the lowest range of energy levels, whereas local behaviors the highest ones (Figure 9.2, right panel). For intermediate levels of energy we found the prevalence of hybrid behaviors.

Details on how the different behavioral attractors effectively amalgamated to accomplish the evolutionary task are given in Montebelli *et al.* [40,41]. What is remarkable is that:

1. As the energy level was left free to follow its natural dynamics, it constituted an effective, self-organized *dynamic action selection mechanism*. The appropriate classes of behavioral attractors that contributed to an effective behavior were locally available to the agent, depending on its energy level. For example (ref. Figure 9.2, right panel), an energy level of 0.7 led to the expression of attractor C3 (in 70% of the replications), C1 (20%) or B1 (10%). The actual selection of the specific attractor depended on the basin of attraction in which the combination of the starting position and the integrated effects of noise induced the system dynamics.
2. The system, although composed of extremely simple, purely reactive components, was able to *integrate information over time*, with no explicit representations or memory needed. We argued [43] that this capacity was derived from the interplay between the slower timescale of the energy level decay and the fast sensory-motor dynamics during the artificial evolution of the system parameters (see also Section 9.5.2).

In sum, we have shown how:

● minimalist non-neural bodily states (e.g., the energy level in our experiment) can modulate the sensory-motor map implemented by an ANN, and thus the general behavior of the simulated robotic agent coupled with its environment;
● this modulation can be exploited as a dynamic action selection mechanism;
● the cooperation between dynamics at different timescales can boost the cognitive potential of the system, in the case of this experiment, endowing a collection of purely reactive components with the capacity to integrate information over time.

## 9.4 The bodily path of anticipation

### 9.4.1 *Anticipatory dynamics*

Operative definitions of anticipatory behaviors stress the role of expectations and belief about the future, on the determination of the current behavior

[44–47]. According to Braitenberg, the anticipatory behavior of an organism is dictated by expectations together with desires, thus underlining the role of a meaningful relation between an agent and its environment [48]. Much has already been said about the adaptive role of anticipation. In particular it can allow faster and smoother action execution, facilitate action initiation, improve information seeking, decision making, predictive attention, and social interaction [46,49–51]. In a recent paper, Butz argues in favor of an anticipatory tension that characterizes biological agents, a natural tendency toward the prediction of the consequences of their own actions and of the dynamics of their environment. This drive might influence both the development of neural structures and bias the agent to anticipatory behavior [52]. In spite of its strongly representationalist orientation (which clashes with some, e.g., enactive, theories of embodied cognition), we consider the author's general hypothesis quite intriguing, and an affinity with the spirit of our own work in its dynamical view on sensory-motor interactions. In a dynamical system perspective, we suggested that [43]:

> ... *a cognitive system settled on its behavioral attractor constitutes an important instance of an implicitly anticipatory system.*

In fact, the engagement with the attractor binds the system to a stable and fully determined dynamic flow. An autonomous and viable dynamic is inherently endowed with anticipatory power. For example, this capacity can be exploited by an agent to navigate in a known environment once all sensory inputs have been discontinued (e.g., see Reference 53).

### 9.4.2   The bodily anticipation hypothesis

In recent work [43] we extended the framework sketched in Section 9.3, introducing a general scheme for an anticipatory architecture. In the previous experiment we have seen how non-neural internal dynamics can influence the current modality of engagement of an agent with its environment, i.e., its current behavioral attractor. On the other hand, the current behavior determines the current non-neural internal dynamics (e.g., an effective behavior that satisfies the experimental task maintains the energy level at high levels). This bidirectional relation is expressed by the arrows connecting the blocks labeled SENSORY-MOTOR FLOW and NON-NEURAL INTERNAL DYNAMICS in Figure 9.3. The former block represents the dynamic of the degrees of freedom relevant to the current sensory-motor engagement between the agent and its environment [11]. Similarly, the latter embeds the relevant non-neural internal dynamics. In parallel, current sensory-motor flow and internal dynamics drive a neural emulator block (labeled ANTICIPATION) that is capable, in virtue of its evolutionary history and/or ontogenetic adaptation, of dynamic anticipation in the sense introduced in the Section 9.4.1. The main practical function of this emulator is to tune to the current sensory-motor flow and dynamically perturb the bodily dynamics with the anticipated consequences of the current interaction.

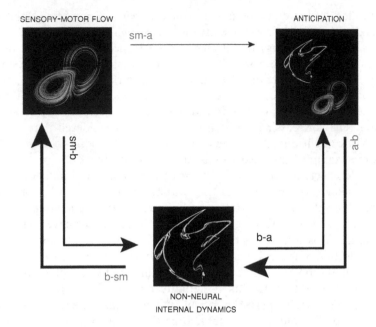

*Figure 9.3*   *Illustration of the bodily anticipation hypothesis. During its daily roaming, our agent gets engaged with a potentially noxious interaction. Neural sensory-motor anticipatory dynamics, here conveniently isolated within the global coupled system (box labeled* ANTICIPATION*), predict the risk by physically perturbing the current non-neural bodily dynamics (*NON-NEURAL INTERNAL DYNAMICS*) through path a-b and from there, indirectly through a further path b-sm, the actual sensory-motor dynamics (*SENSORY-MOTOR FLOW*). Following a quick reorganization of its behavioral attractor, our agent is attuned to face the novel danger thanks to the mediation of its body, without any direct influence of anticipation on the selection of the new behavior [after Reference 43]*

For example, consider a specimen agent, a caveman engaged in a relaxing and innocuous activity, for example, picking berries in a forest. Out of the blue, an emotional stimulus, for example, an apparently hungry, massive dinosaur, loudly enters the scene.[3] The caveman's anticipatory system has no difficulty in predicting the most likely future scenario. The sensory-motor flow

---

[3] The enormous time gap that separates the extinction of dinosaurs and the appearance of the first hominids is part of our example. We want to make sure that our specimen is experiencing a novel situation (therefore, a positivist caveman, who only brings solid scientific arguments to prove the dinosaur's anachronism, would be the perfect candidate for premature exhaustion of his own pedagogical role in our chapter).

correspondent to the ongoing activity (picking berries) must be inhibited and redirected to a more conservative attitude. How will the next viable behavior (e.g., an impulsive fleeing) be selected? In Reference 43 we argued against a direct causal connection with the ANTICIPATION block directly eliciting the appropriate fleeing behavior. The main motivation for this rests on the complexity of the search space that the nervous system should face in building the right chain of correlations. We found a much simpler pathway to select an adequate behavior, passing through the body. We considered that the anticipatory block might directly influence the non-neural bodily dynamics. In our prehistoric example that means that once perceived the ECS, our caveman's body would experience as if actually torn by the fangs and nails of the dinosaur. It is likely that the caveman's evolutionary history and his ontogenesis had already created viable correlations between his dramatic visceral reaction and his fleeing for life, although the specific situation had never been experienced before. This constitutes the essence of our BODILY ANTICIPATION HYPOTHESIS: the selection of the next viable action is off-loaded onto the bio-regulatory dynamics of the body. Destabilized by the anticipated effect of the current interaction, the body reacts as if actually engaged in such sensory-motor experience. The bodily perturbation elicits reactions, already stored in the potential of bodily and neural interactions that tend to pull the system back into viable regions.

### 9.4.3   A minimalist implementation of the bodily anticipation hypothesis

In Reference 43 we tested a minimal implementation of our anticipatory architecture. The previous experimental setup was extended to a *go-no-go* task, loosely inspired by Reference 54. In a first experimental condition nothing changes with respect to the setup described in Section 9.3: the two light sources would determine a steady gradient of environmental luminance (*continuous sensory regime*), correlated with a rewarding area randomly located around one of the lights for each replication. However, during each replication this regime alternated according to a fixed schedule with an *intermittent sensory regime*, where the light sources were obscured every third time step. Under this new condition, the randomly chosen area determined a punishment in the form of an energy leak. As a biological metaphor, this alternation between regimes models the case of a succulent berry whose external pigmentation is different when unripe (and toxic) or ripe (and energizing). Again, the goal consisted in maintaining a positive energy level. We compared the performance of the simple architecture described in Section 9.4.2 (a feedforward ANN with no hidden layers) with a more articulated one, our novel minimalist anticipatory architecture is shown in Figure 9.4.

In the former case, the evolutionary algorithm adapted the ANN's weights and biases on the new task, starting either from the final population evolved in the previous experiment or from a randomly generated population. In the case

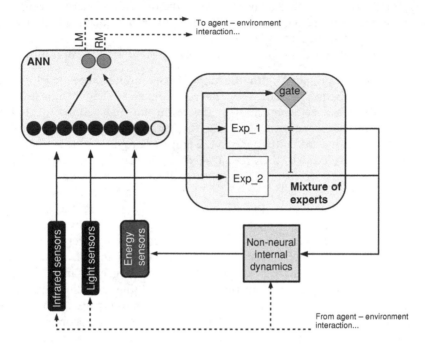

*Figure 9.4   Minimalist anticipatory architecture. The sensory information
(infrared, light, and energy sensors) drives the left and right motors
(LM and RM) through a feedforward ANN with no hidden layers.
The sensory flow is also processed by a mixture of recurrent experts,
preadapted so that each expert is tuned to a specific sensory regime.
The information on the current best expert (corresponding to one of
the two regimes) is given by the gating signal, that selects between
the original energy mechanism of the agent and an overriding one
whose dynamic adapts to the new task [after Reference 43]*

of our anticipatory architecture the original ANN (i.e., the simple ANN, whose
weights and biases were extracted adopting the final population evolved during
the previous experiment) was backed by a preadapted MIXTURE OF RECURRENT
EXPERTS [55–57] that processed the sensory flow. During its preadaptation, each
expert competed with the others to generate the best prediction of the sensory
state at the next time step. By doing so, two different experts became specialized by tuning to the specific dynamic flow of the two different regimes. Of
course the discrimination could be accomplished by much simpler architectures. However, through this design choice we wanted to emphasize the
agent's engagement with the different sensory dynamics in the two sensory
regimes. Crucially, in the new architecture the activation of the expert tuned to
the intermittent sensory regime triggered a new energy mechanism that

overrode the original one. The decay rate of the overriding energy mechanism, rather than hardwired as before, is the one single parameter adapted by an evolutionary algorithm on the new task.

In short (a detailed analysis is available in Reference 43), we found that:

1.  The systems provided with the anticipatory architecture developed an effective dynamic relation with its environment. They demonstrated a straightforward engagement with the rewarding light source during the continuous sensory regime, and a swift disengagement from the penalizing area during intermittent regime (ref. Figure 9.5, right). On the other hand, systems provided with the original ANN architectures (both trained from random weights or from the population at the last generation of the previous experiment) tended to cope with the new task by relying on stereotypical behavioral attractors (Figure 9.5, left). During the continuous sensory regime they engaged in loops containing both light sources, approaching them close enough to enter their potential rewarding areas. During the intermittent regime they simply relaxed their trajectories with respect to the light sources, keeping at a slightly larger distance from them and consequently clear from the critical area, thus avoiding the punishment. This behavior ignores the effect of the recharging area on the energy level, merely relying on light sensor information and geometrical constraints. Such strategy lacks robustness, failing as soon as even minimal geometrical variations affect the environment.

2.  In the case of the anticipatory architecture, the adaptive process for the new task proved easy, as even a random search could immediately generate agents with satisfactory performance. The evolutionary search was much more problematic for the original ANN, evolved from both starting conditions.

## 9.5   Discussion

### 9.5.1   *On the internal/external dichotomy*

We hope to have clarified enough the importance of conceptualizing the phenomenon of cognition as emergent from the coupling of body (with its external morphology and the richness of its internal bio-regulatory mechanisms), nervous system and environment. Within this systemic view, the boundary separating each subsystem is nothing but a useful artifice, functional to the analysis of a complex system dominated by circular relations. Each component participates in the global cognitive process with equal weight. In this sense, even defending the traditional labels of cognitive robotics, where the nervous system would be assimilated the control system, would be problematic. What is controlled? What is doing the controlling? From our example it seems clear enough that different parts of the system mutually influence and are influenced

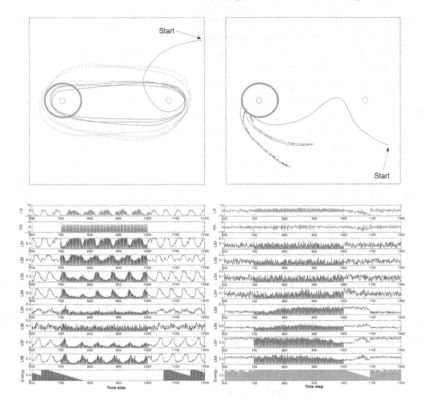

*Figure 9.5*    *Typical spatial trajectories developed by the different architectures during evolutionary adaptation. Top left: Simple feedforward ANNs tend to deploy a stereotypical strategy, i.e., the robot's trajectories systematically engage in exploratory loops between the two light sources (small circles to the left and to the right of the arena), entering the recharging area (large circle to the left of the arena) during the continuous regime (continuous line) and slightly relaxing to avoid its punishment during the intermittent regime (dashed line). Top right: On the other hand, the behaviors that tend to emerge from our minimal anticipatory architecture dynamically engage and disengage with the rewarding/punishing area according to the different sensory regime (again, continuous/dashed lines represent the trajectories during continuous/intermittent sensory regimes). For a better resolution of details in the trajectory, the two pictures zoom in the area of main interest surrounding the light sources. Bottom: The left and right panels exemplify, respectively for a feedforward and an anticipatory architecture, the activation of the two motoneurons (LM and RM), of the light sensors (LS1-8) and of the energy level sensor, during 600 time steps that include a double regime transition (continuous–intermittent–continuous) occurring at time steps 700 and 1000 [after Reference 43]*

by others (e.g., the energy level can modulate the behavior of the sensory-motor map, that in return affects the energy level).

This tight coupling casts a light on an interesting point. What is internal? What is external? Of course we have no difficulty at drawing a line from our distal, anthropomorphic perspective. Nevertheless we can easily argue that a simple agent, even substantially more complex than our elementary model, might find defining such a boundary difficult. We prefer to avoid such dichotomy, as we consider more useful focusing on the global system composed of dynamically interacting parts. At any given time its dynamic balance will be perturbed by stimuli coming from different sources (e.g., the external environment, the agent's regulatory mechanisms, its nervous system). Each perturbation would produce a consonant reaction of the system's trajectory in its phase space. Each time, according to the needs of the analysis, we will have to properly redraw the boundary between input and output, cause and its effect. Parisi suggested objective criteria for partitioning the inside and outside of the body in natural agents, on the grounds of the physical–chemical processes that tend to dominate the two interfaces [27]. Local and specific interactions with fast dynamics, archetypal of physical processes, tend to characterize the interface with the external world. Global and diffused variations with slower timescales, characteristic of chemical processes, tend to take place inside the organisms. Although this is just a generalization, the focus on the different timescales prepares us for the next fundamental observation.

## 9.5.2   On the role of multiple timescales

An obvious objection can be raised against our model. What is it that determines the distinction between neural and non-neural? Could the non-neural internal dynamic be translated into purely neural mechanisms? After all, the work of other groups (e.g., [55,58–60]) seems oriented in that direction.

Rather than taking a defensive stance, we will simply redirect the problem and dissolve it in its abstract computational formalization. What we consider crucial to our model is the interplay of the different timescales that characterize the energy mechanism and the other sensory-motor interactions with the environment. In the experiment reported in Section 9.3, during the artificial evolution of the system, the slower dynamic of the energy level organized the continuous sensory-motor flow in dynamically related events. This endowed the system, composed of purely reactive elements, with the capacity to integrate information over time. On a more general level we formulated the following hypothesis [43]:

> ... *The access to a collection of attuned dynamic sub-systems characterized by intrinsic dynamics at different timescales and the exploitation of such differences, constitutes a powerful mechanism of embodied cognition, widely operating at the different levels of organization of biological cognition. A mechanism providing the cognitive system with the capacity to structure information on events which are*

*relevant to its survival, with no need for explicit representations, mem-*
*ory or consciousness.*

With this in mind we can look at the plethora of bio-regulatory phenomena
with new eyes. The characteristic timescales of non-neural bodily processes, so
different from the normal dynamics of the sensory-motor interactions between
an agent and its environment, might provide exactly that dynamical richness
that we are advocating. The role of multiple timescales is currently attracting
the attention of the scientific community, both in computational neuroscience
[61,62] and cognitive robotics [55,58–60,63–65]. Interestingly, Maniadakis
showed how an evolutionary algorithm can self-organize multiple timescales in
a hierarchy of recurrent neural networks [66].

### 9.5.3    Experimental evidence for the bodily anticipation hypothesis

The paths in the general scheme sketched in Figure 9.3 are actually less arbitrary than they might look at first glance. In the present subsection, we report
some experimental evidence that supports our bodily anticipation hypothesis,
from natural and artificial systems. Our own and related work in cognitive
robotics [40,41,58–60], motivates the arrows representing the relation between
the non-neural internal dynamics and the sensory-motor flow blocks (paths
sm-b and b-sm). The claim that in organisms the internal dynamics of the body
(e.g., a sudden injection of adrenaline) affect the behavior and that behavior
affects the body (e.g., eating or declining the fifth slice of your birthday cake)
should not strike us as bizarre. The capacity of the brain to anticipate sensory-
motor correlates (path sm-a) is currently object of intensive research in neuroscience (e.g., see Reference 47). Examples in cognitive robotics are in
References 55,60. Interestingly, Ziemke *et al.* show how a viable anticipation
does not have to be identical to the anticipated phenomenon [53]. An example
of how anticipation, i.e., a neural event taking place in the nervous system,
might affect the body is given in Reference 30. The case of a professional
musician is reported, who could systematically control her emotional machinery in experimental conditions. Also the seemingly arbitrary switch between the
natural energy dynamic and the overriding energy mechanism taking over
during intermittent sensory regime is inspired by neurophysiological analogs.
False bodily information can sometimes substitute for the actual state, for
example, in the case of endogenously altered nociceptive signals. There is an
obvious advantage for a wounded organism to ignore the pain when it is fleeing from the danger that produced it [31].

### 9.5.4    The bodily path for search-space compression
###             during readaptation

Obviously, our bodily anticipation hypothesis does not deny the possibility of a
purely neural pathway between anticipation and sensory-motor flow (the
missing path a-sm in Figure 9.3). Nevertheless, we point to the fact that our
minimalist anticipatory architecture drastically simplifies the problem of

readapting to a new task. Our proposal focuses on the knowledge that is already embedded in the body after the long history of biological evolution and ontogenesis, and might be exploited during readaptation. The search space during readaptation, characterized by the potentially enormous number of degrees of freedom of an ANN, is reduced by our bodily anticipation hypothesis to the much smaller dimensionality of the bodily neuromodulators (the energy level in our minimalist example). We believe that the bodily anticipation hypothesis could be of help at least in virtue of such drastic compression of the adaptive search space, particularly in circumstances that require, for example, fast, non-deliberated decision making. Rather than searching the massive space of the system's degree of freedom for the proper associations supporting the a-sm pathway, the system can limit its exploration to the subspace of the bodily parameters. Pragmatically, even a random search of the appropriate decay rate of the overriding energy dynamic in our antici-patory architecture can swiftly readapt the system to the new problem, whereas such readaptation proves slow with the original architecture. This is obviously related to Ashby's work on *ultrastable agents*. A random change in the beha-vioral coupling between the agent and its environment is induced whenever a variation of an *essential variable* threatens its survival [8,26].

An opposite hypothesis seems to be brought forth by Damasio, as he introduces the *as-if body loops* [30,31]. The emotional machine, deeply groun-ded in the homeostatic process as we saw in Section 9.2, is in Damasio's theory central even to highly logical functions, for example, decision making [30,67]. Its support can be elicited directly, but after repeated exposure the brain can build consistent causal associations and thus totally bypass the body in the decision process. Nevertheless, Bechara refers to preliminary results showing how in the process of decision making the role of the as-if body loop might be restricted to the most predictable situations (choice under certainty). As the decision drifts from certainty to risk or ambiguity (full uncertainty) the *body loop* mode of operation, where the bodily mechanisms are directly engaged, becomes prominent [67]. We find this observation perfectly tuned with the intuition inspiring our model.

## 9.5.5    *Future work*

We consider our minimal anticipatory architecture as a promising and com-plete illustration of our bodily anticipation hypothesis, although still at its initial stage of development. Nevertheless, together with a few answers, it suggests plenty of supplementary questions. Accordingly, we admit that it needs and deserves further investigation and validation.

Our model might be accused of being an *ad hoc* arrangement, built on the basis of the previous experiment. In other words, it might be suspected that we embed built-in solutions in our minimalist anticipatory architecture: first, for the arbitrary decision to override the original non-neural internal mechanism (although we have demonstrated in the Section 9.5.4 how the same strategy can be found in natural agents); second, for selecting the decay rate of the overriding energy mechanism as critical parameter to be adapted by the evolutionary algorithm. This is a reasonable criticism. Nevertheless, given the extreme simplicity of our current setup, such design choices were necessary. In our model, simplicity constitutes a deliberate preference. For the sake of a detailed analysis, we try to implement the minimal model capable of producing the phenomenon under study. However, we welcome such objection, confident that it can be more easily confuted given a slightly more complex model, both in terms of task and architecture. In particular, future work will specifically address the implementation of more realistic internal dynamics, inspired by natural metabolic systems as well as by the work on prototypical robotic agents endowed with *microbial fuel cells* [68].

## 9.6 Conclusions

This chapter takes on and extends the tradition of a more systemic view of AI research [16,17,40]. Cognition is conceived and analyzed in terms of coupled systems: the body (encompassing both its external morphology and its internal bio-regulatory mechanisms), the nervous system, and the environment constitute a cognitive aggregate. Such interpretation dissolves the internal–external dichotomy into a computational formalization in terms of coordinated multiple timescales. The cognitive role of the body is taken in account with special and novel emphasis on what happens inside of the body. Biological cognition, more than simply inspiring problems and solutions, is seen as the living implementation of the basic organizational principles of intelligence, still mostly to be unraveled.

In a first experiment (Section 9.3) we showed how non-neural internal dynamics, following a slow timescale, can modulate the activity of an ANN and consequently the behavior of an agent coupled with its environment. A traditional evolutionary algorithm self-organized this modulation, implementing a dynamic action selection mechanism. The analysis showed how the coordination of multiple timescales might support the emergence of more sophisticated cognitive capacities, such as the capacity to integrate information over time in a system composed of purely reactive parts. In a second experiment (Section 9.4) we extended the previous system to an anticipatory architecture, offering a minimalist implementation of the bodily anticipation hypothesis presented in this chapter. The novel architecture provided flexible and dynamic engagement of the agent with its environment, as a swift adaptation to a brand new task was accomplished. Crucially, the search for novel behaviors was drastically simplified, as it operated on the limited subspace of the non-neural internal parameters, rather than on the high dimensional space of the ANN. We believe that this work illustrates promising results in terms of basic organizational principles of cognition that can be explored by minimally cognitive architectures.

## Acknowledgements

We thank our colleagues Malin Aktius, Boris Duran, Anthony Morse, Pierre Philippe, Filippo Saglimbeni, Henrik Svensson, and Serge Thill. This work has been supported by a European Commission grant to the project *Integrating Cognition, Emotion and Autonomy* (ICEA, www.iceaproject.eu, IST-027819) as part of the European *Cognitive Systems* initiative.

## References

1.  Bergé, P., Pomeau, Y., Vidal, C.: *Order within Chaos: Towards a Deterministic Approach to Turbulence.* Wiley-Interscience, New York, Toronto, Chichester, Brisbane, Singapore (1984).
2.  Haken, H.: *Synergetics: Introduction and Advanced Topics.* Springer, Berlin, Heidelberg (2004).
3.  Strogatz, S.H.: *Nonlinear Dynamics and Chaos.* Westview Press, Cambridge, MA (1994).
4.  Köhler, W.: *Gestalt Psychology.* Liveright, New York (1947).
5.  Rock, I., Palmer, S.: 'The legacy of gestalt psychology'. *Scientific American*, **263**(6) (1990) 84–90.
6.  Gibson, J.J.: *The Ecological Approach to Visual Perception.* Houghton Mifflin, Boston (1979).
7.  Ashby, W.R.: *An Introduction to Cybernetics.* Chapman & Hall, New York, NY (1957).
8.  Ashby, W.R.: *Design for a Brain: The Origin of Adaptive Behavior.* John Wiley & Sons Inc., New York, NY (1960).
9.  Wiener, N.: *Cybernetics, or Control and Communication in the Animal and the Machine.* MIT Press, Cambridge, MA (1965).
10. Van Gelder, T.: 'The dynamical hypothesis in cognitive science'. *Behavioral and Brain Sciences*, **vol. 21** (2000) pp. 615–628.
11. Kelso, J.A.S.: *Dynamic Patterns: The Self-organization of Brain and Behavior.* MIT Press, Cambridge, MA (1995).
12. Thelen, E., Smith, L.B.: *A Dynamic Systems Approach to the Development of Cognition and Action.* MIT Press, Cambridge, MA (1994).
13. Beer, R.D.: 'Dynamical approaches to cognitive science'. *Trends in Cognitive Sciences*, **4**(3) (2000) 91–99.
14. Maturana, H.R., Varela, F.J.: *The Tree of Knowledge: The Biological Roots of Human Understanding.* Shambhala Publications, Inc., Boston, London (1992).
15. Stewart, J.: 'Cognition = life: Implications for higher-level cognition'. *Behavioural Processes*, **35**(1–3) (1995) 311–326.
16. Ziemke, T., Lowe, R.: 'On the role of emotion in embodied cognitive architectures: From organisms to robots'. *Cognitive Computation*, **1**(1) (2009) 104–117.

17. Froese, T., Ziemke, T.: 'Enactive artificial intelligence: Investigating the systemic organization of life and mind'. *Artificial Intelligence*, **vol. 173** (2009) pp. 466–500.
18. Clark, A.: *Supersizing the Mind: Embodiment, Action, and Cognitive Extension*. Oxford University Press, New York (2008).
19. Chiel, H., Beer, R.D.: 'The brain has a body: Adaptive behavior emerges from interactions of nervous system, body and environment'. *Trends in Neurosciences*, **20**(12) (1997) 553–557.
20. Clark, A.: *Being There: Putting Brain, Body, and World Together Again*. MIT Press, Cambridge, MA (1997).
21. Pfeifer, R., Bongard, J.: *How the Body Shapes the Way we Think: A New View of Intelligence*. MIT Press, Cambridge, MA (2007).
22. Varela, F.J., Thompson, E.T., Rosch, E.: *The Embodied Mind: Cognitive Science and Human Experience*. MIT Press, Cambridge, MA (1992).
23. Ziemke, T., Zlatev, J., Frank, R.M. (eds.): *Body, Language and Mind: Embodiment*, vol. 1. Mouton de Gruyter, Berlin/New York (2007).
24. Clancey, W.J.: *Situated Cognition: On Human Knowledge and Computer Representations*. Cambridge University Press, Cambridge, United Kingdom (1997).
25. Frank, R.M., Dirven, R., Ziemke, T. (eds.): *Body, Language and Mind: Sociocultural Situatedness*, vol. 2. Mouton de Gruyter, Berlin/New York (2008).
26. Di Paolo, E.: 'Organismically-inspired robotics: Homeostatic adaptation and natural teleology beyond the closed sensorimotor loop'. In Murase, K., Asakura, T. (eds.): *Dynamical Systems Approach to Embodiment and Sociality*. Advanced Knowledge International, Adelaide (2003) pp. 19–42.
27. Parisi, D.: Internal robotics. *Connection Science*, **16**(4) (2004) 325–338.
28. James, W.: *The Principles of Psychology*. Macmillan, New York (1980).
29. Damasio, A.: *Descartes' error: Emotion, Reason, and the Human Brain*. G.P. Putnam's Sons, New York, NY (1994).
30. Damasio, A.: *The Feeling of What Happens: Body and Emotion in the Making of Consciousness*. Harvest Books, San Diego, New York, London (2000).
31. Damasio, A.: *Looking for Spinoza: Joy, Sorrow, and the Feeling Brain*. Harcourt, Orlando, Austin, New York, San Diego, Toronto, London (2003).
32. Prinz, J.: 'Embodied emotions'. In Solomon, R.C. (ed.): *Thinking about Feeling: Contemporary Philosophers on the Emotions*. Oxford University Press, New York, NY (2004) pp. 44–59.
33. Prinz, J.J.: *Gut Reactions: A Perceptual Theory of Emotion*. Oxford University Press, New York (2004).
34. Panksepp, J.: 'Affective consciousness: Core emotional feelings in animals and humans'. *Consciousness and Cognition*, **vol. 14** (2005) pp. 30–80.

35.  Lowe, R., Herrera, C., Morse, A., Ziemke, T.: 'The embodied dynamics of emotion, appraisal and attention'. In Paletta, L., Rome, E. (eds.): *Attention in Cognitive Systems*. Springer, Berlin (2008) pp. 1–20.
36.  Lowe, R., Philippe, P., Montebelli, A., Ziemke, T.: 'Affective modulation of embodied dynamics'. In *The Role of Emotion in Adaptive Behavior and Cognitive Robotics: Electronic Proceedings of SAB2008 Workshop* (2008).
37.  Muntean, I., Wright, C.D.: 'Autonomous agency, ai, and allostasis: A biomimetic perspective'. *Pragmatics '&' Cognition*, **15**(3) (2007) 485–513.
38.  Newell, A.: 'Physical symbol systems'. *Cognitive Science*, **4**(2) (1980) 135–183.
39.  Newell, A.: *Unified Theories of Cognition*. Harvard University Press, Cambridge, MA (1990).
40.  Montebelli, A., Herrera, C., Ziemke, T.: 'On cognition as dynamical coupling: An analysis of behavioral attractor dynamics'. *Adaptive Behavior*, **16**(2–3) (2008) 182–195.
41.  Montebelli, A., Herrera, C., Ziemke, T.: 'An analysis of behavioral attractor dynamics'. In Almeida e Costa, F. (ed.): *Advances in Artificial Life: Proceedings of the 9th European Conference on Artificial Life*. Springer, Berlin (2007) pp. 213–222.
42.  Barnes, M.B., Beverly, J.L.: 'Nitric oxide's role in glucose homeostasis'. *American Journal of Physiology – Regulatory, Integrative and Comparative Physiology*, **vol. 293** (2007) pp. 590–591.
43.  Montebelli, A., Lowe, R., Ziemke, T.: 'The cognitive body: From dynamic modulation to anticipation'. In Pezzulo, G., Butz, M.V., Sigaud, O., Baldassarre, G. (eds.): *Anticipatory Behavior in Adaptive Learning Systems: From Sensorimotor to Higher-level Cognitive Capabilities*. Springer, Berlin, Heidelberg (2009).
44.  Rosen, R.: *Anticipatory Systems*. Pergamon Press, Oxford (1985).
45.  Butz, M.V., Sigaud, O., Gérard, P.: 'Anticipatory behavior: Exploiting knowledge about the future to improve current behavior'. In Butz, M.V., Sigaud, O., Gérard, P. (eds.): *Anticipatory Behavior in Adaptive Learning Systems: Foundations, Theories and Systems*. *LNCS*, Springer-Verlag, Berlin (2003) pp. 1–10.
46.  Grush, R.: 'The emulation theory of representation: Motor control, imagery, and perception'. *Behavioral and Brain Sciences*, **vol. 27** (2004) pp. 377–442.
47.  Hesslow, G.: 'Conscious thought as simulation of behaviour and perception'. *Trends in Cognitive Sciences*, **6**(6) (2002) 242–247.
48.  Braitenberg, V.: *Vehicles: Experiments in Synthetic Psychology*. MIT Press, Cambridge, MA (1984).
49.  Butz, M.V., Pezzulo, G.: 'Benefits of anticipation in cognitive agents'. In Pezzulo, G., Butz, M.V., Castelfranchi, C., Falcone, R. (eds.): *The Challenge of Anticipation: A Unifying Framework for the Analysis and Design of Artificial Cognitive Systems*. Springer Verlag, Berlin, Heidelberg (2008) pp. 45–62.

50. Barsalou, L.W.: 'Perceptual symbol systems'. *Behavioral and Brain Sciences*, **vol. 22** (1999) pp. 577–660.

51. Barsalou, L.W.: 'Social embodiment'. In Ross, B.H. (eds.): *The Psychology of Learning and Motivation*. Academic Press, Amsterdam, Boston, Heidelberg, London, New York, Oxford, Paris, San Diego, San Francisco, Singapore, Sydney, Tokyo (2003) pp. 43–92.

52. Butz, M.V.: 'How and why the brain lays the foundations for a conscious self'. *Constructivist Foundations*, **4**(1) (2008) 1–42.

53. Ziemke, T., Hesslow, G., Jirenhed, D.A.: 'Internal simulation of perception: A minimal neuro-robotic model'. *Neurocomputing*, **vol. 68** (2005) pp. 85–104.

54. Schoenbaum, G., Chiba, A.A., Gallagher, M.: 'Orbitofrontal cortex and basolateral amygdala encode expected outcomes during learning'. *Nature Neuroscience*, **1**(2) (1998) 155–159.

55. Tani, J., Nolfi, S.: 'Learning to perceive the world as articulated: An approach for hierarchical learning in sensory-motor systems'. *Neural Networks*, **12**(7–8) (1999) 1131–1141.

56. Haykin, S.: *Neural Networks: A Comprehensive Foundation*. Prentice Hall, New York, Boston, San Francisco, London, Toronto, Sydney, Tokyo, Singapore, Madrid, Mexico City, Munich, Paris, Cape Town, Hong Kong, Montreal (1999).

57. Jacobs, R.A., Jordan, M.I., Nowlan, S.J., Hinton , Geoffrey, E.: 'Adaptive mixtures of local experts'. *Neural Computation*, **3**(1) (1991) 79–87.

58. Tani, J.: 'Learning to generate articulated behavior through the bottom-up and the top-down interaction processes'. *Neural Networks*, **vol. 16** (2003) pp. 11–23.

59. Tani, J., Ito, M.: 'Self-organization of behavioral primitives as multiple attractor dynamics: A robot experiment'. *IEEE Transactions on Systems, Man, and Cybernetics, Part B*, **33**(4) (2003) 481–488.

60. Ito, M., Noda, K., Hoshino, Y., Tani, J.: 'Dynamic and interactive generation of object handling behaviors by a small humanoid robot using a dynamic neural network model'. *Neural Networks*, **19**(3) (2006) 323–337.

61. Kiebel, S.J., Daunizeau, J., Friston, K.J.: 'A hierarchy of time-scales and the brain'. *PLoS Computational Biology*, **4**(11) (2008).

62. Fusi, S., Asaad, W.F., Miller, E.K., Wang, X.J.: 'A neural circuit model of flexible sensorimotor mapping: Learning and forgetting on multiple timescales'. *Neuron*, **54**(2) (2007) 319–333.

63. Yamashita, Y., Tani, J.: 'Emergence of functional hierarchy in a multiple timescale neural network model: A humanoid robot experiment'. *PLoS Computational Biology*, **4**(11) (2008).

64. Paine, R.W., Tani, J.: 'Motor primitive and sequence self-organization in a hierarchical recurrent neural network'. *Neural Networks*, **vol. 17** (2004) pp. 1291–1309.

65. Paine, R.W., Tani, J.: 'How hierarchical control self-organizes in artificial adaptive systems'. *Adaptive Behavior*, **13**(3) (2005) 211–225.

66.  Maniadakis, M., Tani, J.: 'Dynamical systems account for meta-level cognition'. In Asada, M., Hallam, J.C., Meyer, J.A., Tani, J. (eds.): *From Animals to Animats 10: Proceedings of the 10th International Conference on Simulation of Adaptive Behavior*, SAB 2008. Springer, Berlin, Heidelberg (2008) pp. 311–320.

67.  Bechara, A.: 'The role of emotion in decision-making: Evidence from neurological patients with orbitofrontal damage'. *Brain and Cognition*, **vol. 55** (2004) pp. 30–40.

68.  Melhuish, C., Ieropoulos, I., Greenman, J., Horsfield, I.: 'Energetically autonomous robots: Food for thought'. *Autonomous Robots*, **vol. 21** (2006) pp. 187–198.

*Chapter 10*

# The strategic level and the tactical level of behaviour

*Fabio Ruini, Giancarlo Petrosino,
Filippo Saglimbeni and Domenico Parisi*

## Abstract

We introduce the distinction between a strategic (or motivational) level of behaviour, where different motivations compete with each other for the control of the behaviour of the organism, and a tactical (or cognitive) level, where the organism executes the activities aimed at reaching the goal decided at the strategic level. To illustrate and operationalise this distinction we describe three sets of simulations with artificial organisms that evolve in an environment in which in order to survive they need either to eat and drink, or to eat and avoid being killed by a predator. The simulations address some simple aspects linked to the strategic level of behaviour, i.e. the role played by the environment in determining what are the motivations driving an organism and what is the strength of each of these motivations. Other phenomena investigated are the usefulness for the organism's brain to receive information from its own body (e.g., in the form of hunger or thirst), how inter-individual differences among individual organisms may concern both the strategic and the tactical level of behaviour, and how the unsolved competition between very strong motivations can lead to pathological states such as depression.

## 10.1 Introduction

The behaviour of organisms has both a strategic level and a tactical level. The first is the level at which an organism decides the particular activity in which it will be engaged at any particular time. The tactical level is the level at which the organism executes the specific behaviours that implement the activity decided at the strategic level, allowing in this way the organism to reach the goal of the activity. The tactical level is obviously important because it is the actual level of behaviour. Unless the organism is able to generate the appropriate behaviours that will allow it to reach the goal of the activity decided at the strategic level, the organism's survival will likely be compromised. But the strategic level is even more critical because, in order to survive and reproduce, the organism has to accomplish many different activities and generally it cannot be involved in more than one single activity at any given time. Therefore, there must exist some mechanism within the organism for deciding to which specific activity to dedicate. Examples of different activities are eating, drinking, avoiding predators and other dangers, finding a partner for reproduction, insuring the survival of one's offspring, sleeping, reacting appropriately to physical pain, etc. Every time an organism faces such a dilemma, it must decide, for example, to eat rather than drink or sleep, and then execute the specific behaviours leading to the desired goal. The strategic and tactical level of behaviour may be labelled 'motivational' and 'cognitive' respectively. An organism's chances to survive and reproduce depend on both its capacity to appropriately manage the competition between different motivations (choosing the specific one that must govern its behaviour at any given time; strategic or motivational level), and its ability to generate the appropriate sensory-motor mappings (tactical or cognitive level) that constitute the activity aimed at satisfying the current motivation.

One must note that the words 'strategic' and 'tactical' tend to be used in a different sense. If the activity chosen at the motivational level is complex and it involves a hierarchical structure of sub-goals and sub-activities, one may call 'strategic' the higher levels of this hierarchical structure and 'tactical' the lower levels. For example, if I decide to dedicate myself to eating, this activity may involve going out to buy some food, cooking it, and then ingesting the cooked food. This is a complex hierarchy of sub-activities and sub-sub-activities. In terms of our distinction between a strategic and a tactical level of behaviour all the sub-activities and sub-sub-activities that implement the activity of eating belong to the tactical (cognitive) level of behaviour. On top of the tactical or cognitive level there is the strategic (motivational) level where I must decide whether to dedicate myself to the activity of eating or to some different activity instead. This decision arises from an internal competition between motivations that have different strengths. If I am actually eating, this means that the motivation of eating has won the competition against other motivations for the control of my behaviour at this particular time. The study of the hierarchical structure of activities is a classical topic of research in both psychology and artificial intelligence (Miller, Galanter, & Pribram, 1986; Fikes & Nilsson, 1971). Motivation, of course, is another classical topic of psychology. However, little work has been done so far on constructing artificial systems that have different motivations and know how to decide which motivation must control their behaviour at any given time. Attempting to understand how these artificial systems could work may help us to reach a better understanding of real organisms' behaviour.

In this chapter we will illustrate our distinction between the strategic and the tactical levels of behaviour by describing a series of computer simulations in which populations of artificial organisms evolve in various types of environments. The goal of these simulations is to operationalise and articulate the notions of a strategic and a tactical level of behaviour, as well as to address some simple questions concerning the motivational level (for a general discussion on motivation in artificial organisms see Parisi, 1996).

One important question is how motivations win the competition against other motivations for controlling the organism's behaviour at any given time. One common interpretation consists in seeing each motivation as associated with a specific strength. The motivation that wins competition is simply the strongest one. The strength associated to different motivations can be operationally defined by observing the behaviour of an organism in its natural (ecological) or in a controlled (experimental) environment. For example, if an organism is exposed to food and water at the same time and it reacts by approaching food rather than water, we can conclude that its motivation to eat is stronger than its motivation to drink. The strength of any particular motivation may vary among different individuals and may also be different at different times for the same individual. Various factors can explain the current strength of a motivation. Examples are the age of the individual, or the current inputs received by the individual's brain from within the body, or from the external environment, or self-generated by the organism's brain itself (Parisi, 2007).

A second important question concerns the role played by the environment in which an individual lives in determining both the motivations of the organism and their relative strength. The behaviour of an organism is aimed at allowing its survival, in that only if it is able to survive the organism can hope to have offspring and, in this way, to insure the presence of one or more copies of its genes in future generations. But the behaviour exhibited by a particular organism depends on the particular characteristics of the environment in which the organism lives or, more precisely, the environment where the population to which the organism belongs has lived in the past. Behavioural ecology is the discipline that investigates how behaviour reflects the particular environment in which organisms live (Krebs & Davies, 1997; Stephens & Krebs, 1986). But, of course, it is more generally evolutionary biology that studies how organisms become adapted to their environment through the selective reproduction of the best individuals and the constant addition of new variants to the population, mainly resulting from random mutations affecting the inherited genes. Using a simulation framework very similar to that of our simulations, Seth (2002; 2007) has simulated the evolution of the motivational system of artificial organisms living either in a social or a non-social environment. His research has shown how the nature of the environment shapes the motivational system of the organisms.

Evolution can shape both the motivational and the cognitive level of an organism's behaviour in that the inherited genes contain information specifying both what the different motivations and their basic strength are, as well as which behaviours the organism must exhibit in order to satisfy the different motivations. Inherited genes only specify the basic motivations and the basic behaviours or behavioural dispositions of the organism. Especially in humans the particular experiences acquired during the life of the individual and the ability to learn from these experiences are responsible for the appearance of different motivations and behaviours.

A third question is how the motivational and the cognitive level are actually implemented in the organism. Very often the strategic level involves the arrival of information from the organism's body to the organism's brain. For example, if the organism must decide whether to look for food or water it may be useful for the organism's brain to receive information concerning the quantity of energy (hunger) and liquids (thirst) currently present in the body. The brain of an organism may be viewed as made up of two interacting parts or circuits, one sub-serving the motivational level and the other sub-serving the cognitive level. The motivational circuit processes information from the body, while the cognitive circuit processes information gathered from the external environment. Information from the body arrives to the motivational circuit, which acts together with the cognitive circuit to determine the organism's behaviour. However, since the body constitutes an internal environment which, unlike the external environment, co-evolves with the brain in order to insure the organisms' reproductive chances, a system for informing the brain about the current level of energy and liquids may be more useful in certain types of environments. In other cases, however, it is the sensory input from the external environment that may trigger the motivational circuit and cause the organism to decide the activity to be executed. For example, to survive in an environment that contains both food and a predator an organism should look for food when the predator is absent but it should cease looking for food and react appropriately to the predator instead, for example by flying away when the latter appears. In these circumstances it may be important for an organism's brain to include a motivational circuit that, like the cognitive circuit, is activated by sensory inputs coming from the external environment. However, the two circuits still have different roles and they may process different aspects of the sensory input gathered from the environment. At any particular time the environment usually sends many different inputs to the organism's brain, and one crucial role of the motivational circuit is to cause the cognitive circuit to 'pay attention' to one of these inputs while ignoring the others. To say that the organism 'pays attention' to one input while ignoring the others is the same as saying that the organism responds to that input and not to the others. For example, if the organism's body informs the organism's brain that there is little energy but sufficient liquids in the body, i.e., if the organism feels hungry but not thirsty, it will ignore sensory input from water and respond to sensory input from food by approaching and eating it. In a different environment, if sensory input coming from outside the body informs the organism's brain that a predator has appeared in the environment, the motivational circuit will cause the cognitive circuit to ignore sensory input from food and to respond exclusively to the information related to the predator.

Finally, artificial organisms with motivational circuits can be useful to explore inter-individual differences in behaviour and pathologic conditions of the psychiatric (e.g., depression) rather than neurologic (e.g., aphasia) type. Inter-individual differences may be not only differences in ability levels (cognitive) but they may also be differences in character or personality (motivational). Pathological conditions such as depression can be interpreted as due to competition between different but equally strong motivations so that no motivation is able to prevail, leading to maladaptive or pathological behaviours. Individual organisms that are unable to survive and reproduce may belong to two different typologies: individuals that are not very good in properly executing an activity such as approaching food or water, or flying away from a predator (cognitive deficit), and individuals which become 'paralyzed' or behave in other maladaptive ways when they are exposed to two competing motivations that are both very strong (motivational deficit).

## 10.2   Simulations

The simulations we are going to describe have been developed using the Evorobot[1]simulator (Nolfi & Floreano, 2000). Evorobot is a computer simulation tool which permits to carry out experiments using a simulated replica of the Khepera robot (Mondada, Franzi & Guignard, 1999). The robot has a cylindrical body of 55 mm of diameter and 30 mm of height, eight proximity and light sensors, and two DC motors with incremental encoders respectively connected to two independent wheels that make it possible for the Khepera robot to move and rotate its body.

### 10.2.1   Eating and drinking

In this experimental setup (Saglimbeni & Parisi, 2009) a population of organisms has both to eat and drink in order to survive and reproduce. The body of each organism contains a store of energy and one of water. At each time-step they consume a certain amount of both energy and water just to stay alive. If either one of the two stores become empty, the organism dies. In order for the organism to survive, the two stores have to be regularly replenished with new energy and water supplies. The environment used for this setup is a square of 1,000 × 1,000 pixels (px), surrounded by walls. The organism occupies a round space with a diameter of 75 px. The environment contains food (providing energy) and water (providing liquids) tokens, each of them represented by a circle with a 30 px diameter.

To eat a food token or drink a water token, all an organism needs to do is to move over one of these tokens with the centre of its body. The token then disappears, and it is immediately replaced by a new one of the same type located in a random position. Each organism lives alone in its individual copy

---

[1] http://laral.istc.cnr.it/evorobot/simulator.html

of the environment, for ten epochs of 1,500 time-steps each. At the beginning of each epoch, the environment is cleared and a new random distribution of food and water tokens replaces the preceding one.

The organisms can live in one of four possible scenarios. Scenario 1 contains five food tokens and five water tokens, randomly distributed, capable to refill the corresponding store in the organism's body with 0.2 units of either energy or water. Scenario 2 contains more food than water tokens or, in other simulations, more water than food tokens (the ratio is always five-to-one). The different abundance of food and water is kept constant during the ten epochs. Since to survive in the new environment is more difficult than in the previous one, each token refills the corresponding store in the organism's body by an amount equal to 0.4 instead than 0.2 units. Scenario 3 is the same as Scenario 2, but now the distribution of food and water changes seasonally. In one season (epoch) food is more abundant than water but in the next season water is more abundant than food, cyclically. Scenario 4 includes two distinct zones or patches, one containing three food tokens and the other one containing three water tokens. The patches are square areas of 60 px of side and their centres are located at a distance of 600 px from each other. In this case the quantity of energy or water contained in a single token is 0.03 units. The four different environments are also associated to different energy and water consumption rates for the organisms. All of the organisms start each epoch with the maximum amount of energy and water in their bodily stores, i.e. 1.0 unit, and at no time during the simulation this value can exceed the starting level. At each time-step they consume 0.004 water and energy units in Scenarios 1 and 2; 0.0025 in Scenarios 3 and 4. A minimum lifetime is therefore guaranteed even without eating any food or drinking any water: 250 time-steps per epoch in Scenarios 1 and 2, and 400 time-steps in Scenarios 3 and 4.

The organism's behaviour is governed by a feed-forward neural network made of three layers of units (see Figure 10.1(a)). The input layer includes two units encoding the distribution of food tokens currently contained in the organism's visual field, and two units encoding the location of water tokens. The reason for having two units for each type of tokens is that the visual field of the organism is divided into two half-fields, respectively left and right. One unit in each pair encodes the presence of food (or water) tokens in the left half of the visual field, while the other unit does the same for the right half. Assuming $0°$ as the organism's current facing direction, the left visual half-field is centred at $+60°$ and covers an area ranging from $+105°$ to $+15°$; the right visual half-field is centred at $-60°$ and focuses from $-15°$ to $-105°$. The vision range of the organisms is unlimited in terms of extension. The activation values of the input units are calculated according to eq. (10.1).

$$activation\_level = \sum_{i=1}^{n} K\left[A + B^* \log\left(\frac{1}{d_i^2}\right)e^{-\left(\frac{(\phi_i-60)^2}{2\sigma^2}\right)}\right] \tag{10.1}$$

where: $n$ is the number of food or water tokens within the current half-field, $d_i$ and $\phi_i$ respectively identify the distance and the angle (based on the current heading direction) between the organism and the $i$-th token detected, $K$ is a factor that accounts for the number of tokens present in the scenario ($K = 0.75/\text{Log}(N)$ for Scenario 1, 2, and 3; $K = 0.5/\text{Log}(N)$ for Scenario 4, with N equal to 10 for Scenario 1, 6 for Scenarios 2, 3, and 4); A and B are parameters, arbitrarily set to 1.596 and 0.11 respectively, used in order to prevent a sensory unit from being activated with a value greater than one or lesser than zero; $\sigma$ corresponds to half of the angular view capability for each 'eye', which is 45. Linking the input to the output units are four hidden neurons sharing the following sigmoidal activation formula (where $y_i$ is the activation value of the $i$-th neuron and $x$ is the weighted sum of the all inputs received):

$$y_i(x) = \frac{1}{1 + \exp(-x)} \tag{10.2}$$

These four hidden units are fully connected to both the input and output layers. The two output units are respectively linked to the two motors controlling the wheels and therefore the movements of the organism (0: maximum backward speed, 0.5: do not move, 1: maximum forward speed). The organisms' maximum speed corresponds to 8.6 px per time-step. Bias is applied to the

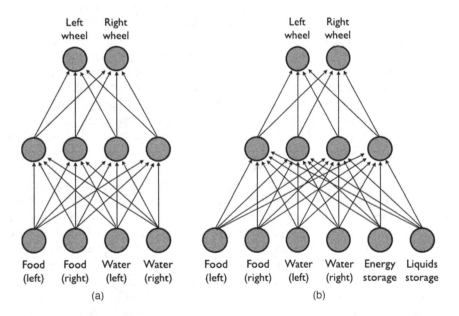

| Left wheel | Right wheel | | | | Left wheel | Right wheel | | | |

| Food (left) | Food (right) | Water (left) | Water (right) | | Food (left) | Food (right) | Water (left) | Water (right) | Energy storage | Liquids storage |

(a)          (b)

*Figure 10.1    The neural network architectures used for the simulations described in section 10.1*

units of both the hidden and output layers. The input units do not have any bias instead and they rely on a simple identity transfer function.

The organisms must respond to sensory input from food and water tokens by approaching and consuming them efficiently and in such a way that they can avoid dying because of lack of energy or water in their body. The simulation starts with a population of 100 organisms whose neural networks have randomly assigned synaptic weights and biases within the range $[-5.0; +5.0]$. Those values are encoded as binary digits sequences in the individuals' genomes. Starting with random genomes implies that at the beginning of the simulation the organisms' behaviour is not very efficient. They are unable to approach food and water tokens efficiently and they may eat and not drink, or vice versa, so that their average life expectancy is quite short. However, the individuals that succeed in eating and drinking sufficiently and in a balanced way, and therefore live longer, have more offspring than other individuals. The evolution towards a population of organisms able to correctly cope with the desired task takes place through a genetic algorithm. At the end of the ten epochs spent inside the simulated environment, all individuals are assigned a fitness value corresponding to the total number of time-steps they have managed to survive. The 20 individuals with the highest fitness values are selected for reproduction. Each of them then generates five offspring that inherit the same connections weights and biases of their (single) parent. The 100 new individuals constitute the population at the next generation. Random mutations are applied to all offspring: each bit of the genome's binary representation has a 0.04 probability of being mutated, thus modifying the corresponding real value. Since offspring inherit the same synaptic weights of their parents with the addition of random mutations, the result is a progressive increase in the effectiveness of the behaviour of the organism as well as a progressive lengthening of their life. The organisms of later generations tend to approach food and water tokens efficiently and in a balanced way so that the energy and water stores in their body are unlikely to become empty at any given time. The evolutionary process is iterated for 1,000 generations and then repeated ten times, each time re-assigning new random connection weights and biases to the members of the starting population. The results coming from the ten experimental replicas are averaged together in order to identify a non-noisy evolutionary trend. For each type of scenario we contrast two different populations of organisms. In one population the neural network controlling the organisms' behaviour has the architecture described above. Therefore in this condition the organisms' behaviour is determined exclusively by the sensory inputs coming from the external environment. In the second population we add two more units to the organisms' neural network (see Figure 10.1(b)). The activation level of these additional units co-varies (linearly between 0 and 1) respectively with the current level of energy and water in the two stores inside the organism's body. The two new units are fully connected to the hidden layer of the network. In this second condition, therefore, the organisms' behaviour is controlled both by the sensory input from the external environment (perceiving food and water tokens) and by the internal input from their own body (level of hunger and thirst). The neural pathway from the sensory input units to the hidden layer simulate the cognitive circuit discussed in the Introduction, while the neural pathway from the two additional units to the hidden units simulates the motivational circuit. The 'brain' of the organisms that possess only the cognitive circuit has no information from within the body but only information from the external environment. These organisms ignore what hunger and thirst are. The 'brain' of the organism endowed with both a cognitive and a motivational circuit receives information both from the external environment and from within the body, which means that the organisms may feel hungry and thirsty. (For simulations in which artificial organisms can be either hungry or thirsty but not both hungry and thirsty at the same time, see Cecconi and Parisi, 1993.)

The results of the simulations show that there are no differences in terms of fitness (as measured by length of life) between the populations with and without the motivational circuit in Scenarios 1 and 2 (Figures 10.2 and 10.3), whereas the population with the motivational circuit has more fitness than the population without the motivational circuit in Scenarios 3 and 4 (Figures 10.4 and 10.5). Why? To answer this question one has to consider that the behaviour of the organisms has to be effective at both the strategic and tactical levels. At the strategic level an organism's behaviour is effective if it makes the organism able to approach food when its body needs energy, and water when its body needs liquids. At the tactical level the organism's behaviour is effective if the organism is able to approach both food and water tokens efficiently, by

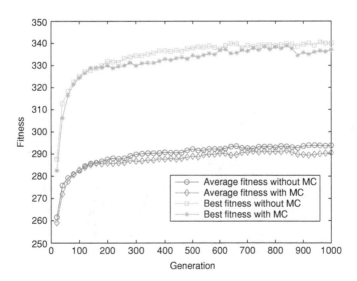

*Figure 10.2    Average and maximum fitness for organisms living into Scenario 1, respectively with and without receiving information about their energy and liquids reserves*

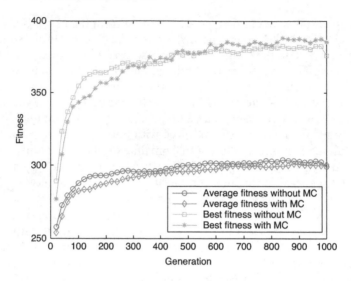

*Figure 10.3    Average and maximum fitness for organisms living into Scenario 2, respectively with and without receiving information about their energy and liquids reserves*

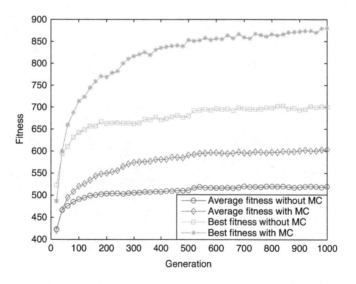

*Figure 10.4    Average and maximum fitness for organisms living into Scenario 3, respectively with and without receiving information about their energy and liquids reserves*

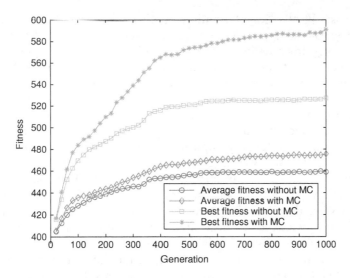

*Figure 10.5    Average and maximum fitness for organisms living into Scenario 4,
respectively with and without receiving information about their
energy and liquids reserves*

going straight and fast towards the desired token. The conditions that make
the behaviour of our organisms effective at the tactical level are identical in all
four scenarios, whether or not their 'brain' contains a motivational circuit. In
contrast, the conditions that make their behaviour effective at the strategic
level vary with the different scenarios, with the motivational circuit playing a
crucial role in Scenarios 3 and 4 and no role in Scenarios 1 and 2.

In Scenario 1 food and water are equally abundant. They are randomly
distributed across the environment and therefore the organisms can be effective
at the strategic level by simply approaching the nearest token, whether it is
food or water. Given the nature of the environment this behaviour will insure a
balanced diet of food and water, with no need for a motivational circuit. A
motivational circuit is also not necessary in Scenario 2 where food is always
more abundant than water (or vice versa). In this scenario the organisms evolve
a tendency to go towards the type of token which is less abundant, unless the
more abundant token is very close. This is an extremely simple behaviour,
nonetheless capable of insuring a balanced diet of food and water even if the
organisms' neural network lacks a motivational circuit. Therefore in Scenarios
1 and 2 there are no difference in terms of fitness between the organisms with
and those without a motivational circuit. This implies that it makes no sense,
from an evolutionary perspective, to invest in the development of a motivational circuit.

The situation is different for Scenarios 3 and 4. Scenario 3 is seasonal, with
more food than water during one season and the opposite during the following
one. Since the organisms ignore what the current season is, to survive they

must necessarily rely on information coming from their body. If they feel hungry they have to approach the nearest food token, while if they feel thirsty they have to go towards the closest water token. For these organisms, therefore, it is necessary to possess a motivational circuit that informs their 'brain' of the current level of energy and water in their body. The same is true for Scenario 4, where food and water tokens are located in separate zones. The organisms without a motivational circuit that happens to end up in the zone with food and without water will eat a lot but it is likely that they will die due to a lack of water. The opposite will happen if they end up in the zone abundant in water but with rare food tokens. These organisms too need a motivational circuit capable of informing their 'brain' about the current level of energy and water in their body in order to leave the food zone to go to the water zone when they feel thirsty and to do the opposite when they feel hungry. This explains why for the organisms living in Scenarios 3 and 4 the existence of a motivational circuit results in higher fitness.

The different adaptive patterns of the distinct populations can be clearly demonstrated if we test an individual organism in a standardised (laboratory) situation in which the organism perceives both a single food token and a single water token. Manipulating the distance between the organism and the two tokens (as well as its levels of hunger and thirst for the organisms with a motivational circuit), we can measure the organism's preferences by examining if it approaches the food or the water token. The organisms evolved in Scenario 1 tend not to have any systematic preference for either the food or the water token when the two tokens are located at the same distance from the organism. They simply tend to approach the closest token, regardless of whether it is food or water. As the distance between the organism and the farthest token increases, this tendency to go towards the closest token strengthens. For these organisms, however, it does not make much of a difference whether they have a motivational circuit or not. This is because their adaptive pattern does not rely on the presence of a motivational circuit and on their ability of respond appropriately to hunger and thirst. The organisms that have adapted to Scenario 2, for example to an environment with food stably more abundant than water, tend to consistently approach the less abundant water token if the two tokens are at the same distance. They approach the food token only if the water token is much more distant than the first one. For these organisms, too, the presence of a motivational circuit has no noticeable influence on their behaviour in the experimental situation. In contrast, for the organisms that have adapted to Scenarios 3 and 4, where the possession of the motivational circuit is critical for evolving an effective behavioural pattern, the behaviour in the experimental situation is clearly influenced by their level of hunger and thirst. If food and water tokens are at the same distance they clearly approach food if they are hungry and water if they are thirsty.

If we define as 'environment' the origin of all the inputs arriving to an organism's brain, we can distinguish between an external environment, which consists in what lies outside the organism's body, and an internal environment, which consists in what lies inside its body. On the basis of the simulations described in this Section we can conclude that organisms having two distinct needs to satisfy (but the results can probably be generalised to numbers higher than two) have to evolve a behaviour which in some evolutionary scenarios depends only on the external environment, and in other scenarios depends on both the external and the internal environments. In Scenario 1 the internal environment is not important because the equal abundance of food and water and their random distribution allow the organisms to behave effectively without having to choose to approach food or water or, as we may also say, without deciding whether to pay attention to food and ignore water or vice versa. In Scenario 2 the internal environment is also irrelevant because of the greater abundance of food with respect to water (or the opposite in different simulations), which allow the organisms to always prefer the less abundant type of tokens, thus developing a tendency to ignore the more abundant one. In Scenarios 3 and 4, on the contrary, the internal environment becomes critical. The organisms can survive much better if they respond to the internal environment (to their hunger and thirst) when deciding to approach food and ignore water or to approach water and ignore food.

## 10.2.2 Eating and escaping from predators

In the simulations described in the preceding section, the motivational circuit has its origins within the internal environment, reflecting the level of energy and water currently present in the organism's body. But a motivational circuit can support the strategic level of behaviour even if the event that triggers the circuit lies outside the body. In this Section we describe simulations that illustrate how the strategic level of behaviour can rely on information gathered from the external environment (Petrosino & Parisi, in preparation). The following description will only highlight the elements of novelty contained in the new simulations compared to the preceding ones. All the non-specified details should be considered the same as in the simulation described in Section 10.2.1.

A population of organisms lives in an environment containing 24 randomly distributed food tokens. In order to survive and reproduce the individuals have to approach and eat these food tokens. The environment is bigger compared to the one described in the previous paragraph and its size is now $1,500 \times 1,500$ px. The organisms have a store of energy in their body, part of which they consume at each time-step to stay alive. The starting level of energy for all the organisms is 1.0 unit (level that cannot be exceeded) and the organisms consume 0.0028 energy units at each time-step. Therefore they have to eat regularly in order to replenish their store of energy and avoid dying. The food tokens have a circular shape with a diameter of 68 px. Each of these contains 12 energy units. When the centre of the body of an organism moves on a token, the diameter of the latter is reduced by 4 px, while the energy contained in the individual's body is increased by 0.014. The token disappears when its diameters becomes less than 20 px. The organisms have also to face a second problem however. At random intervals a predator (represented by a circle with 70 px diameter) appears and approaches the organism. Each organism lives for five epochs made of 2,000 time-steps each, while the predator can appear during any step between zero and *PredatorAppearance*.[2] If the predator reaches the organism, the latter is killed and no further epochs are evaluated. Therefore, to stay alive the organisms have to both approach and eat the food tokens when the predator is absent, and to ignore food and fly away from the predator when it appears. The predator remains into the environment for 150 time-steps before disappearing (and reappearing again during the next epoch). Therefore, in order to remain alive the organism must be able to avoid being reached by the predator until it disappears. Unlike the organisms' behaviour, the predator movements are hardwired rather than evolved. When it appears, at each time-step the predator approaches the organism in the best possible way, by minimising the Manhattan distance from the two agents.

The neural network that controls the organisms' behaviour includes two input units providing information about the location of the nearest food token as well as two neurons encoding the location of the predator (when it is present into the environment). In order for the organism to see a food token or the predator, these need to be within a distance of 260 px from the centre of the organism's body. The input units are fully connected to a layer of internal units, which is in turn fully connected to the output units controlling the displacements of the organism in the environment (Figure 10.6(a)). The initial population of organisms has randomly assigned connection weights and biases. The organisms evolve their behaviour in a succession of generations, with the individuals able to live longer having more offspring than the individuals with a shorter life. Living longer depends on both the ability to eat food when the predator is absent and the ability to escape from the predator when present. The fitness of an individual is measured as the total number of food units collected during its entire lifespan. At the end of each generation, a genetic algorithm selects the 20 fittest individuals for reproduction. Elitism is applied so these 20 individuals are copied to the next generation without any modification. Each parent generates four offspring inheriting its connection weights and biases. Their genomes are randomly mutated by switching the value of each of the binary digits with probability 0.02.

Two different versions of the simulation have been run. In the first one the organisms' neural network is as we have described it: the input units encode the perceptions of food and predator, the output units control the organism's movements, and the internal units provide a link between the input and the output units. In the second version a new internal unit is added to the organisms' neural network. This unit receives connections only from the input neurons encoding the presence and location of the predator and it sends its

---

[2] Various simulations have been run, using different values for *PredatorAppearance* (150, 200, 250, and 300 respectively).

connections to the output units (see Figures 10.6(b)), bypassing the hidden layer. This additional unit is like all other internal units, except for just one difference. While an ordinary internal unit has an intrinsic and constant bias, which always activates the unit together with the values arriving from the input units, the bias of the additional (motivational) unit is only present when the predator is present and perceived by the organism. This implies that when the predator is absent or not perceived by the organism the motivational unit has an activation value equal to zero, thus not having any role in determining the organism's behaviour. The neural pathway which goes from the input units encoding the predator to the motivational unit and then to the output layer is what we define as the 'motivational circuit'. In contrast, the 'cognitive circuit' is the neural pathway linking all the input units to the ordinary internal units and then to the output units.

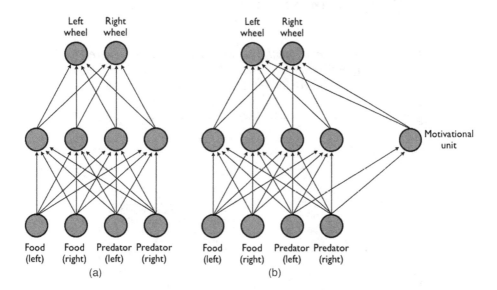

*Figure 10.6    The neural network architectures used for the simulations described in Section 10.2.2*

The results of these simulations (Figures 10.7 and 10.8) show that the addition of the motivational circuit to the organisms' neural network allows the individuals to reach higher levels of fitness compared to the organisms lacking this circuit. How can we explain this difference in fitness? To understand these result it is important to look at the organisms' behaviour when the predator is absent and when it is present. When the predator is absent, both the organisms with and those without the motivational circuit appear to be only interested in food. They approach the nearest food token, eat it, and then move to the next one. However, if we test the organisms in a controlled environment

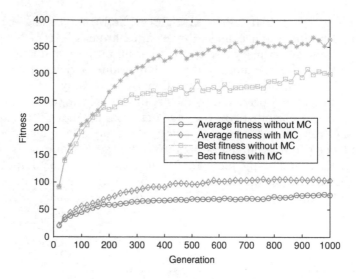

*Figure 10.7    Average and maximum fitness for the organisms evolved in the simulation described in section 10.2.2, respectively relying or not on a motivational circuit (Predator Appearance = 250)*

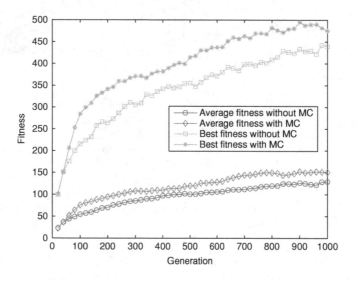

*Figure 10.8    Average and maximum fitness for the organisms evolved in the simulation described in section 10.2.2, respectively relying or not on a motivational circuit (Predator Appearance = 300)*

containing only food and no predator, we see that the organisms with the motivational circuit are better in approaching the food tokens than the organisms lacking the motivational circuit. This therefore appears to be one of the reasons why the organisms endowed with the motivational pathway obtain a higher fitness than the organisms without this circuit.

But the more interesting difference in behaviour between the two categories of organisms can be seen if we look at what happens when the predator appears. In this situation, what is expected from the organisms is that they ignore the food tokens and exclusively focus on trying to escape from the predator. Both the organisms with the motivational circuit and those without it are able to do so but, again, the organisms with the motivational circuit are better than the organisms lacking it. This is another reason why they have a higher fitness. The greater ability of the organisms with the motivational pathway to escape from the predator is observable both in a controlled (experimental) and in an ecological condition. In the controlled condition the organisms are tested inside an environment containing only the predator and no food. In the ecological condition the predator appears at regular intervals within an environment containing a certain amount of food tokens. In the controlled condition the organisms with the motivational pathway appear to be better able to escape from the predator. But the more interesting results concern how the two types of organisms react when the predator appears in their natural environment containing food, forcing them to suddenly starting to ignore food in order to dedicate themselves solely to flying away from the predator. If we observe the behaviour of the organisms when they first perceive the predator we see that an organism endowed with the motivational circuit tends to react 'impulsively' by running at the maximum speed along its current direction of movement, whatever that direction is. Only when the predator gets very close to the organism, the latter reacts in a more reasoned way by moving in a direction opposite to the one from which the predator is coming. This type of complex behaviour is typically observed in the organisms endowed with the motivational circuit while it is rarely present in the organisms lacking it. The organisms without a motivational circuit tend instead to adjust their speed according to the distance between themselves and the predator. Rather than reacting impulsively, running at the maximum speed possible when the predator appears, they progressively increase their speed the more the predator gets close to them.

If we examine what happens in the neural networks of the two types of organisms we may find an explanation for why they react differently at the first appearance of the predator. We first measure the activation level of the two output units determining the organism's motor behaviour when the predator is at different distances from the organism, from very distant to very close. Then we compare the activation level of the same two units both when food is normally present in the environment and when food is experimentally removed from it. Notice that, from a neural point of view, the degree to which the organisms ignore (i.e., do not pay attention to) the food sensory input when the predator is present is indicated by how similar the activation level of the two motor output units is when food is present but is not attended to by the organism (ecological condition) and when food is actually absent (experimental condition). The organisms whose neural networks include a motivational circuit are better able to ignore food and react to a suddenly appearing predator than the organisms lacking the motivational pathway. This is indicated by the fact that the activation level of their two output units, and therefore their motor behaviour, tends to be very similar both when (1) the predator has just appeared but is very distant and there is no other sensory input because there is no food present (experimental condition) and when (2) the predator has just appeared and there is input from food (ecological condition) also. In fact, the similarity between the activation level of the output units in the two conditions can be considered as the neural basis for ignoring food when the predator appears in the ecological condition. However, as the predator approaches the organism and it is perceived as nearer by the organism, the difference between the activation level of the two motor output units in presence and in absence of food first increases and then it decreases until it reaches almost zero. This indicates that, when the predator is very close to the organism, food is completely ignored by the organism although the sensory input from food continues to arrive to the organism's sensors. (A further analysis of this phenomenon, which appears not to depend on the 'cognitive capability' of the organism, but on the presence of the motivational circuit instead, can be found in Ruini & Parisi, 2008.)

Returning to the organisms' behaviour we can describe the behaviour of the organisms with the motivational circuit in response to the predator as made up of two successive stages. When the predator is first perceived and it is still distant from the organism, the organism responds with an 'impulsive' or 'emotional' and not well reasoned behaviour: the organism runs very quickly in any possible direction, which is not necessarily the one maximising its distance from the predator. However, this emotional response allows the organism to better ignore food and to pay attention to the predator only. And in fact, among the organisms endowed with a motivational circuit, the individuals that display this type of emotional response to the first appearance of the predator tend to have more fitness than the organisms that do not exhibit this emotional reaction (see Figure 10.9).

One can better understand the respective roles played by the motivational and the cognitive pathways if we alternatively lesion these two circuits and then we examine the activation level of the motivational pathway's internal unit when the predator is standing at different distances from the organism. As we have already said, when the predator is absent or it is not perceived by the organism because of the distance, the internal unit of the motivational pathway has zero level of activation and therefore the motivational pathway itself has no influence on the organism's behaviour. However, as soon as the predator is perceived by the organism, this internal unit is activated and cooperates with the cognitive pathway in determining the proper behaviour to adopt. If we measure the activity level of the internal unit of the motivational pathway when the predator is perceived by the organism, but is at various distances, we see that this value is very high when the predator is distant from the organism but it quickly lowers and

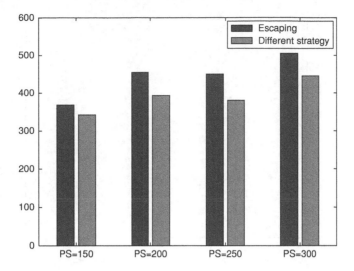

*Figure 10.9   Fitness values for organisms endowing the motivational circuit but showing different behaviours when the predator appears (left: moving at a maximum speed along a random direction; right: following a different strategy). On the X axis different values of Predator Appearance are represented*

almost reaches a zero level when the predator gets closer to the organism (see Figure 10.10). This appears to show that the motivational pathway plays a role in determining the organism's behaviour especially when the predator first appears but it rapidly becomes less important when the predator approaches the organism. When the predator first appears the motivational circuit is activated and it has two types of influences on the organism's behaviour: it cancels the role of the sensory input from food in determining the organism's behaviour so that the behaviour turns out to be determined only by the sensory input from the predator and it causes in the organism an emotional reaction which, as we have already said, consists in moving very fast but not necessarily away from the predator. On the other hand, when the predator gets closer to the organism, the internal unit of the motivational circuit quickly reduces its activation level, which implies that it is the cognitive pathway which takes control of the organism's behaviour, inducing a more reasoned behaviour of moving away from the predator. This analysis is confirmed by the fact that if we lesion the motivational pathway, the organism is less able to avoid the predator in the experimental condition in which only the predator is present and there is no food. In contrast, if we lesion the cognitive pathway, what is observed is that, when the predator approaches the organism, the latter tends to be unable to avoid being killed.

Our conclusion is that, for the organisms endowed with a motivational circuit, evolution has created a sort of division of labour between the motivational and the cognitive circuit. This division of labour may explain why the

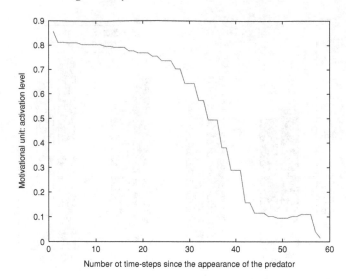

*Figure 10.10*    *This graph has been obtained by deploying six individuals*
*(exhibiting the behaviour of moving at the maximum speed*
*possible when the predator appears), one by one, at the centre*
*of an empty environment. Cutting their connections to the output*
*layer they have been made unable to move. Then a predator has*
*been put into the environment, letting it hunt the organism for*
*60 steps. For each organism, the test is repeated five times with*
*the predator appearing in different positions and the measures*
*averaged. What this plot shows is the average activation value of*
*the motivational unit with the passing of the time-steps*

organisms with a motivational circuit are better than organisms lacking this circuit not only at avoiding being killed by the predator but also at eating food when the predator is absent. In fact, since the motivational circuit responds to the presence of the predator and to the distance of the predator from the organism, while the cognitive circuit responds only (or mainly) to the direction from which the predator is approaching the organism, by having a less heavy information load the cognitive circuit can better process information from food (when the predator is absent) thus responding more effectively to it. In the organisms lacking the motivational pathway this division of labour is impossible and this may explain why they are both less able to avoid being killed by the predator and to eat food efficiently.

### 10.2.3    Entering or not entering the predator's zone

In this Section we describe a third set of simulations similar to the ones described in Section 10.2.2. The main difference consists in the fact that the predator can now only appear inside a specific area of the environment and cannot exit from there. The environment is a bit smaller than the previous one, given its size of 1,200 × 1,200 px. The zone where the predator lives is a circular area of 300 px of diameter. 18 food tokens, each of these having a diameter of 35 px, are randomly distributed across the environment. Food is present both outside and inside the predator's zone (11 food tokens are outside the predator's zone and seven inside) but the food contained in the predator's zone is more energetic than the food situated outside. Therefore, the organism is caught in a conflict between remaining outside the predator's zone but not eating the more energetic food present there and penetrating the predator's zone to eat the more energetic food that can be found there but risking being killed by the predator. The organisms do not have any motivational pathway in their neural network. Two different simulations have been run: one in which the neural network controlling the organisms' behaviour includes an additional sensory input unit encoding the information that the organism has penetrated the predator's zone (say, a sensory unit encoding an odour associated with the predator) and another version in which the neural network lacks this additional sensory unit (see Figure 10.11). The sensory unit is connected to the neural network's internal layer in the same way as all the other inputs. The other input units respectively encode the location of the nearest food token, and the

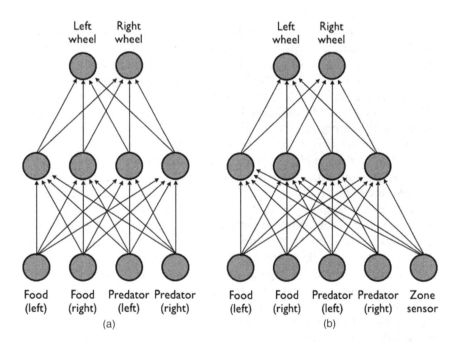

*Figure 10.11   The neural network architectures used for the simulations descri-bed in Section 10.2.3*

location of the predator. The latter units only get activated when both the organism has penetrated the predator's zone and the predator is present there.

Furthermore, in both versions of the simulation we have varied two parameters: the dangerousness of the predator (as measured by its speed of movement) and the energetic value of the food present inside the predator's zone, which is always more energetic than the food outside the predator's zone but can, in different simulations, be twice or five times richer. The results obtained (summarised in Table 10.1) show that the organisms' behaviour is sensitive to these experimental manipulations.

*Table 10.1    Fitness values (calculated on the entire populations as average of the last 20 generations) in the different experimental conditions*

|  | S5/E2 | | S5/E5 | | S9/E2 | | S9/E5 | |
|---|---|---|---|---|---|---|---|---|
|  | **BF** | **AF** | **BF** | **AF** | **BF** | **AF** | **BF** | **AF** |
| Without zone sensor | 483 | 200 | 557 | 228 | 270 | 110 | 316 | 123 |
| With zone sensor | 530 | 217 | 596 | 238 | 329 | 133 | 318 | 127 |

*Note*: S, predator's speed; E, energetic value of the food; BF, best fitness; AF, average fitness.

The organisms are more likely to enter the predator's zone when the predator is less dangerous than when it is more dangerous, as well as when the food available inside the predator's zone is more energetic than the food outside. These results apply to both the organisms endowed with the additional sensory unit telling them that they have entered the predator's zone and the organisms lacking this additional sensory unit. Another result is that the organisms with the odour sensory unit reach higher level of fitness than the organisms without. This is particularly true when the predator is very dangerous and when the food in the predator's zone is highly more energetic than the food outside the zone. When the predator is more dangerous or the food in the predator's zone is only slightly more energetic than the food outside, it is not very useful or sensible for the organisms to enter the 'hazard zone'. Therefore the organisms endowed with the odour sensory unit tend to exit the predator's zone as soon as this unit tells them they have entered it. This possibility is not available to the organisms without the odour sensory unit, and this explains why these organisms have less fitness than the organisms possessing the additional unit. On the other hand, when the predator is not very dangerous or the food in the predator's zone is much more energetic than the one outside, it makes more sense to enter the predator's zone, even at risk of being killed. In all cases the organisms endowed with the odour sensory unit can rely on an entirely reliable clear-cut (yes or no) information that they are inside the predator's zone, which they can adaptively exploit. In contrast, the organisms lacking the odour sensory units do not know if they have entered the predator's zone and they can only rely on the appearance of the predator and respond to this volatile information. This is another possible explanation for why the organisms with a motivational circuit score higher fitness values than the organisms lacking the circuit.

## 10.3   Inter-individual differences in behaviour and the appearance of pathological behaviour

Inter-individual differences in behaviour are an important aspect of living organisms. Of course, all kinds of inter-individual differences play a crucial role in biology since evolution is based on inter-individual variability, i.e., on the reproduction of some individuals against others and the constant addition of new variability through genetic mutations. But the study of inter-individual differences might also be useful to better understand the complex mechanisms underlying behaviour. Inter-individual differences may in fact result from the different functioning of the various components of the neural and, more generally, bodily systems underlying behaviour, therefore allowing us to identify what these components are and how they exactly work. Artificial organisms endowed with both a motivational and a cognitive system are located in a larger space of possible inter-individual variability than organisms only relying on a cognitive system. Our simulation experiments make possible to start the exploration of this larger space of inter-individual variability.

The simulations described in this chapter use a genetic algorithm to evolve the organisms in their environments. The use of an evolutionary algorithm implies that in each generation some individuals do reproduce while other individuals do not (or some individuals have more offspring than others). Since evolutionary algorithms tend to be used in order to obtain some desired behaviour, the researcher is typically interested in the individuals that reproduce and therefore approximate the desired behaviour rather than in the individuals that do not reproduce. But if one is interested in inter-individual differences in behaviour, the individuals that do not reproduce may be more interesting than the ones that do reproduce. In our simulations the individuals that do reproduce tend to be organisms functioning effectively at both the motivational and cognitive levels. On the other hand, the individuals that do not reproduce can be individuals that function well at the motivational level but not at the cognitive level, or vice versa. In fact, if in our simulations we look at the individuals that do not reproduce, we see that the reason behind their inability to reproduce may reside in their motivational or in their cognitive system, but not necessarily in both. If in our simulations with both food and predator we test in controlled conditions (with either food or the predator present but not both) various individuals that do not reproduce, we may find that some of them are good at approaching food in absence of the predator and at flying away from the predator in the absence of food, but they have trouble in avoiding being killed by the predator in their natural environment (in which the predator appears when food is also present). These individuals tend to be distracted by food and, for this reason, they may easily be killed by the predator. These are individuals good at the cognitive level but not at the motivational level. On the other hand, other individuals that do not reproduce may present the opposite pattern: they may be good at ignoring food when the predator appears, but not very effective at approaching food in absence of the predator. These individuals are good at the motivational level but not at the cognitive level. The existence of these two types of individuals confirms the usefulness of our distinction between the strategic and the tactical level of behaviour.

Similar results have been obtained in the simulations in which the predator can only appear in a particular zone of the environment. Among the individuals that do not reproduce some are very good at approaching food outside the predator's zone. But these same individuals behave inefficiently when they happen to enter the predator's zone. When they enter the predator's zone these individuals tend to remain at the border of the zone because they are attracted by the more energetic food present inside but at the same time they are 'afraid' of the predator. Therefore they are unable to eat both the food inside the predator's zone and the food outside. These individuals can be classified as not good at the motivational level. On the other hand, there are other non-reproducing individuals that are good at the motivational but not at the cognitive level. When they happen to enter the predator's zone they either try to eat some of the more energetic food present in that region or they quickly exit the predator's zone becoming interested in the food present outside. However, these individuals are not very good at approaching food, either inside or outside the predator's zone, and this low level of functioning at the cognitive level is sufficient to prevent them from reproducing.

The behaviour of the first type of organisms (motivational deficit) can be considered as pathological and, more specifically, exemplifying (at a very basic level, of course) the clinical condition of depression. The individual is caught between two strong motivations and it is unable to decide which one to pursue at any given time. The consequence is that the individual does nothing, which of course implies that the individual will probably not reproduce. It is interesting to note that this type of pathological behaviour can be more easily observed when the two motivations are both strong. As we have observed, different motivations compete for the control of behaviour based on their strength. The strength of a motivation can be an intrinsic and therefore constant property of the motivation or it may vary with time and circumstances. For example, in the simulations with equal abundance of food and water in the environment we observe the pathological condition of doing nothing when the organism is exposed to both a food token and a water token put at the same distance if the organism is both very hungry and very thirsty.

## 10.4    Conclusions

In this Chapter we have described some simulations that illustrate our distinction between the strategic (motivational) level and the tactic (cognitive) level of the behaviour of organisms. This has allowed us to investigate some simple phenomena involving the strategic level and its interaction with the cognitive level. To survive and reproduce organisms must be able to satisfy many different motivations. Since they usually can pursue only one motivation at a time they need a mechanism that allows them to decide which motivation attempt to satisfy during each moment. This is the strategic level of behaviour. Once they have decided which motivation to pursue, they must be able to execute the activity which allows them to reach the goals of the winning motivation. This is the tactical level of behaviour. At the strategic level different motivations compete for the control of the organism's behaviour and the motivation that wins the competition is the one which currently has the highest strength. Using simple computer simulations of artificial organisms evolving within artificial environments, we have shown that the motivations of an organism (and their strength) depend both on the particular environment in which the organism lives and on the current state of its own body. In some scenario, but not in all of them, it is useful that the body sends information about its current state to the organism's brain so that the organism can take into account this information together with information coming from the external environment in order to decide which motivation to pursue. Our simulations have shown that, in these scenarios, organisms endowed with an information channel going from the body to the brain and transmitting information about the level of energy and liquids currently contained into the body reach higher levels of fitness than organisms that do not have this information channel available. Like real organisms, artificial organisms can spontaneously evolve such information channels. For example, organisms that have to cease any type of activities in order to recover more quickly and safely when their body is damaged, will autonomously evolve a pain sensory unit in the neural network controlling their behaviour. This unit will tell their brain that some damage has occurred at the physical level (Acerbi & Parisi, 2007).

While the cognitive level of behaviour mainly uses information originated in the external environment, the motivational level tends to use inputs that originate within the organism's body. However, as we have seen in our simulations with predators, sensory input from the external environment can activate motivations and therefore it can indirectly control the behaviour of an organism by passing through the organism's motivational system. Another role of the external environment that we have not explored is that sensory input coming from outside the body can augment the current value of some motivations, leading in this way a particular motivation to win the fight against other competitors and take control of the organism's behaviour. In psychology this type of sensory inputs from the environment are called incentives. For example, passing in front of a restaurant can increase the current value of my motivation to eat, and as a result I can decide to dedicate myself to satisfy it by interrupting any other activities aimed at a different motivation. This shows that the play of different motivations in my brain can not only motivate me to engage in one activity rather than in a different one but it can also cause the interruption of the current activity.

## 10.5   Discussion

In this chapter we have distinguished two basic systems that determine the behaviour of organisms, the motivational and the cognitive system. Although the two systems are distinct and are implemented by mostly different neural circuits, the motivational and the cognitive levels of behaviour influence each other in a number of ways. The motivational system influences the cognitive system in that it directs the attention of the organism to only one of the many inputs arriving to the organism's sensors from the environment. This is an important direction of research that can be pursued using our simulation framework. For example, some experiments carried out on human subjects have shown that stressful conditions increase the selectivity of attention (Chajut & Algon, 2003), and one might replicate these results with artificial organisms in order to formulate explicit models of this effect. On the other hand, the activities that the organism executes to pursue its current motivation may generate inputs influencing the motivational system and thus causing the organism to shift to another motivation. This also is a very interesting direction of future research.

One aspect that has not been investigated throughout this chapter is how the strategic level of an organism's behaviour could be affected by the emotions he experiences. Emotions are clearly related to motivations, but the strategic level of behaviour concerns motivations, not emotions. Emotions can be defined as the arrival of inputs to an organism's brain that are generated inside the organism's body (or perhaps even within the brain itself). These inputs, of course, play a crucial role in deciding which motivation will control the organism's behaviour. However, we think that motivation is the most basic phenomenon, so the one that is interesting to investigate. Motivations control in fact the behaviour of an organism even when the organism does not feel any particular emotion. Emotion is part of the motivational system but it is a particular phenomenon which occurs in particular circumstances, that is, when one motivation wants to 'raise its voice' in order to be heard and to win the competition with other motivations. (For the influence of emotion on the behaviour of artificial organisms, see Canamero, 2005; Perez, Moffat & Ziemke, 2006; Ziemke, 2008.)

We have also seen that constructing artificial organisms endowed with a motivational level of behaviour makes it possible to explore a larger space of inter-individual differences that are observed in real organisms. If an organism has just one single goal or motivation, an individual can only differ from another individual in that one is better than the other at reaching that particular goal. This is a quantitative difference in ability. But real organisms differ among them not only in a quantitative but also in a qualitative way, because they have many different motivations. Differences in personality or character are qualitative, not quantitative differences. If one constructs artificial organisms with many different motivations and a motivational level of behaviour it becomes possible to study organisms which differ from one another not only in level of ability but also in character or personality, where differences in character and personality are mainly differences in motivations and in their strength. One can even construct individual profiles of artificial organisms based on a number of different characteristics referring to both their abilities and their motivations.

In our artificial organisms the motivational level of behaviour is implemented as a special pathway in the neural network that controls an organism's behaviour. This pathway is extremely simple and this simplicity contrasts sharply with the richness and complexity of the neural systems that underlie the motivational level of behaviour in real organisms. Another important direction of research is to use what we know about the brain and the brain's interaction with the rest of the body to endow the neural network of our artificial organisms with more realistic neural pathways that capture how motivations control an organism's behaviour (French & Canamero, 2005). The motivational system may reside in different parts of the brain with respect to the cognitive system (e.g., sub-cortical vs. cortical) and may have different structural and functional properties (more rapid and more prolonged action). More generally, it may turn out that while the cognitive system of an organism can be simulated by remaining at the cellular level (a neural network's unit corresponding to a neural cell or neuron) and by considering only the organism's brain, the organism's motivational system can only be simulated if one goes one step down and reaches the molecular level, and if one considers not only the organism's brain but also the interactions of the brain with the rest of the organism's body (Parisi, 2004).

## References

Acerbi, A. & Parisi, D. (2007). 'The evolution of pain'. *In Advances in Artificial Life: Proceedings of the 9th European Conference on Artificial Life*, Lisbon, Portugal, pp. 816–824.

Canamero, L. (2005). 'Emotion understanding from the perspective of autonomous research'. *Neural Networks*, vol. 18, pp. 445–455.

Cecconi, F. & Parisi, D. (1993). 'Neural networks with motivational units'. In *From Animals to Animats 2: Proceedings of the 2nd International Conference on Simulation of Adaptive Behavior*, Honolulu, HI, pp. 346–355.

Chajut, E. & Algon, C. (2003). 'Selective attention under stress: implications for theories of social cognition'. *Journal of Personality and Social Psychology*, vol. 85, pp. 231–248.

Davis, D.N. (2000). 'Agents, emergence, emotion and representation'. *In Proceedings of the 26th Annual Conference of the IEEE Industrial Electronics Society*, Nagoya, Japan, vol. 4, pp. 2577–2582.

Davis, D.N. (2008). 'Linking perception and action through motivation and affect'. *Journal of Experimental and Theoretical Artificial Intelligence*, vol. 20, pp. 37–60.

Fikes, R. & Nilsson, N. (1971). 'STRIPS: A new approach to the application of theorem proving to problem solving'. *Artificial Intelligence*, vol. 2, pp. 189–208.

French, L.B. & Canamero, L. (2005). 'Introducing neuromodulation to a Braitenberg vehicle'. *In Proceedings of the 2005 IEEE International Conference on Robotics and Automation*, Barcelona, Spain, pp. 4199–4204.

Krebs, J. & Davies, N.B. (1997). *Behavioural Ecology: An Evolutionary Approach*. Oxford, Blackwell.

Miller, G.A., Galanter, G., & Pribram, K. (1986). *Plans and the Structure of Behaviour*. New York, Holt.

Mondada, F., Franzi, E., & Guignard, A. (1999). 'The development of Khepera'. *In Proceedings of the 1st International Khepera Workshop*, Paderborn, Germany, pp. 7–14.

Nolfi, S. & Floreano, D. (2000). 'Evolutionary Robotics'. *The Biology, Intelligence, and Technology of Self-Organizing Machines*. Cambridge, MIT Press.

Parisi, D. (1996). 'Motivation in artificial organisms'. In Tascini, G., Esposito, V., Roberto, V., Zingaretti, P. (eds.) *Machine Learning and Perception*, pp. 3–19. Singapore, World Scientific.

Parisi, D. (2004). 'Internal robotics'. *Connection Science*, vol. 16, pp. 325–338.

Parisi, D. (2007). 'Mental robotics'. In Chella, A., Manzotti, R., *Artificial Consciousness*, pp. 191–211.

Perez, C.H., Moffat, D.C. & Ziemke, T. (2006). 'Emotions as a bridge to the environment: on the role of body in organisms and robots'. In From Animals to Animats 9: *In Proceedings of the 9th International Conference on Simulation of Adaptive Behavior*. Rome, Italy, pp. 3–16.

Petrosino, G. & Parisi D. (in preparation). 'Deciding whether to look for food or to avoid predators'.

Ruini, F. & Parisi D. (2008). 'Selective attention in artificial organisms (abstract)'. *In Artificial Life XI: Proceedings of the 11th International Conference on the Simulation and Synthesis of Living Systems*, Winchester, UK, p. 799.

Saglimbeni, F. & Parisi D. (in press). 'Input from the external environment and input from within the body'. *In Proceedings of the 2009 IEEE European Conference on Artificial Life*, Budapest, Hungary.

Scheutz, M. (2004). 'Useful roles of emotions in artificial agents: a case study from Artificial Life'. *In Proceedings of the 19th National Conference on Artificial Intelligence*, San Jose, CA, pp. 42–48.

Seth, A.K. (2002). 'Competitive foraging, decision making, and the ecological rationality of the matching law'. *In From Animals to Animats 7: Proceedings of the 7th International Conference on Simulation of Adaptive Behavior*, Edinburgh, UK, pp. 359–368.

Seth, A.K. (2007). 'The ecology of action selection: insights from artificial life'. *Philosophical Transaction of the Royal Society*, vol. B, pp. 1545–1558.

Stephens, D. &, Krebs, J. (1986). *Foraging Theory*. Princeton, Princeton University Press.

Ziemke, T. (2008). 'On the role of emotion in biological and robotic autonomy'. *BioSystems*, vol. 91, pp. 401–408.

*Chapter 11*

# Sequencing embodied gestures in speech

*Juraj Simko and Fred Cummins*

## Abstract

The embodied character of cognitive motor systems that has greatly influenced the understanding of their constitution and function is reflected in many recent models. Embodiment conditions that system behaviours must take appropriate account of energy expenditure and metabolic costs that are unavoidable in a physically realised medium. We here consider that optimisation, presumed to result from both phylogenetic and ontogenetic processes, can be used to constrain the space of potential degrees of freedom of a system, ensuring that the resulting action is efficient and smooth. To understand the emerging adaptations, it is necessary to factor in the properties of the physical and physiological substrate that anchor the system's goal-oriented performance.

However, the embodied nature of speech production has been disregarded by most phonological research to date. This leads to a failure in providing a coherent phonological grounding of a wide range of phenomena extensively documented by experimental phoneticians, in particular those associated with the relative timing of gestures (also called *gestural phasing*) in connected speech and its variability as found in different manners of speech. Existing phonological theories of sequencing rely on essentially external system-wide rules and principles or explicit dynamical constraints governing phasing to account for various suprasegmental properties and prosodic parameters of an utterance. We introduce here a new and highly abstract modelling platform developed to investigate the embodied character of speech. The physically instantiated, second-order dynamic nature of the system allows us to define and exactly evaluate various cost functions, which we hypothesise to play a role in efficient gestural sequencing. We investigate the general dynamical properties of the system, and identify a set of its high-level, intentional parameters linked to the cost functions associated with its goal-oriented performance. We show that the phenomena accompanying gestural sequencing, coarticulation, fluency and prosodic modulation emerge as consequences of a non-trivially formulated efficiency constraint, thus providing a principled phonological account of phonetic reality.

## 11.1   Introduction

Speech production is a well-rehearsed, sequential cognitive activity. Speaking brings about the precise coordination of multiple effectors belonging to a highly complex physical system stretching from diaphragm to lips. As with any form of skilled action, mastery of speech involves learning how to coordinate these diverse parts while respecting the constraints imposed by functional requirements, i.e. communication.

In his papers on Emergent Phonology (Lindblom, 1983, 1999), Lindblom put forward an idea that several general motor action and cognitive principles can, when mapped appropriately into the speech production and perception domain, shed a novel light on known phonological explanations and phonetic phenomena. Rather than being postulated as a system of external and representational laws, phonological phenomena *emerge* as consequences of these basic principles.

The basic constraint that shapes any cognitive embodied skilled motor activity is a requirement of *efficiency*. As we understand it, an *efficient* system is one that displays an energetically optimal trade-off between the conflicting demands of minimising effort in movement and maximising perceptual clarity for the perceiver of speech. By adhering to this principle alone, the speech production cognitive system curtails the complexity of the task of generating and sequencing production primitives so that a continuous stream of speech sounds is produced. In other words, this principle helps to reduce, dramatically, the number of degrees of freedom associated with a redundant motor action in general and with speech production in particular.

In this view, the cognitive system is not seen as an autonomous disembodied controller acting *on* a physiological substrate and governed by abstract rules, but rather as containing the substrate, inseparable from it, acting *in accordance* with the physical constraints imposed by the environment, adapted to them, and taking full advantage of substrate's properties in assembling functionally determined coordinative structures for performing given tasks.

The usefulness of this approach for phonological research has been demonstrated, for example in Dispersion-Focalization Theory (Schwartz *et al.*, 1997), which provides an account of the distribution of individual vowels in a potentially continuous space of vocalic primitives. Schwartz and his collaborators re-interpreted stability criteria postulated in Stevens' Quantal Theory (Stevens, 1989) and perceptual distinctiveness criteria put forward in Lindblom's Dispersion Theory (Liljencrants and Lindblom, 1972) as complementary production and perception cost functions and used a simple optimisation to derive vowel distributions that closely matched various natural vowel systems. This result shows that the efficient global patterns pinpointed by the constrained optimisation have their real counterparts in existing phonologies. During the acquisition of their mother tongue, the speakers can take advantage of the existence of these low-energy attractors in the production dynamics. In our work, we extend this approach to another dimension of speech production: that of the sequencing of gestures in time. We hypothesise that efficiency requirements, arguably influencing the distribution of speech primitives, also play a crucial role in the way that these primitive actions are strung together in time when uttering a connected stream of speech.

Producing an utterance is to a large extent a *sequencing* task. It involves a precise phasing of the execution of primitive articulatory actions. The manner in which the actions are organised into patterns determines the content and quality of the acoustic output. The rules governing the ordering, the precise relative timing and overlaps of actions and the high-level parameters of their execution provide a descriptive framework for many types of phonological variation.

The phonetic manifestation of these sequencing variations encompasses the phenomena generically described as *coarticulation*. Coarticulation refers to the context-dependent manifestation of speech segments or gestures that arise as a direct result of the co-production of adjacent or nearby segments/gestures. Prosodic variation influences coarticulation patterns in a wide variety of ways that have received much attention. In this work, we shall focus primarily on the variations elicited by speaking rate manipulation.

Thomas Gay and his collaborators (Gay *et al.*, 1974; Gay, 1981) noted that, in general, 'the duration of segmental units, the displacement and velocity of articulatory movements, and the temporal overlap between individual segments undergoes non-linear transformations during changes in speaking rate'. The non-linear effects of rate changes were reported in different forms for both speech and non-speech motor actions. For example, they showed that consonants get shortened proportionately less than vowels in fast speech.

Nittrouer *et al.* (1988) and Nittrouer (1991) investigated the influence of rate changes on the phasing details of utterances. They showed that these changes result in changes of the *relative* phasing structure of an utterance. For utterances of $/C_1V_1C_2V_2C_3/$ form, the onset of the inter-vocalic consonantal gesture for $C_2$ – bilabial in (Nittrouer *et al.*, 1988) and alveolar in (Nittrouer, 1991) – starts relatively earlier in the underlying vocalic cycle $V_1$–$V_2$ for fast speaking rate than for slow speaking rate. In both cases, the consonantal gesture also occupies a greater portion of the cycle at fast rates.

On the other hand, Cummins (1999) refined this general principle by observing that nonlinearities accompanying speaking rate changes are not manifested uniformly across the duration of an utterance. In fact, the relative durational relationships between actions grouped within a suprasegmental unit (e.g. a syllable) remain more stable than those observed across unit boundaries. Can these and similar variations be accounted for as natural consequences of adaptations of the cognitive system acting in accordance with the embodied neuromuscular system, our vocal tract, towards *efficient* production of the information-carrying stream of speech?

### 11.1.1    Background

To address this question, we present a developing modelling paradigm in which many of the superficial complexities associated with speech production are finessed, while many basic principles relevant to efficient sequencing and coordination in real time in a physically instantiated, embodied structure are respected.

We draw inspiration for capturing the fundamental properties of an embodied articulatory system from three major spheres of research: the theory of optimality principles, motor action theory and task dynamics. Phonologically, we ground our model of sequencing in the theory of Articulatory Phonology (AP) and its Task Dynamic (TD) implementation.

Optimality principles play a vital role in the performance of skilled cognitive sensorimotor actions. Their appeal 'lies in their ability to transform a parsimonious performance criterion into elaborate predictions regarding the behaviour of a given system' (Todorov, 2004; see also for an overview of the recent trends in this field). Both evolution and ontogenetic development craft behavioural systems under the constraint of efficient production as costs associated with both producing and perceiving a message are always present. To account for gestural patterns resulting from the requirement of motor efficiency, we thus need to include a measure of energy expenditure associated with the system's performance in our model, i.e. we must have modelling access to the magnitudes of forces driving the movement of system constituents. We therefore have to build our model as a system embodied in the physical world, possessed of masses and subject to physical constraints of continuity, impermeability, inertia and the like.

Recent modelling attempts, e.g. Anderson and Pandy (2001), have explored a wide variety of individual cost functions relevant to different tasks. However, capturing the essential features of behavioural data typically requires the inclusion of a number of distinct cost terms. We shall suggest a combination of three well-motivated terms in our cost function and show how the weights representing their prominence elicit the required lawful variation in speech production at multiple rates.

The low-cost patterns of speech production seem to play an important role in shaping many aspects of phonological structure. In the words of a leading proponent of this research paradigm, Björn Lindblom, phonological patterns can be seen as 'products of cultural evolution adapted to universal biological constraints on listening, speaking and learning' (Lindblom, 2000). Our aim is to propose an interlinked system of such constraints balancing the production-oriented and listener-oriented influences. To be able to do that – in particular on the production-oriented side of the trade-off equation – we must identify the right level of analysis allowing us to quantify the cost reflecting the elusive concept of articulatory ease.

Kinematic and dynamic properties of limb (and speech articulator) movement – such as velocity curves, changes in movement duration under different conditions of rate and extent – have been extensively studied, both experimentally (Ostry *et al.*, 1987) and theoretically (Kelso, 1995; Ostry, 1986). Cooke (1997) suggested a second-order dynamics (akin to a damped driven spring with mass) as a suitable approximation for modelling muscle–joint structures, and he showed that continuous change of its high-level physical parameters, e.g. *stiffness*, lead to qualitative changes in the organisational form exhibited by the motor action systems.

This approach has been extended by the school inspired by the work of the great Russian physiologist Nicolai Bernstein which has developed the Equilibrium Point hypothesis of muscular action (Ostry and Feldman, 2003; Latash, 2008). The central idea behind the hypothesis is that muscular action is determined by shifts in the equilibrium position of the muscle dynamics. Equilibrium shift means the equivalent of a change in the resting length of a damped spring after which the load attached to the spring moves to a new position driven by forces dependent on the spring stiffness, damping and the load mass. Although the proponents of this approach explicitly stress the deficiencies of the simple mass-spring dynamics for modelling the highly non-linear characteristics of muscular action (Feldman and Latash, 2005), they nevertheless admit the methodological convenience of the global dynamic parameters such as stiffness and damping in accounting for some broad patterns of the form of movement. Being aware of the simplifications introduced by modelling low-level muscular action this way, we presume that capturing the basic properties of speech articulators as essentially damped mass-springs with adjustable stiffness is sufficient for our main aim of demonstrating the role of efficiency principles for task sequencing in an embodied system.

A related methodology for describing the behaviour of complex motor systems has been introduced by Haken, Kelso and Bunz (1985) and by Kelso and Saltzman as Task Dynamics (Saltzman and Kelso, 1987). They introduced an abstract space of tasks performed by the system where the patterns of the task accomplishment (and not the underlying neuromuscular system) are captured by the second-order dynamics. The overall dynamics of the production system is thus determined functionally, by the behaviour towards the achievement of a given higher-level goal, e.g. constriction of the vocal tract at a specific location. The task generates a synergy between the system's primitive components, groups them into a coordinative structure and, in effect, reduces the number of degrees of freedom of the unconstrained system. The dynamical patterns of the task and their interaction with a physically instantiated articulatory space account for many observed phenomena. This approach has been successfully adapted for speech production modelling.

AP, proposed by Catherine Browman and Louis Goldstein (1991, 1992), postulates functionally defined, physically real dynamical events called *gestures* to be both fundamental units of information, i.e. phonological contrast and primitive units of action, i.e. speech production. Every gesture imposes a set of goals (formation and release of a constriction at some place of the vocal tract) on the vocal tract to produce the desired phonological event.

Within AP, the behaviour of gestural primitives is captured in a top-down manner in three parallel levels of description: a *gestural score* level characterised by activation variables as functions of time, a vocal tract *task* level captured by abstract tract variables and an articulatory (physically real, embodied) level described by model *articulator* variables. The gestural score represents organised patterns (constellations) of gestures participating in an utterance's production, and in particular the precise timing of their onsets and offsets relative to each other – *gestural phasing*. Each tract variable represents the degree to which the goals associated with a gesture are achieved over time. Finally, an articulator variable shows the position of every articulator participating in the gestural target accomplishment. The degree to which a gesture has achieved its targets is captured by several (one or two) tract variables. Each tract variable is in turn associated, not necessarily exclusively, with several model articulators. For example, the tongue tip constriction location (TTCL) tract variable is linked to behaviour of tongue tip, tongue body and jaw, while the jaw at the same time affects the values of the lip protrusion (LP) tract variable. Therefore, the mapping between elements of the articulatory layer (model articulator space) and those of the task layer (tract variable space) is of a many-to-many nature. This mapping is, however, seen as a relatively straightforward transformation between two coordinate systems.

In the standard implementation model of AP, TD proposed by Saltzman and colleagues from Haskins Laboratory (Saltzman and Munhall, 1989; Saltzman, 1991), the dynamics of speech production is derived exclusively from the uncoupled dynamics of *tract variables* in the vocal tract task space. It is the motion of a gesture's tract variables, *not* the motion of the associated individual articulators, which is characterised dynamically. Given a gestural activation pattern and targets for each gesture, the precise manner in which these targets are achieved in time is the solution of a second-order dynamical system where each tract variable is modelled as a mass-spring with arbitrary (unit) mass. This solution is simply recast into the model articulator coordinate system, thereby specifying the kinematics of the individual articulators participating in the target accomplishment. Crucially, the dynamics of the articulatory layer play no role whatsoever in determining the behaviour of this motor action system. Despite its dynamic nature and its emphasis on the importance of production constraints for a phonological theory, this approach models the speech production system in a top-down manner. The physically real properties of articulators, e.g. their masses, are not represented in the TD implementation, which makes it impossible to track the force-dependent quantities, such as energy expenditure, that are presumed to underwrite the efficient (energetically optimal) operation of the system.

The researchers participating in the AP project have managed to provide a coherent description of many phonetic and phonological phenomena. In particular, the theory successfully accounts for motor action robustness to external perturbations, naturally explains coarticulation as co-production (the result of multiple gestures with total or partial activation overlap vying for control over articulatory system), or the existence of hidden gestures (which do not manifest themselves in the acoustic outcome of speech action, but are still present in the underlying production).

Without necessarily committing to all theoretical claims and implications of AP and TD, we will adopt their fundamental views and terminology: functionally defined gestures, the interacting levels of description and the inherently dynamical nature of speech production.

One area where we take a different stance to AP and its TD implementation is in the sequencing and phasing of individual gestures and, in particular, the observed dependency of the gestural score for an utterance on the manner of its production, i.e. the dependency of the phasing relations among the gestures on the speaking rate and other sundry prosodic properties. Browman and Goldstein (2000) propose an essentially grammatical way for determining phase relationships between gestures participating in an utterance. Byrd *et al.* (2000) and Byrd and Saltzman (2003) proposed a more comprehensive theory allowing for flexibility in the phase relationships (windows of relative timing) and an additional layer of the so-called $\pi$ gestures that capture some prosodic properties of an utterance.

Both of these approaches thus deal with the problem of determining the phasing of gestures by specifying essentially explicit descriptions and adding external rules governing phasing for various suprasegmental properties and prosodic parameters of an utterance. When seen as a cognitive system, the speech production apparatus is thus supposed to deal with the vast number of degrees of freedom associated with gestural phasing by a complex, representational structure without exploiting the physical and physiological environment in which it is embodied.

The main aim of our research is to identify constraints imposed on such an embodied articulatory system by its dynamics and efficiency requirements. That way we can investigate the general properties of the embodied articulatory space, describe its dynamics and propose a set of high-level, intentional parameters which are at the disposal of a cognitive system and which elicit the qualitative changes in the organisational form characteristic of speech production at a range of rates.

In this chapter, we propose to exploit the inherent dynamics of the articulatory layer to investigate the space of gestural phasing. This space is delimited by physiology, physics and efficiency principles (arising from both phylogenetic evolution and ontogenetic development), and has to be mastered by and is at the disposal of a cognitive system participating in speech production. Speech conceived as an embodied motor action allows us to bring concepts such as cost, production efficiency and optimality into the discourse on speech production. These then facilitate the identification of high-level production parameters associated with the intentional control of speaking rate and production precision, which in turn help us to talk more precisely about elusive notions in phonetics and phonology such as fluency, trade-offs between precision and rate, etc.

## 11.1.2    Speech production models

The human vocal tract is a very complex physiological system with over 60 muscles and several bone structures engaged in online shaping of its components, the *articulators*, to attain required aerodynamic shapes and movements. It consists of several more or less independent and loosely defined articulatory subsystems: tongue and palate, jaw and lips, velum and vocal folds, to name a few. Within each of these subsystems, articulators are relatively strongly linked by the anatomical organisation of the tract.

To produce a velar stop, the back part of the tongue body (dorsum) must achieve a contact with the rear portion of palate (soft palate). This movement obviously impacts the behaviour of all articulators grouped within the tongue and palate subsystem: it strongly limits the possible shapes of the entire tongue body and its positions relative to the palate and to a lesser extent also the absolute position of the (anatomically quite flexible) tongue tip and its position relative to, e.g. the alveolar ridge or teeth.

On the other hand, articulators from distinct subsystems, e.g. tongue body and vocal folds, can act quite independently: a velar stop can be produced with vocal folds vibrating (producing a voiced consonant, e.g. /g/) or with vocal folds relaxed (producing a voiceless /k/). This flexibility is exploited by the speech production system: the relative functional independence of subsystems allows for a combination of their activity patterns resulting in various phonologically contrasting speech sounds. With some caution, it might be said that during speech production the articulators within each subsystem are coupled strongly by anatomy and relatively weakly functionally, while articulators from different· subsystems are strongly coupled functionally and comparatively weakly linked by anatomy.

There are several physiologically motivated, articulatory models of the vocal tract (Maeda, 1982; Boersma, 1998; Iskarous *et al.*, 2003). These models focus on the very complex task of articulatory speech synthesis, i.e. the accurate translation of changing shapes and movements of the vocal tract into acoustic space. The input to these models is a sequence of parameters defining the shapes and positions of model articulators, which are then used to synthesise the acoustic output. The kinematics of the articulators is fully determined by the input sequence. The models thus abstract away from the dynamical properties of their components, the masses and forces acting on the model articulators: the vocal tract they deal with is thus disembodied with respect to its dynamical properties.

These models are then traditionally used in modelling higher-level characteristics of speech production and perception (Guenther, 1995; Howard and Huckvale, 2005). As mentioned in the case of AP in Section 11.1.1, although the connected theories are successful in accounting for many matters related to speech production, acquisition and even in a projection of speech-related processes into human brain operation, they inevitably stop short of explaining phenomena linked to the embodied nature of speech production, e.g. articulatory efficiency, high-level parameterisation of speaking rate and articulatory precision control, all associated with the emergent lawfulness of gestural phasing.

In recent years, several detailed models of the entire vocal tract or its subsystems taking the embodiment of vocal tract seriously have been proposed (Perrier *et al.*, 2000). These models aim to get hold of the enormous complexity of physics and physiology behind the real vocal tract and its ability to produce speech. As a consequence, the synthesis of an acoustic output or detailed stream of vocal tract configurations for even a short utterance takes very long time even on the fastest computers available to date. Much as they are useful for testing the predictions of current production theories, these models cannot be realistically employed for our task of describing the properties of the entire physically relevant *space* underlying speech production. To do that we need to be able to find optimal, efficient gestural constellations, which in practice means exploration of a search space by running thousands of slightly varying productions of each utterance.

## 11.2   Abstract model

Our approach is thus to build a physiologically and physically motivated model of the speech production system, or, more precisely, of its supraglottal subsystem in the above sense. Instead of trying to capture the superficial details of human vocal tract anatomy and the acoustic principles behind soundwave generation, we focus exclusively on the essential dynamical properties of an embodied motor action system incorporating only high-level features of the human vocal tract that constrain speech production. The primitive constituents of the model must be relevant for our aim to capture adaptations that result in efficient, low-energy sequencing of motor actions.

Therefore, we decided, first, to leave out the synthesis of acoustic output. Our model generates purely articulatory output – kinematic traces of the model articulators. Although some important perception phenomena are related to non-linearities of the articulatory-to-acoustic mapping, we presume that the perceptual quality of the system's output is directly related to the articular precision with which it approaches a given target. Second, we leave out the details of anatomical aspects of vocal tract. The articulators of our model are, for now, purely abstract. They do not map directly onto specific articulators (tongue tip, lips, ...), but instead they capture high-level dynamical properties we are interested in. For a given modelling task, we can adjust the details of our model to reflect particular properties of and relations between speech articulators relevant for the given circumstances.

At some level of abstraction, the vocal tract can be seen as a (particularly complex) system driven by intentionally imposed muscular force impulses. As suggested by research in motor action, such systems can in principle be modelled using an appropriate (presumably, again, very complex) set of second-order differential equations. Our approach is to turn this premise the other way around: take a relatively simple system with second-order dynamics (yoked pendula) and investigate to what extent its behaviour can shed light on the constraints underlying human speech production.

As we aim to account for phenomena related to gestural sequencing and phasing, one of the high-level features of speech production we need to capture by our modelling approach is a different nature of production of vowels and consonants.

As experimentally first documented by Öhman (1966) and subsequently conceptualised by, e.g. Fowler (1983), vowels and consonants fall into two distinct natural articulatory classes. Vowels, in contrast to consonants, are produced by relatively slow, configurational movements of articulators. The vocalic gestures have broader targets and they engage to a large extent only three supraglottal articulators: tongue dorsum, jaw and lips. The consonantal gestures (or at least stop consonants), on the other hand, are generated by rapid, ballistic movements of articulators and involve their mutual collisions forming appropriate full constrictions of the vocal tract. These differences often lead to postulating two separate articulatory tiers; during production, the consonantal tier is superimposed on the vocalic one. In Section 11.3, we will show how we can account for the facts underlying this hypothesis and also what properties of the system are related to the emergence of the separate articulatory tiers. Rather than offering a full, exhaustive model of the speech apparatus, we aim to provide a modelling paradigm which allows us to build a succession of simple setups capturing various tangible phonological and phonetic phenomena. As described in Section 11.2.3, the approach allows for expanding and refining the basic setup to accommodate diverse hypotheses, test them and to serve as an intuition pump for thinking about sequencing and fluency.

The articulators are represented by pendula driven by torsion springs. Their kinematics is given by the solution of the following non-linear differential equation:

$$m\ddot{\theta} + b\dot{\theta} + k(\theta - \theta_0) = 0 \tag{11.1}$$

where $m$ is the moment of inertia of the pendulum, $k$ is the torsion spring coefficient, $b = 2\sqrt{mk}$ is the critical rotational damping parameter and $\theta_0$ is the resting angular deflection of the pendulum. The dynamics prescribed by (11.1) is equivalent to the dynamics of damped mass-spring behaviour; the reason for choosing the torsion spring–driven pendula instead of springs is to enclose their action in a compact space by warping a sequence of springs around in a circle. Because the effect of this decision can be seen as a purely geometrical transformation, for the sake of compatibility with the tradition of approximating the neuromuscular action with damped springs, we shall use the spring equivalents when referring to the coefficients of (11.1), i.e. we

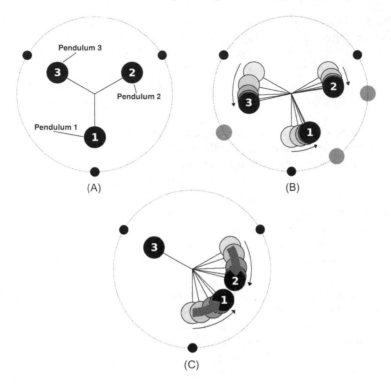

*Figure 11.1   Basic setup (A) and a production of vocalic (B) and consonantal (C) gestures*

shall call *m*, *b* and *k*, mass, damping and stiffness, respectively. Similarly, we shall use the term 'force' when talking about the torque driving the pendulum action. The basic setup of our abstract model is shown in a rest state in Figure 11.1A. It consists of three torsion spring-driven pendula hung on massless rods from a common point. Each pendulum has its own mass $m_i$, stiffness $k_i$, angular equilibrium position $\theta_{0i}$, $i = 1, 2, 3$ (in the above sense). The pendula are critically damped through the damping coefficients $b_i$, to avoid overshoot and oscillation (the critical damping coefficient is calculated during an utterance production to dampen the forces elicited by all active gestures). Each pair of bobs exhibits mutual repulsion force $R_{ij}$ when the bobs approach one another. The force $R_{ij}$ is negligible for all but very small angular distances and infinite for zero distance.

For various setups, it is possible to impose further second-order dynamical constraints on the system. Two pendula can, e.g. be joined (linearly coupled) by a spring of a given length and stiffness to simulate a partial anatomical link between the two abstract articulators (e.g. as in the case of tongue tip and tongue dorsum in the human vocal tract).

The pendula are the abstract articulators of our vocal tract model, presumably belonging to a single anatomically constrained subsystem. They can be acted upon by both configural (vocalic) and ballistic (consonantal) gestures and there exists a one-to-many relation between articulatory goals and articulators. Their movements are subject to physical constraints, and we can specify the dynamics governing their movement in a relatively simple fashion.

The system's rest dynamics describes a *speech ready* state of the articulatory system. Rather than representing an idle, purely anatomical resting position of the articulators induced by, e.g. gravity (the tongue lying flat against the jaw, the lips closed and teeth clinched together), our speech ready equilibrium position is the state in which the articulatory system is 'receptive', pre-configured for action elicited by speech gestures. This state is presumably established during a speaker's native vowel space acquisition and fine-tuning and is the state from which the vowel targets could be reached most economically (Barry, 1998). Phonetically, the resting equilibrium corresponds to the configuration of the vocal tract producing schwa /ə/.

Vocalic gestures (syllabic nuclei) are implemented within the system as sets of additional absolute equilibrium positions involving one or more pendula.

To produce a 'vowel' defined as an absolute articulatory configuration, extra centres of attraction force induced in the same way as the resting equilibria can be switched on around the system at angular positions different to those for the resting position, each acting on one pendulum. We call these attractor tuples (one attractor position for each pendulum) the *attractor sets*. The magnitude of the attractive forces is larger than that of the resting attractor forces, so that pendula get displaced from the resting position towards the vowel configuration as illustrated in Figure 11.1B. This is achieved by setting a proportionally higher value of the stiffness coefficient than the value for the resting attractor, i.e. by multiplying the resting stiffness of each engaged articulator by a vocalic gesture gain. If an attractor set is switched on, after some time and with a velocity proportional to the stiffness and mass of each pendulum involved, the engaged pendula move to stable positions, with their angular deflections close to those of the attractor set. (Not exactly the same, as the default resting position attractor set is still acting on them, although with a relatively weaker force.) When the attractor set is then switched off, the pendula slowly return to their respective resting positions.

Consonantal gestures are represented in a distinctly different fashion. Each is initiated by mutual attraction forces – forming a *consonantal attractor set* – between a pair of pendula, proportional to their distance and depending on each model articulator stiffness. The pendula thus move towards each other (Figure 11.1C), until their mutual repelling forces equal the driving forces. This collision of two pendula constitutes an articulatory target ('closure'). The system remains in this 'closure' state until the attractor set is deactivated. Again, the path taken to reach closure and move away from it, and the path taken during the closure duration will depend on past state and future goals.

## 11.2.1 Formal definition

In this section, we shall present a formal, mathematical description of our model. The resting dynamics of the system can be expressed by a system of three differential equations. The vector $\ddot{\theta}_{\partial}$ describing the acceleration of the model articulators is given by the equation:

$$\ddot{\theta}_u = \mathbf{M}^{-1}\left(-\mathbf{B}\dot{\theta} - k\mathbf{K}_0(\theta - \theta_0) + \mathbf{R}(\theta) - \sum_z k\kappa_z \mathbf{C}_z^*(\mathbf{C}_z\theta - \theta_{z0})\right) \quad (11.2)$$

where $\theta = (\theta_1, \theta_2, \theta_3)^T$ is a vector of the three articulator positions, $\theta_0 = (\theta_{10}, \theta_{20}, \theta_{30})^T$, the vector of their resting equilibria, $\mathbf{M} = \text{diag}(m_1, m_2, m_3)$ is the diagonal mass matrix with pendula bob masses on its diagonal, $\mathbf{B} = \text{diag}(b_1, b_2, b_3)$ is the diagonal damping matrix with the damping coefficients on its diagonal (as mentioned earlier in the text, this element is calculated to dampen all forces active at time, not only the resting one) and $\mathbf{R}(\theta) = (r_{ij}(\theta))$ is the repulsive force matrix, each $r_{ij}(\theta)$ being the magnitude of the repulsive force $R_{ij}$ for the angular deflections $\theta$, $r_{ii}(\theta) = 0$, for all $i, j = 1, 2, 3$. Each optional $z$ in the leftmost sum element imposes a possible linear coupling (spring) between two articulators, with stiffness $k_z = \kappa_z k$ (generally much smaller than any $k_i$) and length $\theta_{z0}$. If $z$ is a coupling between $i$th and $j$th articulators $i < j$, the $1 \times 3$ matrix $\mathbf{C}_z = (c_{z1}\ c_{z2}\ c_{z3})$ is defined as follows: $c_{zj} = 1$, $c_{zi} = -1$ and the third element is set to 0. This matrix represents a projection from the space of our three articulators into the simple task space containing the only task of keeping the $\mathbf{C}_z\theta = \theta_j - \theta_i$ constant (equal to $\theta_{z0}$). The $3 \times 1$ matrix $\mathbf{C}_z^*$ is the Moore–Penrose pseudoinverse of $\mathbf{C}_z$ mapping the status of the task achievement back to the articulatory space. This approach is analogous to the mapping between task space and articulatory layer used by the TD implementation of AP (Saltzman and Munhall, 1989).

The expressions $k\mathbf{K}_0$ and $k\kappa_z$ capture an important design decision incorporated in our model. To reduce the number of the parameters of the optimisation process, the system's various stiffness parameters (the rest dynamics stiffness $k_i$ of the $i$th articulator, each coupling spring stiffness $k_z$, and gestural stiffness described later in the text) are all related to each other in a linear fashion. In other words, each $k_i = \kappa_i k$ and each $k_z = \kappa_z k$, where $k$ is the overall system-wide stiffness; $\kappa_i s$ and $\kappa_z s$ are stiffness *coefficients* of the pendula and coupling springs, respectively. The matrix $\mathbf{K}_0 = \text{diag}(\kappa_1, \kappa_2, \kappa_3)$ is a matrix containing the rest dynamics' stiffness coefficients on its diagonal. These coefficients are parameters of a given setup and remain constant for the given modelling task. It is possible, however, to adjust the overall system stiffness $k$ and thus control how swiftly the pendula react to the application of the resting forces and forces induced by gestural activations.

An active vocalic attractor set $v$ imposes an additional acceleration on articulators

$$\ddot{\theta}_v = -\mathbf{E}_v \mathbf{M}^{-1} \kappa_{\text{voc}} k \mathbf{K}_0 (\theta - \theta_v) \tag{11.3}$$

where $\mathbf{E}_v = \text{diag}(e_1, e_2, e_3)$ is a diagonal matrix prescribing which articulators are engaged in the production of the vowel $v$ ($e_i$ equals 1, if the $i$th attractor is involved, and 0 otherwise), and $\theta_v$ is the vector of the vocalic attractor equilibrium positions, and $\kappa_{\text{voc}}$ is the vocalic stiffness gain.

Each attractor set represents a context-free vowel target, but the approach into and path from the configuration will be context sensitive, depending on the past states of the system and future articulatory goals.

The consonantal collision gestures introduce a target-driven linear coupling between pendula, equivalent to the coupling elicited by the optional anatomically inspired springs discussed in the previous section. Thus, the acceleration imposed by a consonantal gesture $c$ acting on $i$th and $j$th pendula ($i < j$) can be formally expressed as:

$$\ddot{\theta}_c = -\mathbf{M}^{-1} \kappa_{\text{con}} k \mathbf{K}_0 \mathbf{C}_c^* \mathbf{C}_c \theta \tag{11.4}$$

The task-articulator mapping matrix $\mathbf{C}_c$ is defined as the projection matrix $\mathbf{C}_c$ in the previous section, i.e. $\mathbf{C}_c = (c_{c1}\ c_{c2}\ c_{c3})$ where $c_{cj} = 1$, $c_{ci} = -1$ and the third element is set to 0, $\mathbf{C}_c^*$ is its Moore–Penrose pseudoinverse, and $\kappa_{\text{con}}$ (set to 4 for the models presented in this chapter) is the relative stiffness coefficient for consonantal gestures. The equilibrium distance for the consonantal task is 0.

For example, for the consonantal gesture $c_{13}$ between pendula 1 and 3, $\mathbf{C}_{c_{13}} = (-1\ \ 0\ \ 1)$. The pseudoinverse $\mathbf{C}_{c_{13}}^* = (-\frac{1}{2}\ \ 0\ \ \frac{1}{2})^T$ and the product

$$\mathbf{C}_{c_{13}} \mathbf{C}_{c_{13}}^* = \begin{pmatrix} \frac{1}{2} & 0 & -\frac{1}{2} \\ 0 & 0 & 0 \\ -\frac{1}{2} & 0 & \frac{1}{2} \end{pmatrix}$$

The gestural dynamics imposed on the pendula 1 and 3 is then

$$m_1 \ddot{\theta}_{c1} = -\frac{1}{2} \kappa_{\text{con}} \kappa_1 k (\theta_1 - \theta_3) \tag{11.5}$$

$$m_3 \ddot{\theta}_{c3} = -\frac{1}{2} \kappa_{\text{con}} \kappa_3 k (\theta_3 - \theta_1) \tag{11.6}$$

The pendulum 2 is not influenced by the consonantal gesture in this case. Unlike the vocalic gestures defined by (11.3), the consonantal gestures introduce a coupling between the articulators. This task-oriented coupling (as opposed to the coupling reflecting anatomical constraints introduced in (11.2)) exemplifies an important difference in our approach to modelling vowels and consonants. While vowels are seen as *absolute* positional configurations of model articulators, the consonantal gestures impose a mutual coordination between pair of articulators acting in synergy.

## 11.2.2 Activation functions

In principle, each gesture can be triggered independently of any other and multiple gestural activation patterns can partially or totally overlap. Because production targets are defined in different ways for consonants and vowels, co-production of gestures is possible.

Each (vocalic or consonantal) gesture defined for the model can be switched on and off during model's execution. These gestural activation patterns are modelled via activation time functions. For a given gesture $g$, the value of the activation function $a_g(t)$ at time $t$ is set to 1 if the gesture $g$ is active at time $t$, and to 0 if it is not active. The step-wise shift from 0 to 1 in the activation function $a_g$ marks the onset of the gesture's $g$ production, the step-wise shift from 1 to 0, its offset. The ensemble of all activation functions (one per each gesture defined for the model) is equivalent to the gestural score of AP. The activation patterns are illustrated in the top part of Figure 11.2. The overall behaviour of our model is thus given by the equation:

$$\ddot{\theta} = \ddot{\theta}_{\mathfrak{d}} + \sum_v a_v(t)\ddot{\theta}_v + \sum_c a_c(t)\ddot{\theta}_c \qquad (11.7)$$

incorporating (11.2), (11.3) and (11.4). Both the summation expressions range over all vocalic and consonantal gestures defined for the model, respectively. The rest attractors (speech ready) are always on, while the gestures influence the system's dynamics according to their activation functions.

Our model is in many respects similar to the TD implementation of AP described in Section 11.1. The important difference, however, is that the dynamics in the presented model originates in the embodied articulatory layer, i.e. the model's behaviour reflects the physical properties of its constituents (masses, stiffness coefficients, etc.). As we have already argued, this aspect is vital for our main intention of modelling the influence of efficiency require-ments on speech dynamics in general and on the gestural phasing in particular.

## 11.2.3 Model parameters, inputs and outputs

Figure 11.3 shows a schematic representation of possible model setups. They map, tentatively, to a 'linguo-palatal' and 'linguo-labial' articulatory subsystems of the human vocal tract, respectively. In the 'linguo-palatal' setup (Figure 11.3A), the pendula 2 and 3 play the role of a tongue tip and tongue dorsum respectively; they are linked by a spring with relatively low stiffness, and pen-dulum 1 represents the top of the mouth, its left-hand side, the velum and its right-hand side, the alveolar ridge. There are two vocalic attractor sets, each involving two pendula: the set labelled $\iota$ with target angular deflections of 0 and

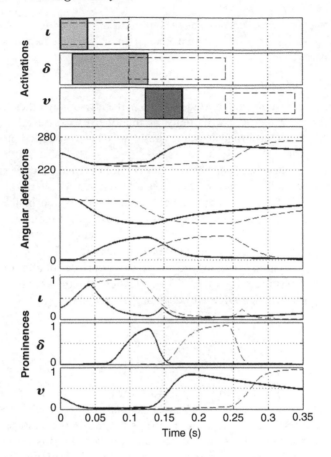

*Figure 11.2    Initial (dashed lines) and optimal (full lines, grey boxes) gestural constellations, and the resulting angular deflections and prominence charts for a sequence ι–δv generated on the model setup illustrated in Figure 11.3*

220 degrees of pendula 1 and 3 respectively, and one labelled $v$ with target angular deflections of 0 and 280 degrees of the same two pendula. The two consonantal gestural targets are labelled $\delta$ (collision of 'tongue tip' pendulum 2 with 'alveolar' side of pendulum 1) and $\gamma$ (collision of 'tongue dorsum' pendulum 3 with the 'velum' side of pendulum 1).

For the sake of comparability of simulation results, both the presented setups share their quantitative characteristics, and differ only in the definition of their vocalic attractors. The 'linguo-labial' setup (Figure 11.3B) exhibits less constrained relationship between the vocalic and consonantal gestures. The pendula 1 and 2 represent lower and upper lip, respectively, and pendulum 3 simulates the tongue dorsum. The two vowels defined for this setup are determined by the position of

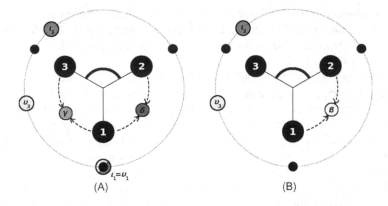

*Figure 11.3*   *'Linguo-palatal' (A) and 'linguo-labial' (B) setups of our model.*
*The filled circles on the perimeter represent target equilibrium*
*positions for given vowels and given pendula, e.g. $v_3$ represent the v*
*target for pendulum 3. The filled circles placed between pendulum*
*rods represent consonantal targets. For both setups, the masses of*
*pendula 1, 2 and 3 are 100, 40 and 50 g, respectively, their stiffness*
*coefficients are 3, 1.8 and 2.5, respectively. The resting length of the*
*spring joining pendula 2 and 3 is 140 degrees, its stiffness coefficient*
*is 0.2. The definitions of vocalic and consonantal attractors are given*
*in the text*

pendulum 3 only. The only consonant, labelled $\beta$ (collision of the two labial pendula) models a bilabial stop.

The masses of pendulum bobs (matrix $\mathbf{M}_0$), the stiffness coefficients ($\mathbf{K}_0$, $\kappa_Z$s, $\kappa_v$s and $\kappa_c$s), the equilibrium vectors $\theta_0$ and $\theta_v$s, the coupling spring lengths $\theta_z$s and vocalic gesture engagement matrices $\mathbf{E}_v$s are all *parameters* of a given model setup. They represent the physiological properties of the vocal tract as well as the articulatory properties of a given phonetic space. They remain constant not only during each single simulation of an utterance production, but also during our entire exploration of the phasing behaviour of the model. The remaining two parameters of the model act as independent inputs to the model. The first one is the collection of activation functions $a_g(t)$ for each gesture $g$ defined in the model, i.e. a multidimensional input stream containing the activation patterns of pre-defined vocalic and consonantal attractor sets. As we have already mentioned, it is also possible to modulate the stiffness parameters of all constituents of the model via overall system stiffness $k$. (At the current stage of the model's development, the overall stiffness is kept constant during an utterance production.) These inputs and their components have to be coordinated to elicit speech-like organisation and kinematics.

The solution of the differential equations describing the model setup parameterised by the inputs is obtained by numerical approximation (implemented in
Matlab) and yields angular positions of pendula in time (see Figure 11.2) representing the articulatory behaviour.

These functions are then used to calculate the degree to which the vocalic or
consonantal phonetic segments (targets of the gestures) are achieved at any
given moment. As mentioned earlier in the text, the output of the system
described by (11.7) is restricted to the traces of pendulum bob positions in time;
there is no actual sound production. For vowels, the distance of the present state
from a given attractor set, absolute or relative, is inversely proportional to the
*prominence* of targeted vowel. Similarly, the prominence of a consonant is
inversely proportional to the mutual distance of two pendula engaged in the
consonant's production. Prominence ranges from 0 to 1, with 1 representing
perfect achievement of the target configuration. The prominence trace $p_g(t)$ of
the gesture $g$ in time plotted for given inputs can be interpreted as a vocal tract
variable as employed by AP and TD. As mentioned earlier in the text, the state
of these 'tract variables' capturing gestural target achievement is in our model
derived from the dynamics of the articulatory layer of description and not the
other way round, as in TD. The prominence $p_g(t)$ does not express the degree of
achievement of the task prescribed by an independent task dynamics, but rather
it captures the degree of realisation of the given segment *derived* from the
overall behaviour of the system and its embodied dynamics parameterised by
activation patterns and overall system stiffness. Unlike TD tract variables, prominence traces are not solutions of an uncoupled second-order differential system. After evaluating all prominence functions for defined gestures, the phonetic
segment with the highest prominence is deemed to be produced at any given
instant.

## 11.2.4    Cost evaluation

The relative simplicity of our abstract model's design allows us to track details
of various physical quantities involved in its speech-like activity: duration of
gestures, groups of gestures and entire utterances, kinematic characteristics
(velocities, acceleration, etc.) of the movement of model articulators and,
importantly, the forces eliciting the dynamics of the system. These and similar
measures are closely linked to the natural notion of *cost* associated with motor
action, which, in turn, plays a crucial role when talking about efficiency, fluency
and, as argued in Section 11.2.5, intentional control of high-level behaviour of
the system. Currently, we are using our model to evaluate three types of cost
naturally related to speech production: *force expenditure cost E* – the overall
physical effort involved in utterance production, utterance *duration cost D*
influencing speaking rate and phonological relevance (salience) of the utterance's suprasegmental entities and *parsing cost P* linked to the receiver's effort
needed to parse the produced utterance. These three cost components allow us to
quantitatively model the trade-off between articulatory ease (represented here by

*E* and *D*) and perceptual clarity (*P*). As we shall show later in the text, all these expenditures are defined as functions of the inputs of our model, gestural activation functions and overall system stiffness.

### 11.2.4.1 Force expenditure

The force expenditure cost *E* is related to the concept of articulatory ease. It has, by far, the most elaborate definition of the three cost components and will be described first. Being a naturally occurring dynamical motor action driven by muscular activity, speech production is shaped by the requirement to minimise, at least in long term, the overall force used by its physiological components during speech production. This force needs to be computed.

The best approximation available for the overall force actively driving an utterance production is the value of what we call the *absolute gestural impulse* **J**: the integral over the duration of the utterance of the sum of magnitudes of all forces that act on model articulators with the exception of repulsive forces, which serve merely to simulate the solidity of pendula, and forces elicited by purely anatomical coupling between the articulators. That is, for production starting in time 0 and completing in time *T*,

$$
\mathbf{J} = \int_0^T \left( \begin{array}{l} |\mathbf{B}\dot{\theta}| + |k\mathbf{K}_0(\theta - \theta_0)| + \sum_v a_v(t)|\mathbf{E}_v \kappa_{\text{voc}} k \mathbf{K}_0(\theta - \theta_v)| \\ + \sum_c a_c(t)|\kappa_{\text{con}} k \mathbf{K}_0 \mathbf{C}_c^* C_c \theta| \end{array} \right) dt \quad (11.8)
$$

is the vector $(J_1, J_2, J_3)^T$ of force expenditures per articulator. The overall absolute gestural impulse is then:

$$
J = J_1 + J_2 + J_3 \tag{11.9}
$$

This way of evaluating the expenditure of force in a motor action system reflects our outlook that behind *every* active force (positively or negatively) influencing the dynamics of the production system there is an active muscular structure. For example, even though most of the time the always-on speech ready dynamics acts against the forces elicited by active gestures, we presume that there is an anatomic structure (possibly one of the pair of agonist/antagonist muscles, or even a subsystem of their fibres) engaged in the pull towards the resting equilibrium. Also, we presume that there is a similar muscular structure engaged in generating the critical damping forces. During the development of our model, we experimented with various definitions of the overall force expenditure. The hypothesis presented here is supported by the insight we gained during this phase.

Further, we presume that the system increases its overall stiffness *k* by increasing the tension (tone) of muscles acting on each articulator in opposite directions with magnitude proportional to $k$. We approximate the force required to maintain the required system's stiffness by the stiffness value $k$ itself. Thus, in our model, the force expenditure cost is defined simply as:

$$E = J + k \tag{11.10}$$

Being a measure of the articulatory effort, this cost function has presumably a universal influence on the speech production. The remaining two cost concepts are more directly related to our aim of finding the efficiency requirements responsible for phenomena associated with speaking rate adjustments and related sequencing variations.

### 11.2.4.2  Parsing cost

One of the strategies people adopt when they speak fast (while maintaining or even reducing the effort, the force expended during speech production) is target undershoot (Lindblom, 1963). The speakers compromise the precision with which some or all of the required gestural targets are achieved, in particular those associated with vocalic segments: they hypoarticulate.

Presumably, however, this target undershoot increases the *receiving* party's cost associated with parsing the utterance.

Speech is a social activity, so its production and perception aspects are inseparable – mere articulation is only a part, albeit important, of an utterance production; the utterance must be successfully parsed, to be fully realised by the receiving party. More undershoot generally implies higher parsing cost. Therefore, we include a measure of the receiver's parsing cost in the overall cost of utterance production. The parsing cost $P$ is thus defined as a sum of undershoots over all relevant phonetic segments produced in an utterance, i.e. distances of their peak prominences from the ideal, perfect prominence value of 1.

The requirement of minimising the force expenditure cost and the parsing cost pose conflicting constraints on the speech production: lowering the force expenditure increases the cost of parsing the resulting utterance and vice versa. Efficient production of a given utterance is thus the result of a weighted compromise between these conflicting demands. This observation and its connection to phonetic variation is the basis of Lindblom's Hyper-Hypoarticulation Theory (Lindblom, 1990). Here we provide a quantitative expression of his predominantly theoretical outlook.

### 11.2.4.3  Duration cost

When choosing the speaking rate and other prosodic aspects of speech, overall utterance duration, or the duration of some well-defined functional suprasegmental units (e.g. syllables) seems to be a natural parameter to control. The first approximation of the duration cost value $D$ is the overall duration $T$ of the utterance.

The same sequence of phonetic segments can be often uttered in two distinct ways, consider for example two utterances /di—did/ and /did—id/, where the dash marks a syllabic boundary. The requirement of salience of the differently organised syllables in these two utterances poses stricter temporal constraints on the inter-gestural phasing of the segments lying within the same syllable than on those lying across the syllabic unit boundary. The phasing patterns governing the production of a syllable are more stable than inter-syllabic ones (Byrd, 1996). We hypothesise that this suprasegmental salience constraint emerges as a consequence of the uneven distribution of the duration cost function over utterance production. To test this hypothesis, we designed the duration cost function $D$ to reflect the required suprasegmental (e.g. syllabic) structure of an utterance.

The first modelling approximation of duration cost function conceived this way is an integral

$$D = \int_0^T \pi(t)dt \tag{11.11}$$

where the step function $\pi$ is defined in the following way: $\pi(t)$ is set to 1 if $t$ falls between onset and offset of a suprasegmental unit, e.g. a syllable, and to a constant value $\pi_c$, $0 \leq \pi_c \leq 1$, otherwise.[1] Thus, the duration of periods when $\pi$ has less than its maximal value 1 'counts less' then the duration of the intra-segmental periods. In the simulations presented in this chapter, the onset and offset of the CV syllables is associated with the moment of maximal prominence of C and the moment of maximal prominence of V, respectively, and *not* the onsets and offsets of the underlying gestures. Thus, the function $\pi$ reflects the surface structure of the utterance, and therefore influences the gestural phasing indirectly.

All cost components $E$, $D$ and $P$ defined here are functions of a given utter-ance (sequence of gestures) in general, and of the manner of its production in particular, i.e. of the precise gestural constellation details, plus the overall sys-tem's stiffness related to system's ability to attain the required gestural targets. We do not claim that the three cost measures introduced here provide an exhaustive list of constraints behind efficient speech production. On the contrary, we are fully aware that there are many other possible candidates for cost func-tions, which can be used to account for phonetic and phonological phenomena, e.g. jerk (maximal acceleration of model articulators) or work, to name just a few. As argued in the rest of this chapter, our selection nevertheless seems to be the right one to shed novel light on the aspect of speech production under scrutiny here, i.e. fluent, efficient gestural sequencing and its dependency on intentionally adjusted speaking rate and articulatory precision.

---

[1] This approach to imposing a suprasegmental structure upon an utterance is inspired by Byrd's and Saltzman's prosodic $\pi$-gestures mentioned in Section 11.1.1.

## 11.2.5　Optimisation

The central hypothesis behind our work is that the requirement of optimal behaviour, i.e. the drive towards the minimisation of all the three cost measures, $E$, $P$ and $D$, reveals general properties of the space of efficient, natural gestural constellations.

Having a vector of three cost measures to be minimised simultaneously, we are presented with a multiobjective optimisation problem. We approach it in a standard way – a weighted sum strategy – and convert it into a scalar optimisation problem by considering a weighted sum of all objectives:

$$C_\alpha = \alpha_E E + \alpha_P P + \alpha_D D$$

where $\alpha = (\alpha_E, \alpha_P, \alpha_D)$ is a vector weighting the cost components, and hence instantiating a specific trade-off between production and perceptual costs. The small changes in gestural constellation and overall stiffness value which lower the value of one of the cost functions under consideration generally cause an increase of the value of one or both remaining cost measures. For example, a smaller undershoot (decrease of the parsing cost $P$) can be achieved by increasing the overall stiffness of the system (increasing the force expenditure cost $E$) or by lengthening the duration of gestural activation (in effect increasing the durational cost $D$). This important property of the selected system of cost functions guarantees that, for a given weight distribution $\alpha$, there exists a 'compromise' solution – a gestural constellation and overall stiffness value – of the given optimisation problem of minimising the overall cost $C_\alpha$.

The vector $\alpha$ expresses the biases in this trade-off game between the cost components. For example, the requirement to speak faster is directly linked to an increase in the value of $\alpha_D$, which makes shorter versions of a given utterance relatively 'cheaper'. The cost 'saved' this way can be offset against a proportional increase of one or both of the other two components. The speaker can increase the undershoot (if she's confident that the listener will be able to parse the utterance anyway) or the stiffness (and the force expenditure, e.g. in a noisy environment). This choice is again reflected in the ratio of the other two weight coefficients $\alpha_P$ and $\alpha_E$.

The weight coefficients neither prescribe any details of gestural phasing nor, indeed, straightforwardly determine the value of system stiffness; rather, they are high-level, intentional parameters of the physically embodied speech production system. It is not our claim that the speakers use a cost components' weight distribution (akin to $\alpha$) as parameters of the associated minimisation processes for *online* production of utterances. Rather, we believe that these trade-offs play a vital role during the long-term development of speech as a skilled human activity, and are thus reflected in the phonological laws underlying the speech production (see, Dispersion-Focalization Theory described in Section 11.1.1). In other words (paraphrasing Lindblom), speech is adapted to be spoken. During speech acquisition and the accompanying fine-tuning of our own production dynamics, we take advantage of these low-energy patterns, which act as attractors in vastly high dimensional space of all possible productions.

We use a simplified simulated annealing method[2] to identify those model input streams that minimise the overall cost $C_\alpha$. The compound function of the pendulum model of the vocal tract and the overall cost function, mapping the gestural constellations and system stiffness parameters to a single numerical efficiency measure, is used as an objective function of the optimisation problem. For a given model setup, a (non-optimal) initial configuration of input (a collection of gestural activation functions producing the desired utterance and a value of overall stiffness), parameters of the suprasegmental function $\pi$ and an assignment of cost weight distribution $\alpha$, the optimisation procedure searches the space of possible inputs (adjusting the gestural onset and offset points and the value of overall stiffness) until it finds a gestural constellation/ stiffness pair minimising the total cost value $C_\alpha$. To guarantee that the optimal input actually produces the given utterance, a very high additional cost is assigned to those inputs that do not produce the required sequence of segments.

Figure 11.2 illustrates the optimisation procedure. It shows initial (dashed lines) and optimal (full lines) gestural phasing, computed angular deflections of pendula and resulting prominence functions for the 'linguo-palatal' model setup illustrated in Figure 11.3A for a gestural sequence $\iota$–$\delta v$. The initial constellation has been designed so that the subsequent gesture is triggered as soon as the preceding gesture sufficiently approaches its target – reaches a prominence of 0.95 (see the dashed lines on the Activation and Prominence charts). The starting overall stiffness (torsion spring coefficient) of the system was set to 15 Nmrad$^{-1}$. These initial input streams are then used as a starting point of the optimisation procedure, along with the model setup parameters and the cost components weight distribution vector $\overrightarrow{\alpha} = (1, 60, 100)$. For this particular setup, the (experimentally tested) meaningful values of weight component $\alpha_D$ range between 20 and 200, the values of $\alpha_P$ range from 10 to 1000.

The full lines show the resulting optimal gestural constellation (grey boxes), the associated angular deflection traces and the prominence charts. The final overall stiffness (not plotted) was calculated as 17.58 Nmrad$^{-1}$. Due to the particular details of this model's setup and relatively low durational cost weight, the activation intervals of participating gestures overlap only minimally. The partial overlap for the gestures $\iota$ and $\delta$ is a consequence of production synergies between these two gestures. The maximal reached prominence of each gesture (vocalic ones in particular) has decreased for the resulting constellation reflecting the relatively low precision requirements allowing greater undershoot.

---

[2] Despite the intended simplicity of our model and cost definition, the objective function $C\alpha$ is still fairly complex with plenty of local minima where simplex (or a hill descent method) tends to 'get stuck'.

## 11.3   Simulation results

Figure 11.4 shows a 'slice' of the efficient production space for the sequence /ɪ–δv/ produced by the 'linguo-palatal' model setup. The top graph shows a sequence of

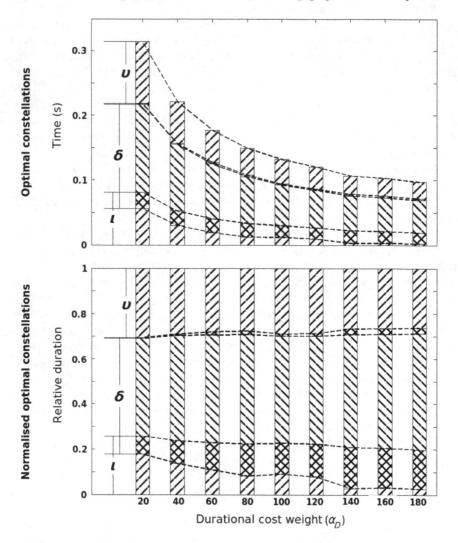

*Figure 11.4   Optimal productions of a sequence ɪ–δv by the 'linguo-palatal'
model setup. The top two charts plot a series of the optimal
gestural constellations (absolute and normalised gestural activation
intervals respectively) for increasing durational cost weight
αD plotted on the x axis. Each bar represents an optimal
constellation, its striped boxes correspond to activation intervals
of gestures, in a bottom-up order, ɪ, δ and v, respectively*

(simplified) optimal gestural constellations ($y$ axis) and an overall trend (interpolation of gestural onsets and offsets) for fixed force expenditure and precision cost weights $\alpha_E = 1$ and $\alpha_P = 100$ and for duration cost weight $\alpha_P$ ranging from 20 to 180 ($x$ axis). The second pane shows the same constellations with their overall duration normalised to see the relative activation durations and overlaps of gestures. (The optimal gestural constellation and the accompanying charts plotted in Figure 11.3 show the production details for one instance, that of $\alpha_D = 60$, of Figure 11.4.)

Several expected consequences of increasing the weight $\alpha_D$ of the durational cost can be observed in this simulation example: the shortening of the overall duration of the utterance, increase of the overall system's stiffness and increase of target undershoot by participating gestures (not plotted). This pattern is consistent with human speakers increasing their speaking rate (Ostry *et al.*, 1987; Lindblom, 1963).

More interestingly, the normalised constellation plot captures two less straightforward consequences of speaking rate increase observed by phoneticians.

First, it shows that the relative gestural activation durations do not remain constant as the rate increases – the relative duration of the consonantal gesture $\delta$, e.g. increases with the rate. This is consistent with Gay's result mentioned in Section 11.1.1. Moreover, the phasing of gestural onsets and offsets changes in a non-linear fashion: the relative activation overlap of gestures $\iota$ and $\delta$ gets larger as the overall duration of the utterance shortens. So, our model correctly reproduces the qualitative behaviour of sequencing variations as reported, e.g. in Nittrouer *et al.* (1988) and Nittrouer (1991) and discussed in Section 11.1.1 of this chapter.

Second, the relative phasing of gestures $\delta$ and $\upsilon$ which are declared to belong to one suprasegmental unit (syllable) by the durational step function $\pi$ remains more or less constant with no or little overlap. On the other hand, the activation overlap between gestures $\iota$ and $\delta$ across a boundary is more flexible and participates in overall shortening of the utterance with increasing rate – with the cost defined using the function $\pi$, to expand the lag between the gestures is 'cheaper' across the boundary than it is within a syllable. This is again consistent with observed influence of speaking rate increase on relative gestural timing (see Cummins, 1999 and our interpretation in Section 11.1.1).

As mentioned earlier in the text, when faced with the task of phasing the gestures which impose competing targets on shared articulators, the optimisation procedure generates a gestural constellation with no or small overlaps of the gestures' activation intervals. What happens when subsequent gestures impose less mutually interacting constraints on the articulators involved in an utterance production; i.e. when the sets of articulators engaged in the subsequent motor actions are comparatively weakly anatomically coupled, as for example in the case of vowels overlaid with bilabial stops? Can we expect an emergence of a strong phasing relationship between similar gestures sharing the common articulators (e.g. vowels) and a separate layer of anatomically quite independent gestures (e.g. bilabial consonants) functionally phased relative to the underlying (vocalic) gestures?

To answer to this question, we used our 'linguo-labial' setup described in Section 11.2 and illustrated in Figure 11.3B.

Figure 11.5 shows the initial (dashed line) and optimised (full line) constellations for a sequence $\iota\text{–}\beta v$ realised in this adapted model setup; the initial overall system's stiffness was set to 15 Nmrad$^{-1}$, the obtained resulting one was 18.3818 Nmrad$^{-1}$.

The optimal constellation shows the tight sequential phasing of the vocalic gestures $\iota\text{–}v$ (marked by a vertical dotted line and an asterisk) interleaved with the consonantal gesture $\beta$. The gesture $\delta$ is phased with respect to the vocalic layer

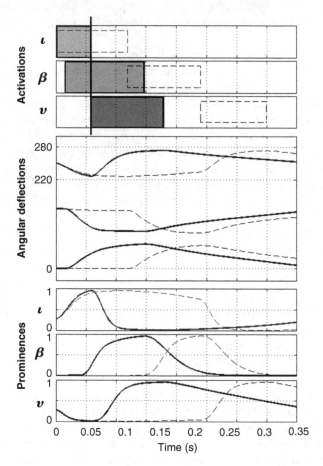

*Figure 11.5    An optimal constellation for a gestural sequence $\iota\text{–}\beta v$ by the adapted model setup (see the text) with vocalic attractors relatively weakly anatomically linked with a consonantal ($\beta$) attractor. For the description of figure's components, see Figure 11.2*

according to the *functional* requirement of the correct order of prominence peaks achieved by subsequent gestures - see the prominence panel of Figure 11.5. This is an emergent phenomenon: the initial constellation was designed by the same procedure as in the previous case; the gestures were initially phased in a simple sequence.[3]

The presence of the two separate layers, one carrying vocalic and the other one consonantal gestures, is analogous to the existence of functional tiers postulated by several phonological theories, mentioned in Section 11.2; (Fowler, 1983; Browman and Goldstein, 1991; see also Keating, 1990). Using our modelling approach and, crucially, taking embodiment of the speech production cognitive system seriously, this phenomenon *emerges* as a result of an interaction between sets of constraints that are best captured on two parallel levels of description: the functional constraints formulated on the abstract *task level* and efficiency requirements set down on the embodied *articulatory level.*

## 11.4  Discussion

We have argued that some phenomena of speech production and perception that have traditionally been postulated and described in a representational, grammatical fashion by various phonological and phonetic theories (e.g. separation of functional tiers, non-linearity of gestural sequencing with regards to speaking rate, etc.) emerge as consequences of the dynamical properties of a physically instantiated production system, together with the requirement of efficiency in production.

As far as we are aware, our modelling paradigm is the first one to provide a platform for capturing these vital dynamical properties of the speech production cognitive system in a simple and intuitive fashion. It allows us to describe and explain some of the known production patterns as examples of behaviour of an embodied motor action system, and account for them in the language of intentionally motivated high-level parameters linking the system's dynamics, cost functions and efficiency without the need of bringing in any additional external phonological postulates. Some of the phenomena, in particular those associated with efficient speech production and sequencing of primitive actions can thus be treated as emergent properties in the sense of Lindblom (1999). The intended simplicity of our modelling approach gives an additional support to our claim that it is the character of the task in hand and the *nature* of second-order dynamics of an embodied system enforced by the cost efficiency principle (rather than complex details of phonological rules and of vocal tract physiology) that on their own bring about important phenomena accompanying speech production. We do not claim that an account of an embodied articulatory system is the only 'correct' level of description on which it is possible to talk meaningfully about

---

[3] Although Fig. 11.5 shows only one gestural constellation, the same pattern has emerged for an entire series of constellations with varying durational and precision cost weights $\alpha_D$ and $\alpha_P$ similar to one represented in Fig. 11.4.

phonological and phonetic phenomena. We agree, e.g. with the school of AP that there are at least three such informative levels, each one designed to best describe constraints imposed on speech production by its linguistic, functional and physiological aspects, respectively. Identifying any of these constraints and, as seen in the second example in the previous section, the cross-level interdependencies between them, helps us to better understand the cognitive processes underlying the production and perception of speech. The model setups presented in this chapter reflect only very high-level organisational characteristics of the end effectors of the human vocal tract. In follow-up work, we have studied models that more closely capture the anatomical and functional properties of the speech production system. In these models, the task-oriented and embodied aspects of model articulators are defined hierarchically at interconnected levels of description closely related to the architecture used in the TD implementation of AP (Saltzman and Munhall, 1989). To be able to use these models of the vocal tract in conjunction with the optimality paradigm presented here, we adapted the TD definition of their behaviour so that it takes into account the embodied character of speech articulation.

The results of simulations performed on these instances of the general abstract modelling paradigm presented here provide further insights into emergent patterns of coarticulation and functional tier separation. Moreover, as reported elsewhere (Simko and Cummins, 2009), the details of the optimal articulatory behaviour generated by these models are in a considerable agreement with the gestural sequencing patterns manifested by human speakers (Browman and Goldstein, 1988; Löfqvist and Gracco, 1997, 1999, 2002).

This chapter has presented a novel mathematical model in some detail. It is, perhaps, worthwhile to summarise the overall account provided herein. We begin with many of the same assumptions as underwrite the Articulatory Phonological framework and its TD implementation: we assume that gestures are primitives of linguistic organisation, that they are crucially sequenced in time and that their evolution is constrained in lawful fashion by a task-specific dynamic regime. In contrast to previous approaches, we choose to define tasks in the space of physically instantiated articulators, and we make use of the inertial properties of these articulators to shed light on the sequencing of gestures in real time, and in dependence on such high-level speech properties as speaking rate. We find that it is possible to operationalise the conflicting constraints of articulatory ease and perceptual clarity within a single additive cost function. We then demonstrate, through simulation, that this cost function and the overarching concept of efficiency can, indeed, suitably constrain the organisation of gestures in time.

# Acknowledgements

This work has been funded by Principal Investigator Grant number 04/IN3/I568 from Science Foundation Ireland to Dr. Fred Cummins, UCD, Dublin.

# References

Anderson, F.C. and Pandy, M.G. (2001). 'Dynamic optimization of human walking'. *ASME Journal of Biomechanical Engineering*, vol. 123, pp. 381–390.

Barry, W.J. (1998). 'Time as a factor in the acoustic variation of schwa'. In *Proceedings of ICSLP-1998*, Sydney, Australia.

Boersma, P. (1998). *Functional Phonology: Formalizing the Interactions between Articulatory and Perceptual Drives*. Ph.D. dissertation, University of Amsterdam.

Browman, C.P. and Goldstein, L. (1991). 'Tiers in articulatory phonology, with some implications for casual speech'. In Kingston, J. and Beckman, M.E. (eds.), *Papers in Laboratory Phonology I: Between the Grammar and the Physics of Speech*. Cambridge, U.K.: Cambridge University Press; pp. 341–376.

Browman, C.P. and Goldstein, L. (1992). 'Articulatory phonology: an overview'. *Phonetica*, vol. 49, pp. 155–180.

Browman, C.P. and Goldstein, L. (2000). 'Competing constraints on intergestural coordination and self-organization of phonological structures'. *Bulletin de la Communication Parlee*, vol. 5, pp. 25–34.

Browman, C.P. and Goldstein, L.M. (1988). 'Some notes on syllable structure in articulatory phonology'. *Phonetica*, vol. 45, pp. 140–155.

Byrd, D. (1996). 'Influences on articulatory timing in consonant sequences'. *Journal of Phonetics*, vol. 24, pp. 209–244.

Byrd, D., Kaun, A., Narayanan, S., and Saltzman, E. (2000). 'Phrasal signatures in articulation'. In Broe, M. and Pierrehumbert, J. (eds.), *Papers in Laboratory Phonology V: Acquisition and the Lexicon*. London: Cambridge University Press; pp. 0–87.

Byrd, D. and Saltzman, E. (2003). 'The elastic phrase: modeling the dynamics of boundary-adjacent lengthening'. *Journal of Phonetics*, 31(2):149–180.

Cooke, J. (1979). 'The organization of simple, skilled movements'. In *Motor Learning and Control*. Senanque, France: NATO Advanced Study Institute; pp. 199–212.

Cummins, F. (1999). 'Some lengthening factors in English speech combine additively at most rates'. *Journal of the Acoustical Society of America*, 105(1):476–480.

Feldman, A. and Latash, M. (2005). 'Testing hypotheses and the advancement of science: recent attempts to falsify the equilibrium point hypothesis'. *Experimental Brain Research*, 161(1):91–103.

Fowler, C.A. (1983). 'Converging sources of evidence on spoken and perceived rhythms of speech: cyclic production of vowels in sequences of monosyllabic stress feet'. *Journal of Experimental Psychology: General*, vol. 112, pp. 386–412.

Gay, T. (1981). 'Mechanisms in the control of speech rate'. *Phonetica*, vol. 38, pp. 148–158.

Gay, T., Ushijima, T., Hirose, H., and Cooper, F.S. (1974). 'Effects of speaking rate on labial consonant-vowel articulation'. *Journal of Phonetics*, vol. 2, pp. 47–63.

Guenther, F.H. (1995). 'Speech sound acquisition, coarticulation, and rate effects in a neural network model of speech production'. *Psychological Review*, **102**(3):594–621.

Haken, H., Kelso, J.A.S., and Bunz, H. (1985). 'A theoretical model of phase transitions in human hand movements'. *Biological Cybernetics*, vol. 51, pp. 347–356.

Howard, I.S. and Huckvale, M.A. (2005). 'Training a vocal tract synthesizer to imitate speech using distal supervised learning'. In *Proceedings of the SPECOM 2005*. Patras, Greece. pp. 159–162.

Iskarous, K., Goldstein, L.M., Whalen, D., Tiede, M., and Rubin, P. (2003). 'Casy: the haskins configurable articulatory synthesizer'. In *Proceedings of the 15th International Congress of Phonetic Sciences*, Barcelona, Spain. Universitat Auto'noma de Barcelona.

Keating, P.A. (1990). 'The window model of coarticulation: articulatory evidence'. In Kingston, J. and Beckman, M. (eds.), *Papers in Laboratory Phonology I: Between the Grammar and Physics of Speech*. Cambridge: Cambridge University Press; pp. 451–470.

Kelso, J.A.S. (1995). *Dynamic Patterns: The Self-Organization of Brain and Behavior*. Cambridge, Massachusetts, London, England: MIT Press.

Latash, M. (2008). *Synergy*. New York: Oxford University Press.

Liljencrants, J. and Lindblom, B. (1972). 'Numerical simulation of vowel quality systems: the role of perceptual contrast'. *Language*, **48**(4):839–862.

Lindblom, B. (1963). 'Spectrographic study of vowel reduction'. *Journal of the Acoustical Society of America*, **35**(11):1773–1781.

Lindblom, B. (1983). 'Economy of speech gestures'. In MacNeilage, P.F. (ed.), *The Production of Speech*. New York: Springer-Verlag.

Lindblom, B. (1990). 'Explaining phonetic variation: a sketch of the H&H theory'. In Hardcastle, W.J. and Marchal, A. (eds.), *Speech Production and Speech Modelling*. Dordrecht: Kluwer Academic Publishers; pp. 403–439.

Lindblom, B. (1999). 'Emergent phonology'. In *Proceedings of the 25th Annual Meeting of the Berkeley Linguistics Society*. University of California, Berkeley.

Lindblom, B. (2000). 'Developmental origins of adult phonology: the interplay between phonetic emergents and the evolutionary adaptations of sound patterns'. *Phonetica*, 57(2–4):297–314.

Löfqvist, A. and Gracco, V.L. (1997). 'Lip and jaw kinematics in bilabial stop consonant production'. *Journal of Speech, Language, and Hearing Research*, vol. 40, pp. 877–893.

Löfqvist, A. and Gracco, V.L. (1999). 'Interarticulator programming in VCV sequences: lip and tongue movements'. *Journal of the Acoustical Society of America*, vol. 105, pp. 1864–1876.

Löfqvist, A. and Gracco, V.L. (2002). 'Control of oral closure in lingual stop consonant production'. *Journal of the Acoustical Society of America*, **111**(6): 2811–2827.

Maeda, S. (1982). 'A digital simulation method of the vocal-tract system'. *Speech Communication*, vol. 1, pp. 199–229.

Nittrouer, S. (1991). 'Phase relations of jaw and tongue tip movements in the production of VCV utterances'. *Journal of the Acoustical Society of America*, **90**(4):1806–1815.

Nittrouer, S., Munhall, K., Kelso, J., Tuller, B., and Harris, K.S. (1988). 'Patterns of interarticulator phasing and their relation to linguistic structure'. *Journal of the Acoustical Society of America*, vol. 84, pp. 1653–1661.

Öhman, S.E.G. (1966). 'Coarticulation in VCV utterances: spectrographic measurements'. *Journal of the Acoustical Society of America*, **39**(1):151–168.

Ostry, D.J. (1986). 'On viewing motor behavior as a physical system'. *Journal of Phonetics,* vol. 14, pp. 145–147.

Ostry, D.J., Cooke, J.D., and Munhall, K.G. (1987). 'Velocity curves of human arm and speech movements'. *Experimental Brain Research*, vol. 68, pp. 37–46.

Ostry, D.J. and Feldman, A. (2003). 'A critical evaluation of the force control hypothesis in motor control'. *Experimental Brain Research*, vol. 221, pp. 275–288.

Perrier, P., Perkell, J., Payan, Y., Zandipour, M., Guenther, F., and Khalighi, A. (2000). 'Degrees of freedom of tongue movements in speech may be constrained by biomechanics'. In *Proceedings of the 6th International Conference on Spoken Language Processing, ICSLP 2000*, Beijing, China.

Saltzman, E.L. (1991). 'The task dynamic model in speech production'. Chapter 3. In Peters, H.F.M., Hulstijn, W., and Starkweather, C.W. (eds.), *Speech Motor Control and Stuttering*. Amsterdam: Elsevier Science.

Saltzman, E.L. and Kelso, J.A.S. (1987). 'Skilled actions: a task-dynamic approach'. *Psychological Review*, **94**(4):84–106.

Saltzman, E.L. and Munhall, K.G. (1989). 'A dynamical approach to gestural patterning in speech production'. *Ecological Psychology*, **1**(4):333–382.

Schwartz, J.-L., Boe, L.-J., Vallee, N., and Abry, C. (1997). 'The dispersion-focalization theory of vowel systems'. *Journal of Phonetics*, vol. 25, pp. 255–286.

Simko, J. and Cummins, F. (2009). 'Sequencing of articulatory gestures using cost optimization'. In *Proceedings of the Interspeech Conference*, Brighton, UK.

Stevens, K.N. (1989). 'On the quantal nature of speech'. *Journal of Phonetics*, vol. 17, pp. 3–45.

Todorov, E. (2004). 'Optimality principles in sensorimotor control'. *Nature Neuroscience*, vol. 7, pp. 907–915.

*Chapter 12*

# A neural network model for the prediction of musical emotions

*Eduardo Coutinho and Angelo Cangelosi*

## Abstract

This chapter presents a novel methodology to analyse the dynamics of emotional responses to music in terms of computational representations of perceptual processes (psychoacoustic features) and self-perception of physiological activation (peripheral feedback). The approach consists of a computational investigation of musical emotions based on spatio-temporal neural networks sensitive to structural aspects of music. We present two computational studies based on connectionist network models that predict human subjective feelings of emotion. The first study uses six basic psychoacoustic dimensions extracted from the music pieces as predictors of the emotional response. The second computational study evaluates the additional contribution of physiological arousal to the subjective feeling of emotion. Both studies are backed up by experimental data. A detailed analysis of the simulation models' results demonstrates that a significant part of the listener's affective response can be predicted from a set of psychoacoustic features of sound – tempo, loudness, multiplicity (texture), power spectrum centroid (mean pitch), sharpness (timbre) and mean STFT flux (pitch variation) – and one physiological cue, heart rate. This work provides a new methodology to the field of music and emotion research based on combinations of computational and experimental work, which aid the analysis of emotional responses to music, while offering a platform for the abstract representation of those complex relationships.

## 12.1  Introduction

The ability of music to stir human emotions is a well-known fact (Gabrielsson & Lindström, 2001), and accumulated evidence leaves little doubt that music can at least express emotions. This association is so profound that music is often claimed to be the 'language of emotions' and a compelling means by which we appreciate the richness of our affective life. Music gives a 'voice' to the inner world of emotions and feelings, which are often very hard to communicate in words (Langer, 1942).

In spite of the acknowledgment of the emotional power of music, supported by scientific and empirical evidence, the manner in which music contributes to those experiences remains uncertain. One of the main reasons is the large number of syndromes that characterise emotions. Another, is its subjective and phenomenological nature. The emotion created by a piece of music may be affected by memories, the environment and other situational aspects, the mood of the person who is listening, individual preferences and attitudes, cultural conventions, among others (a systematic review of these factors and their possible influence in the emotional experience can be found in Scherer and Zentner, 2001). Nonetheless, a considerable corpus of literature on the emotional effects of music in humans has consistently reported that listeners often agree rather strongly about what type of emotion is expressed in a particular piece or even in particular moments or sections (a review of accumulated empirical evidence from psychological studies can be found in Juslin & Sloboda, 2001; see also Gabrielsson & Lindström, 2001). Naturally this leads to a question: can the same music stimulus induce similar affective experiences in all listeners, somehow independently of acculturation, context, personal bias or preferences?

A very important aspect of perception of emotion in music is that it is only marginally affected by factors such as age, gender or musical training (Robazza, Macaluso, & D'Urso, 1994), since it indicates that the general neurophysiological mechanisms that process emotional stimuli are involved in the perception of emotion in music. Such an idea is supported by the finding that the ability to recognise discrete emotions is correlated with measures of 'Emotional Intelligence' (Resnicow, Salovey, & Repp, 2004). Furthermore, Peretz, Gagnon, and Bouchard (1998) presented even more compelling evidence showing that the perceptual analysis of the music input can be maintained for emotional purposes, even if impaired for cognitive ones. Peretz *et al.* (1998) suggest the possibility that emotional and non-emotional judgements are the products of distinct neurological pathways. Some of these pathways were found to involve the activation of subcortical emotional circuits (Blood & Zatorre, 2001; Blood, Zatorre, Bermudez, & Evans, 1999), which are also associated with the generation of human affective experiences (e.g., Damasio, 2000; Panksepp, 1998) that can operate outside an individual's awareness. Panksepp and Bernatzky (2002) even suggest that a great part of the emotional power derived from music may be generated by lower subcortical regions, where basic affective states are organised (Damasio, 2000; Panksepp, 1998).

One cannot ignore the fact that most listeners appreciate music through a diverse range of cortico-cognitive processes, which rely upon the creation of mental and psychological schemas derived from exposure to the music in a given culture (e.g., Meyer, 1956). Nevertheless, the affective power of sound and music suggests that it may be related to the deeper affective roots of the human brain (Zatorre, 2005; Panksepp & Bernatzky, 2002; Krumhansl, 1997). Certain basic neurological mechanisms related to motivation/cognition/emotion automatically elicit a response to music in the receptive listener. This gives rise to profound changes in the body and brain dynamics, and to interference with ongoing mental and bodily processes (Panksepp & Bernatzky, 2002; Patel & Balaban, 2000). This multimodal integration of musical and non-musical information takes place in the brain (Koelsch, Fritz, Cramon, Müller, & Friederici, 2006), suggesting the existence of complex relationships between the dynamics of musical emotion and the perception–action cycle response to musical structure. This is also the belief of several researchers who imply the existence of causal relationships between musical features and emotional response (e.g., Gabrielsson & Lindström, 2001).

The basic perceptual attributes involved in music perception (also referred to as psychoacoustic features) are loudness, pitch, contour, rhythm, tempo, timbre, spatial location and reverberation (Levitin, 2006). While listening to music, our brains continuously organise these dimensions according to different gestalt and psychological schemas. Some of these schemas involve further neural computations on extracted features which give rise to higher order musical dimensions (e.g., meter, key, melody, harmony), reflecting (contextual) hierarchies, intervals and regularities between the different music elements (e.g., Levitin, 2006). Others involve continuous predictions about what will come next in the music as a means of tracking structure and conveying meaning (Meyer, 1956). In this sense, the aesthetic object is also a function of its objective design properties, and so the subjective experience should be (at least partially) dependent on those features (Kellaris & Kent, 1993).

Within such framework, researchers usually focus on the similarities between listeners' affective experiences to the same music expecting to find relationships with the nature of the stimulus. Such observations have already a long history (back to Socrates, Plato and Aristotle), but they gained particular attention after Hevner's studies during the 1930s (Hevner, 1936) that are amongst the first to systematically analyse which musical parameters (e.g., major vs. minor modes, firm vs. flowing rhythm, direction of melodic contour) are related to the reported emotion (e.g., happy, sad, serene). Since then, a core interest amongst music psychologists has been the isolation of the perceptible factors in music, which may be responsible for the many observed effects, and a fairly regular stream of publications have attempted to clarify this relationship (for a review of past studies please refer to Gabrielsson & Lindström, 2001). We have now strong evidence that certain music dimensions and qualities communicate similar affective experiences to many listeners. The results of more than 100 studies clearly show that listeners are consistent and reliable in their judgements of emotional expression in music (see Juslin & Laukka, 2004), and there is also evidence of the universality of music affect (Balkwill & Thompson, 1999; Fritz *et al.*, 2009).

## 12.1.1   Emotional responses and the perception of music structure

By focusing on the music stimulus and its features as important dynamic characteristics of affective experiences, many studies suggested the influence of various structural factors on emotional expression (e.g., tempo, rhythm, dynamics, timbre, mode, harmony, among others; see Gabrielsson & Lindström, 2001; Schubert, 1999a for a review). Unfortunately, the nature of these relationships is complex, and it is common to find rather vague and contradictory descriptions, especially when the music structural factors are considered in isolation (Gabrielsson & Lindström, 2001).

Many researchers have used qualitative descriptions to describe the perceived changes in the music, often supported by discrete scales considering only two extreme levels (e.g., fast and slow in the case of tempo). Such an approach leaves aside all the intermediate levels of these variables assuming that they behave linearly within the extreme categories defined, thus neglecting a wide range of musical possibilities and complexity of interactions between variables. Another problem of such an approach arises when we consider that music can elicit a wide range of emotions in the listener. A piece of music is characterised by changes over time, which are a fundamental aspect of its expressivity. The dynamical changes over time are perhaps the most important ones (Dowling & Harwood, 1986), especially if we consider that musical emotions may exhibit time-locking to variations in psychological and physiological processes, consistent with a number of studies that show temporal variations in affective responses (e.g., Goldstein, 1980; Nielsen, 1987; Krumhansl, 1997; Schubert, 1999a; Korhonen, 2004a).

Much of the research in this area has focused on general emotional characterisations of music (e.g., identification of basic emotion, lists of adjectives or affective labels), by controlling parameters that can show some degree of stability throughout a piece (e.g., tempo, key, timbre, mode). In some studies, sets of specially designed stimuli have been used (e.g., probe tone test), while other studies were based on a systematic manipulation of real music samples (e.g., slow down tempo, changing instruments). More recently, following the claim that music features and structure are characterised by emotionally meaningful changes over time (e.g., Dowling & Harwood, 1986), new methodological paradigms using real music and continuous measurements of emotion emerged (e.g., Schubert, 2001). Nevertheless, only a few studies have focused on the analysis of temporal patterns of emotional responses to music, concentrating on the perceived sound.

## 12.2   Temporal patterns of emotional responses to music and the perception of music structure

### 12.2.1   Measures of emotional responses to music stimuli

The majority of studies involving the measurement of human emotions make use of three main classes of quantities (Berlyne, 1974): physiological responses (e.g., heart rate (HR), galvanic response), behavioural changes or behaviour preparation (e.g., facial expressions, body postures) and the subjective feeling component (self-report of emotion: e.g., checklists and rating scales). The two modalities most often measured in music and emotion research are subjective feelings and physiological arousal. One is associated with the feeling of emotion during music listening, and the way the music may trigger or represent a specific emotion or representation of it. The other investigates patterns of physiological activity associated with music quantities (such as the relationships between tempo and HR), as well as the peripheral routes for emotion production (how physiological states relate with entire emotional experience).

### 12.2.1.1  Subjective feelings

The subjective feeling of emotion refers to its experienced qualities, based on the understanding of the felt experiences from the perspective of an individual. While the perception of emotional events leads to rapid (some automatic and stereotyped) emotional responses, feeling states have a slower modulatory effect in cognition, and ultimately also in behaviour and decision-making (according to the nature and relevance of the eliciting stimulus). More generally the feeling of emotion can be seen as a (more or less diffuse) representation that indexes all the main changes (in the respective components) during an emotional experience. This is the compound result of event appraisal, motivational changes and proprioceptive feedback (from motor expression and physiological reactions) (Scherer, 2004). These conscious 'sensations' are an irreducible quality of emotion, unique to the specific internal and external contexts, and to a particular individual (Frijda, 1986; Lazarus, 1991).

Schubert (2001) proposed the use of continuous measurements of cognitive self-report of subjective feelings, using a dimensional paradigm to represent emotions on a continuous scale. According to Wundt (1896), differences in the affective meaning among stimuli can succinctly be described by three pervasive dimensions (of human judgement): pleasure ('lust'), tension ('spannung') and inhibition ('beruhigung'). They can be represented in a three-dimensional space, with each dimension corresponding to a continuous bipolar rating scale: pleasantness–unpleasantness, rest–activation and tension–relaxation. This model has received empirical support from several studies that have shown that a large spectrum of continuous and symbolic stimuli can be represented using these dimensions (see Bradley & Lang, 1994). Other studies have provided evidence that the use of only two dimensions is a good framework to represent affective responses to linguistic (Russell, 1980), pictorial (Bradley & Lang, 1994) and music stimuli (Thayer, 1986). These dimensions are labelled as Arousal and Valence. Arousal corresponds to a subjective state of feeling activated or deactivated. Valence stands for a subjective feeling of pleasantness or unpleasantness (hedonic value; Russell, 1989).

Schubert (1999a) has applied this concept to music creating the Emotion-Space Lab experimental software. While listening to music, participants were asked to continuously rate the emotion 'thought to be expressed' by music.

Each rating would correspond to a point in a two-dimensional emotional space (2DES) formed by the Arousal and Valence dimensions. This approach overcomes some of the drawbacks related to other techniques which do not take into consideration changes in emotion during music (Sloboda, 1991), and it also supports the study of the interaction between quantifiable and meaningful time-varying features in music (psychoacoustic dimensions) and emotion (Arousal and Valence). Several other authors (e.g., Grewe, Nagel, Kopiez, & Altenmuller, 2005; Korhonen, 2004a) have also used this model to find interactions between ratings of Valence and Arousal and the acoustic properties of different pieces. They also permit the representation of a very wide range of emotional states, since they describe a continuous space where each position has no label attached (the only restriction is the meaning of each scale). This is particularly important in the context of musical emotions. Its simplicity in terms of psychological experiments and good reliability (Scherer, 2004) have consistently promoted its use in emotion research.

### 12.2.1.2   Physiological arousal

Currently, neurobiological models of emotion recognise not only the importance of higher neural systems on visceral activity (top-down influences) but also the influences in the opposite direction (bottom-up) (see Berntson, Shafi, Knox, & Sarter, 2003 for an overview). While the top-down influences allow cognitive and emotional states to match the appropriate somato-visceral substrate, the bottom-up ones are suggested to serve to bias emotion and cognition towards a desired state (e.g., guiding behavioural choice, Bechara, Damasio, & Damasio, 2003). Modern conceptualisations propose that a stimulus (appraised via cortical or sub-cortical routes) triggers physiological changes, which in turn facilitates action and expressive behaviour. In this way, together with other components of emotion, physiological activation contributes to the affective feeling state. Moreover, it has been suggested (Dibben, 2004) that individuals may implicitly use their body state as a clue to the Valence and intensity of the emotion they feel.

Recent studies using physiological measurements have provided consistent evidence about the relation between affective states and bodily feelings (e.g., Harrer & Harrer, 1977; Khalfa, Peretz, Blondin, & Manon, 2002; Krumhansl, 1997; Rickard, 2004). Krumhansl (1997) measured the widest spectrum of physiological variables (e.g., spectrum of cardiac, vascular, electrodermal and respiratory functions) and some emotion quality ratings (e.g., happiness, sadness, fear, tension), reported by participants on a second-by-second basis. Krumhansl supports the idea that distinguishable physiological patterns are associated with different emotional judgements. In another series of studies (Witvliet & Vrana, 1996; Witvliet, Vrana, & Webb-Talmadge, 1998), among other variables, researchers investigated the effect of music on skin conductance and HR. Like others (e.g., Iwanaga & Tsukamoto, 1997; Khalfa *et al.*, 2002; Rickard, 2004), these studies have shown that HR and skin conductance response (SCR)[1] increase with arousing or emotionally powerful music.

Another important aspect is to locate the participant's responses within the mechanisms of emotion. Participants are asked to either focus on the experienced or the perceived emotions (Gabrielsson, 2002). In any case, their own feelings of emotion (emotion 'felt'), or the emotion known to be represented by the music, are the reported ratings. In this way the framework focuses on the subjective feelings of emotion, without considering, at least explicitly, other components of the emotional experience. Some studies have shown that these dimensions may also relate to physiological activation. Lang and other colleagues (Greenwald, Cook, & Lang, 1989; Lang, Bradley, & Cuthbert, 1998) have recently compiled a large database indicating that cardiac and electrodermal responses, and facial displays of emotion, show systematic patterns in affect as indexed by the dimensions of Arousal and Valence ('pleasure'). Feldman (1995) also suggests that the conscious affective experience may be associated with a tendency to attend to the internal sensations associated with an affective experience.

Although evidence of an emotion-specific physiology was never found (Ekman & Davidson, 1994; Cacioppo, Berntson, Larsen, Poehlmann, & Ito, 1993), research in peripheral feedback provides evidence that body states can influence the emotional experience with music (Dibben, 2004; Philippot, Chapelle, & Blairy, 2002). Research on emotion has delivered strong evidence that certain patterns of physiological activation are reliable references of the emotional experience. Peripheral feedback has also been considered to be able to change the strength of an emotion even after this has been generated in the brain (Damasio, 1994).

## 12.2.2 Modelling continuous measurements of emotional responses to music

Within a continuous measurements framework, only two models using psychoacoustic variables and time series analysis were proposed (Schubert, 2004; Korhonen, 2004a). In both studies, the authors attempted to model the interaction between psychoacoustic signals and self-report of emotion, taking into account the variations in emotion as a piece of music unfolds in time. Schubert proposed a methodology based on the combinations of time series analysis techniques to analyse the data and to model such processes. Korhonen proposed the use of system identification (Ljung, 1986).

Schubert (2004) applied an ordinary least squares stepwise linear regression model (using sound features as predictors of the emotional response) and a first-order autoregressive model (to account for the autocorrelated residuals and providing the model with a 'memory' of past events) to his experimental data. For each piece, Schubert created a set of models of emotional ratings

---

[1] Measure of the net change in sweat gland cells activity in response to a stimulus or event.

(Arousal and Valence) and selected musical features (melodic pitch, tempo, loudness, frequency spectrum centroid and texture). Each sound feature was also lagged (delayed from the original variable) by 1–4 s. Schubert's assumption was that the emotional response will occur close to or a short time after the causal musical event, and the choice of a 4 s lag was based on preliminary exploration and on other continuous response literature (Krumhansl, 1996; Schubert, 2001; Sloboda & Lehmann, 2001).

The modelling technique used by Schubert suffers from a number of drawbacks. First, it assumes that the relationships between music and emotional ratings are linear. This is a very optimistic view taking into account the nature of the neural processes involved in sound perception. Second, the relationships between sound features and emotional response are considered to be mutually independent. This factor is particularly restrictive, since it discards altogether the interactions between sound features. This is an oversimplification of the relationships between sound features and emotional response, and an acknowledged limitation in music and emotion studies (Gabrielsson & Lindström, 2001). The interactions between variables are a prominent factor in music. As also concluded by Schubert, more sophisticated models that can account for more detailed descriptions of the relationships between the dynamic qualities of music structure and emotions are required to better understand the nature of this process. Another limitation of Schubert's work is lack of prediction of the emotional responses to novel music, since he created separate models for each piece and each affective dimension. Moreover, the relationships found between sound features and affective response for the different piece were piece specific (sometimes even contradictory), compromising the validation of model.

In the second study, Korhonen (2004a) extended these experiments and addressed some of Schubert's study limitations. The sound features space and the musical repertoire were increased, in order to incorporate more music and psychoacoustic (sound) features. The modelling techniques include all the music variables in a single model and the generalisation to new music is also tested. System identification describes a set of mathematical tools and algorithms that create models from measured data. Typically, these models, are either based on predefined (although adjustable) model structures (e.g., state–space models), or with no prior model defined (e.g., neural network). Korhonen considered state–space models, ARX (autoregressive model with exogeneous input) and MISO (multiple input single output) ARX models (the last using delays estimated automatically from the step response (see Korhonen, 2004a for further details). The best model with highest generalisation performance explains 7.8 per cent of Valence and 75.1 per cent of Arousal responses. The performance for Valence is very poor, even though only one piece at a time was used for generalisation. Comparing with Schubert's results, Korhonen's models showed worst performance for some pieces and better for others (particularly worst for Valence). It also used more sound features including tempo, loudness, texture, mean pitch and harmony.

In a follow-up study, Korhonen, Clausi, and Jernigan (2004) also assessed the contribution of feed-forward neural networks[2] with input-delay elements, and state–space models. Each model was again evaluated by the average performance of six models testing the generalisation for each piece (using five of the pieces to estimate the parameters of the model). State–space models were again unsuccessful at modelling Valence (suggesting that linear models may not be appropriate to estimate Valence). The neural network model used improved the model performance only for Valence, explaining 44 per cent of the response. Again, the generalisation is only for one piece at a time. Additionally, the interactions between sound features and the contribution of non-linear models are issues still not addressed.

### 12.2.3   A novel approach: spatio-temporal neural networks

Schubert and Korhonen studies constitute a starting point for our computational investigations. The use of continuous measurements is motivated by the idea that musical emotions may exhibit time-locking to variations in psychological and physiological processes, consistent with a number of studies that show temporal variations in affective responses (e.g., Goldstein, 1980; Nielsen, 1987; Krumhansl, 1997; Schubert, 1999a; Korhonen, 2004a). Because the static attributes of music are only partially responsible or indicative of emotional response to music, which can be intense and momentary (e.g., Dowling & Harwood, 1986), the study of its dynamics in the context of time series analysis needs to be explored. The dynamic aspects of music are perhaps amongst its most important qualities since a piece is characterised by constant changes over time.

In this chapter we propose a novel methodology to analyse the dynamics of emotional responses to music consisting of spatio-temporal neural networks sensitive to structural aspects of music. The computational studies are backed up by experimental data, such as the models that are trained on human data to 'mimic' human affective responses to music and predict new ones. Our hypothesis is that the spatio-temporal patterns of sound convey information about the nature of human affective experiences with music. Our intention is to understand how the psychoacoustic properties of music convey information along two pervasive psychological dimensions of affect: Arousal and Valence.

## 12.3   A spatio-temporal neural network model of musical emotions

The intention of creating a computational model is not only to represent a desired system, but also to achieve a more refined understanding of its underpinnings. In that way, the knowledge to be extracted from the model should

---

[2] Type of artificial neural network in which the information moves in only one direction: forward – from the input nodes, through possible hidden nodes and then to the output nodes. There are no cycles or loops involved in the information processing of the network.

reveal information about the nature of the underlying mechanisms, permitting to generate new hypotheses and to make predictions in novel scenarios. In this context, we will consider the contribution of spatio-temporal connectionist models, i.e., models that include both a temporal dimension (e.g., the dynamics of musical sequences and continuous emotional ratings) and a spatial component (e.g., the parallel contribution of various music and psychoacoustic factors), to model continuous measurements of musical emotions.

A spatio-temporal connectionist model can be defined as 'a parallel distributed information processing structure that is capable of dealing with input data presented across time as well as space' (Kremer, 2001, p. 2). Similar to conventional connectionist models, spatio-temporal connectionist networks (STCN) are equipped with memory in the form of connection weights. These are represented as one or more matrices, depending on the number of layers of connections in the network. This memory extends back past the current input pattern to all the previous input patterns. Accordingly, it is usually referred to as long-term memory. Once a connectionist network has been successfully trained, it remains fixed during the operation of the network. A distinctive characteristic of STCNs is the inclusion of a short-term memory. This memory that allows STCNs to deal with input and output patterns that vary across time as well as space. While conventional connectionist networks compute the activation values of all nodes at time $t$ based only on the input at the current time step, in STCNs the activations of some of these nodes is computed based on previous activations. Unlike the weights (long-term memory) which, as explained, remain static once the training period is completed, the short-term memory is continually re-computed with each new input vector in both training and operation.

An important class of STCNs are recurrent neural networks, a type of artificial neural networks which propagate data from input to output (like feed-forward neural networks), but also the data from later processing stages to earlier stages. These models involve various forms of recurrence (feedback connections), and through these some of the information at each time step is kept as part of the input to the following computational cycle. By allowing feedback connections, the network topology becomes more flexible, and it is possible to connect any unit to any other, including to itself. Due to this flexibility, various proposals and architectures for time-based neural networks (see Kremer, 2001 for a review) making use of recurrent connections in different contexts have been proposed. Examples of these models are the Jordan Network (Jordan, 1990) and the Elman neural network (ENN) (Elman, 1990). Jordan and Elman neural networks are extensions of the multilayer perceptron, with additional context units, which 'remember' past activity. These units are required when learning patterns over time (i.e., when the past computations of the network influence the present processing).

The approach taken by Jordan (1990) involves treating the network as a simple dynamic system in which previous states are made available to the system as an additional input. During training, the network state is a function of the input of the current time step, plus the state of the output units of the previous time step. By contrast, in the Elman network, the network's state depends on the current inputs, plus the model's internal state (the hidden units activations) of the previous cycle. This is achieved through an additional set of units (memory or context units), which provide (limited) recurrence. These units are activated on a one-for-one basis by the hidden units, copying their values: at each time step the hidden units activation are copied into the context units. In the following cycle, the context values are combined with the new inputs to activate the hidden units. The hidden units map the new inputs and prior states to the output. Because the hidden units are not trained to respond with specific activation values, they can develop representations in the course of learning. These encode the temporal structure of the data flow in the system: they become a 'task-specific' memory (Elman, 1990). Because they themselves constitute the prior state, they must develop representations that facilitate this input–output mapping. Recurrent neural networks have been extensively used in tasks where the network is presented with a time series of inputs, and are required to produce an output based on this series. Some of the applications of these models are the learning of formal grammars (e.g., Lawrence, Giles, & Fong, 2000; Elman, 1990), spoken word recognition (McClelland & Elman, 1986), written word recognition (Rumelhart & McClelland, 1986), speech production (Dell, 2002) and music composition (e.g., Mozer, 1999).

For our computational studies, the ENN was chosen. The basic functional assumption of this connectionist model is that the next element in a time series sequence can be predicted by accessing a compressed representation of previous hidden states of the network and the current inputs. If the process being learned requires that the current output depends somehow on prior inputs, then the network will need to 'learn' to develop internal representations which are sensitive to the temporal structure of the inputs. During learning, the hidden units must accomplish an input–output mapping and simultaneously develop representations that systematic encoding of the temporal properties of the sequential input at different levels (Elman, 1990). In this way, the internal representations that drive the outputs are sensitive to the temporal context of the task (even though the effect of time is implicit). The recursive nature of these representations (acting as an input at each time step) endows the network with the capability of detecting time relationships of sequences of features, or combinations of features, at different time lags (Elman, 1991). This is an important feature of this network because the lag between music and affective events has been consistently shown to vary in the order of 1–5 s (Schubert, 2004; Krumhansl, 1996; Sloboda & Lehmann, 2001). Another important aspect is that ENNs have very good generalisation capabilities. This technique has been extensively applied in areas such as language (e.g., Elman, 1990) and financial forecasting systems (e.g., Giles, Lawrence, & Tsoi, 2001), among others.

The remaining of this chapter describes two computational studies. The first study ('Model 1') describes a new neural network model of musical emotions, which accounts for the subjective feeling component of emotions. The neural network is trained to predict affective responses to music, based on a set of psychoacoustic components extracted from the music stimuli. In the second study ('Model 2'), we develop a new computational investigation for the extension of 'Model 1', to include physiological cues. The aim is to verify if physiological activity has meaningful relationships with the affective response. The final part of this chapter summarises the research, and discusses its implications and contributions to the field of music perception, emotion and cognition.

### 12.3.1   Modelling subjective feelings

This simulation experiment will consist of the training of an ENN to learn to predict the subjective feelings of emotion from the input of musical excerpts. Following the modelling stage and recurring to a set of analytical techniques, a special emphasis will be put on the study of the relationships between sound features and affective responses (as represented by the model). The experimental data was obtained from a study conducted by Korhonen (2004a), which includes the emotional appraisals of six selections of Western Art ('classical') music (see Table 12.1), obtained from 35 participants (21 male and 14 female). Using a continuous measurement framework, emotion was represented by its Arousal and Valence dimensions, using the EmotionSpace Lab (Schubert, 1999b). The emotional appraisal data was collected at 1 Hz (second-by-second). The music excerpts are encoded as sound (psychoacoustic) features, and emotions are represented by its Arousal and Valence components. The data is available online (Korhonen, 2004b), courtesy of the author.

#### 12.3.1.1   Psychoacoustic encoding (model input data)

To encode the music as time-varying variables reflecting the variations occurring in the main perceptual dimensions of sound, each piece was encoded into the psychoacoustic space. For this simulation experiment we used the psychoacoustic data extracted from the stimulus material using Marsyas

Table 12.1   *Pieces used in Korhonen's experiment and their aliases for reference in this chapter. The pieces were taken from Naxos's* Discover the Classics *CD 8.550035-36*

| Piece ID | Title and composer | Duration (s) |
|---|---|---|
| 1 | Concierto de Aranjuez (II. Adagio) – J. Rodrigo | 165 |
| 2 | Fanfare for the Common Man – A. Copland | 170 |
| 3 | Moonlight Sonata (I. Adagio Sostenuto) – L. Beethoven | 153 |
| 4 | Peer Gynt Suite No. 1 (I. Morning mood) – E. Grieg | 164 |
| 5 | Pizzicato Polka – J. Strauss | 151 |
| 6 | Piano Concerto No. 1 (I. Allegro maestoso) – F. Liszt | 315 |

(Tzanetakis & Cook, 2000) and PsySound (Cabrera, 1999) software packages.[3] We will use the same time series data sets used by Korhonen (2004a).

Since Korhonen used several variables to quantify the same psychoacoustic dimensions, we optimised the psychoacoustic data from an initial set of 13 variables[4] to a final set of 6 variables (see Coutinho & Cangelosi, 2009 for further details on a previous study). This set includes six major psychoacoustic groups related with the perception of sound:

1. Dynamics: The loudness level (L) represents the subjective impression of the intensity of a sound (measured in sones), as described in Cabrera (1999), and is used to represent dynamics.
2. Mean Pitch: The power spectrum centroid (P) represents the first moment of the power spectral density (PSD) (Cabrera, 1999), and is used to quantify the mean pitch.
3. Pitch Variation (contour): The mean STFT flux (Pv) corresponds to the Euclidian norm of the difference between the magnitude of the short-time Fourier transform (STFT) spectrum evaluated at two successive sound frames[5] (refer to Tzanetakis & Cook, 2000 for further details).
4. Timbre: Timbre is represented using a sharpness (S), a measure of the weighted centroids of the specific loudness, which approximates the subjective experience of a sound on a scale from dull to sharp (the unit of sharpness is the acum[6]). Details on the algorithm can be found in Zwicker and Fastl (1990) (implementation in PsySound).
5. Tempo: The pace of the music was estimated from the number of beats per minute. Because the beats were detected manually, a linear interpolation between beats was used to transform the data into second-by-second values (details on the tempo estimation are described in Schubert, 1999a).
6. Texture: Multiplicity (Tx) is an estimate of the number of tones simultaneously noticed in a sound; this measure was determined using Parncutt's algorithm (as described in Parncutt, 1989, p. 92) included in PsySound.

## 12.3.1.2 Experimental data on subjective feelings (model output data)

Korhonen (2004a) used the EmotionSpace Lab (Schubert, 1999a) to quantify emotion on the dimensions Arousal and Valence. While listening to each of the pieces, participants' emotional appraisal (i.e., pairs of Arousal and Valence

---

[3] Only tempo was calculated manually, using Schubert's (11999a) method (see pp. 274–277).
[4] Although Korhonen used 18 variables, the five sound features representing Harmony variable included in that study are not included here in order to exclude higher level features specific to the music culture and with controversial methods for its quantification.
[5] Although this algorithm is not a specific measure of melodic contour, it has been successfully used as such in music information retrieval applications (Korhonen, 2004a). Nevertheless, in this chapter we refer to this variable as pitch variation because it characterises better the nature of the encoding. Moreover, the relationships between pitch variations and emotion were also the object of some studies (e.g., Scherer & Oshinsky, 1977), as described in Schubert (1999a).
[6] One acum is defined as the sharpness of a band of noise centred on 1 000 Hz, 1 critical-bandwidth wide, with a sound pressure level of 60 dB.

values) was collected at 1 Hz. The use of a sampling rate of 1 Hz follows evidence showing that affective events in response to music consistently follow the stimulus between 1 and 5 s (Schubert, 2004; Krumhansl, 1996; Sloboda & Lehmann, 2001). The assumption is that the emotional response will occur close to or a short time after the causal musical event; therefore, the time series collected are expected to reflect changes in the participants' emotional appraisal following musical events. For this simulation experiment, the average Arousal/Valence values (for each second of music) over the 37 subjects were used to train the network.

### 12.3.1.3    Simulation procedure

The sound features constitute the input to the model. Each of these variables corresponds to a single input node of the network. The output layer consists of two nodes representing Arousal and Valence. Three pieces of music (1, 2 and 5), corresponding to 486 s, were used during the training phase. To evaluate the response to novel stimuli, the remaining three pieces were used: 3, 4 and 6 (632 s of music). Throughout this chapter, the collection of stimuli used to train the model will be referred to as 'Training set', and 'Test set' to the novel stimuli, unknown to the system during training, that test its generalisation capabilities and performance. The pieces were distributed among the sets in order to cover the widest range of values of the 2DES in both sets. The rationale behind the procedure is the fact that it is necessary to train the model with widest range of values possible in order to be able to predict the emotional responses to a diverse set of novel pieces.

The task at each training iteration ($t$) is to predict the next ($t + 1$) values of Arousal and Valence.[7] The target values (aka 'teaching input') are the average Arousal/Valence pairs across all participants in Korhonen's experiments. To adapt the range of values of each variable to be used with the network, all variables were normalised to a range between 0 and 1. The learning process was implemented using a standard back-propagation technique (Rumelhart, Hintont, & Williams, 1986). During training the same learning rate and momentum were used for each of the three connection matrices. The network weights were initialised with different random values. The range of values for each connection in the network (except for the connections from the hidden to the memory layer which are set constant to 1.0) was defined randomly between $-0.05$ and $0.05$.

If the model is also able to respond with low error to novel stimuli, then the training algorithm was able to extract some general rules from the training set that relate musical features to emotional ratings. The maximum number of training iterations and the values of the learning parameters were estimated in order to avoid the over-fitting of the training set data.[8] After preliminary tests

---

[7] In the context of modelling procedure, the term iteration is used to refer to a full input–output activation spreading of the neural network.

[8] In order for the model to generalise, it must not be built around the minimisation of the error in the training data. The ideal point is a compromise between the output errors for both training and test data.

and analysis, the duration of the training was set at 20 000 iterations, using a learning rate of 0.075 and a momentum of 0.0. Testing the model with different numbers of hidden nodes also permitted the optimisation of the size of the hidden layer, which defines the dimensionality of the internal space of representations. The best performance was obtained with a hidden layer of size five (see Coutinho & Cangelosi, 2009 for further details).

### 12.3.1.4   Results

After the set of preliminary simulations described in the previous paragraphs, 37 neural networks (the same as the number of participants in Korhonen experiments) were trained using the network configuration shown in Figure 12.1. The average error (for both outputs) of the 37 networks was 0.050 for the Training set, and 0.076 for the Test set. These values were produced after 20 000 iterations of the training algorithm.

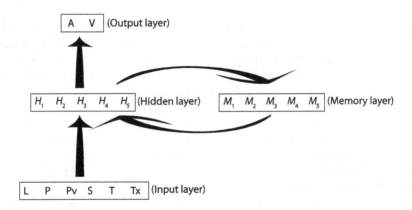

*Figure 12.1   Neural network architecture and units identification for 'Model 1'. Inputs (sound features): loudness level (L), mean pitch (P), pitch variation (Pv), timbre (S), tempo (T), texture (Tx). Outputs: Arousal (A) and Valence (V). The hidden and memory layers contain five artificial neurones: $H_1$–$H_5$ and $M_1$–$M_5$, respectively*

To compare the model output with the experimental data for each piece, the root mean square error (*rmse*) and the linear correlation coefficient (*r*) were used to describe the deviation and similarity between the model outputs and the experimental data. The following analysis will report on the network that showed the lowest average error for both data sets. The *rmse* error and *r* of each output for all the pieces are shown in Table 12.2.

The model performance for Arousal was better for pieces 1, 2, 5 and 6, as shown by the low *rmse* errors: $rmse_1(Arousal) = 0.052$, $rmse_2(Arousal) = 0.040$, $rmse_5(Arousal) = 0.044$ and $rmse_6(Arousal) = 0.052$ (all four values are lower than the mean Arousal for all pieces: $rmse_{all} = 0.056$). Conversely, these same

*Table 12.2    Comparison between the model outputs and experimental data: root mean square (rmse) error and linear correlation coefficient (r) (\*p < 0.0001, \*\*p < 0.001, \*\*\*p < 0.02)*

| Piece ID | rmse | | r | | Set |
|---|---|---|---|---|---|
| | A | V | A | V | |
| 1 | 0.052 | 0.044 | 0.964* | 0.760* | Train |
| 2 | 0.040 | 0.054 | 0.778* | 0.939* | Train |
| 3 | 0.061 | 0.045 | 0.278** | 0.206*** | Test |
| 4 | 0.085 | 0.081 | 0.797* | 0.040 | Test |
| 5 | 0.044 | 0.046 | 0.768* | 0.583* | Train |
| 6 | 0.052 | 0.082 | 0.958* | 0.650* | Test |
| Mean | 0.056 | 0.059 | | | |

pieces show a high $r$ for the Arousal output: $r_1(Arousal) = 0.964$, $r_2(Arousal) = 0.778$, $r_5(Arousal) = 0.583$ and $r_6(Arousal) = 0.650$. Only pieces 3 and 4 had a higher *rmse* error than the mean of all the remaining pieces. Nevertheless, while the model Arousal output for piece 3 shows a low linear correlations with the experimental data, piece 4 achieves a coefficient comparable with the remaining pieces.

Regarding the Valence output, all except pieces 4 and 6 showed an *rmse* above the average of all pieces ($rmse_4(Valence) = 0.081$, $rmse_6(Valence) = 0.082$, $rmse_{all}(Valence) = 0.059$), within a range of values similar to the Arousal output. When looking at the correlations between model and experimental data, only piece 4 shows no significant correlation between the model output and experimental data (the only non-significant correlation), since piece 6 has a significant $r$ of 0.650.

Figure 12.2 shows the Arousal and Valence outputs of the model for two sample pieces, one belonging to the Training set (a) and another to the Test set (b), versus the data obtained experimentally (target values). The model was able to track the general fluctuations in Arousal and Valence for both data sets, whilst the performance varied from piece to piece. The overall successful predictions of the affective dimensions for both known and novel music support the idea that music features contain relevant relationships with emotional appraisals. A visual inspection of the model outputs (for the visualisation of the remaining pieces, please refer to Coutinho & Cangelosi, 2009), confirmed by the *rmse* and *r* measures, also indicates that the model output resembles the experimental data. Furthermore, although the linear correlation coefficients are low for some of the pieces, it can be shown that the model responses to both data sets approximates to a high degree the experimental by means of non-linear measures of similarity between time series (see Coutinho & Cangelosi, 2009 for further details).

It is important to note that the number of pieces used in this simulation experiment is not as relevant as their total length and areas of the Arousal/

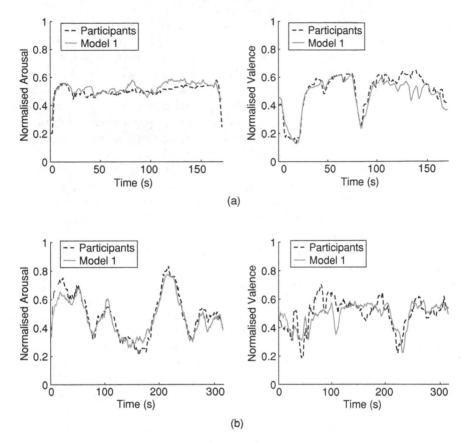

*Figure 12.2    Arousal and Valence model outputs compared to experimental for pieces 2 and 6. (a) Piece 2 (Copland, Fanfare for the Common Man) belongs to the training set and (b) Piece 6 (Liszt, Piano Concerto No. 1) which belongs to the test set*

Valence space covered. Although a set of six pieces seems to be a small sample set, they correspond to a diverse sample of psychoacoustic features and emotional appraisals. The pieces chosen by Korhonen are heterogeneous in terms of instrumentation and style within the gender chosen, i.e., within each piece there is a wide variety of psychoacoustic patterns that vary regularly in the selected pieces throughout the duration of the experiment (for instance, tempo ranges from 0 to 220 bpm, loudness from 0 to 90 phons), and they correspond to a wide range of emotions expressed. Although more pieces could be used with our model, practical limitations related to the maximum time considerable acceptable for listening to music for experimental studies, limit the total length of the pieces to be used to approximately 20 minutes (Madsen, Brittin, & Capperella-Sheldon, 1993). The total length of the pieces in korhonen's experiment was approximately 19 minutes.

## 12.3.1.5   Model analysis

An important observation drawn from the results obtained is the fact that the spatio-temporal relationships learned from the Training set were successfully applied to a new set of stimuli. These relationships are encoded in the network weights, and the flux of information in the internal (hidden) layer of the neural network represents the dynamics of the internal categorisation (or recombination) of the input stimuli, that enables output predictions. One of the advantages of working with an artificial neural network is the possibility of exploring its internal mechanisms, which generate the behaviour and indirectly show how the model processes the information. Aiming at the identification of these relationships between the sound features and the model's predictions it is necessary to identify how the hidden units process the inputs into the outputs. With that information it may be possible to estimate the input–output transformations of the model.

One possibility is to inspect the weights matrices in the model and identify the highest weights. Although simple, this methodology focuses on the long-term memory of the model, which totally discards the dynamics of the model: the temporal structure of the data flow in the system, which, as discussed, behaves as a 'task-specific' memory. Instead, in order to investigate the model dynamics, we analysed the temporal correlations between inputs, hidden units and outputs, in order to obtain the overall relationships between groups of sound features (inputs), hidden units and outputs (Arousal and Valence). To do so, the correlations between inputs, hidden and output units were computed using a canonical correlation analysis (CCA) (Hotelling, 1936), a statistical procedure of analysing linear relationships between two multidimensional variables. A canonical correlation is the correlation of two canonical variables: one representing a set of independent variables, the other a set of dependent variables. The CCA optimises the linear correlation between the two canonical variables to be maximised in the context of many-to-many relationships. There may be more than one linear correlation relating the two sets of variables, each representing a different dimension of the relationship, which explain the relation between them. For each dimension it is also possible to assess how strongly it relates each variable in its own set (canonical factor loadings). These are the correlations between the canonical variables and each variable in the original data sets.

A CCA was performed to assess the relationships between the sequences of input, hidden and output layers activity. In this way it is possible to analyse the contribution of each network layer node or (sets of nodes) to the activity of a different layer and the relationships between input and hidden layers (how the inputs relate with the internal representations of the model), and these with the outputs (which sets of hidden units are more related to the output). In Table 12.3 we show the details of the CCA on the activity of the neural network layers (i.e., the sequence of activations of the neural network hidden layer in response to the changing sound input). The bigger the loading, the stronger the relationships between the original variables (input, hidden, and output units' activity) and the canonical variates.

Considering the relationships between input and hidden layers activities (see left side of Table 12.3), we found that three canonical variables explain 98.3 per cent of the variance in the data. These three dimensions encode the general levels of shared activation in the input and hidden layers. The first pair of variables loads on P, Tx, Ti (inputs set), $H_2$ and $H_5$ (hidden layer). The second, loads only on input D, but it loads on all nodes of the hidden layer, while the third canonical variable loads on C, $H_2$ and $H_4$. The CCA of the hidden and output layers activity has shown instead two canonical variables, which explain all the variance in the data (see right side of Table 12.3). The first root is correlated strongly with Arousal, and the activity in hidden units $H_1$ and $H_2$. The second pair of canonical variables correlates with both Valence (positive) and Arousal (negative), and with the activity in units $H_3$ to $H_5$.

*Table 12.3* *Canonical correlation analysis (CCA): the canonical correlations (the canonical correlations are interpreted in the same way as the Pearson's linear correlation coefficient) quantify the strength of relationships between the extracted canonical variates, and so the significance of the relationship. To assess the relationship between the original variables (input, hidden and output units activity) and the canonical variables, the canonical loadings (the correlations between the canonical variates and the variables in each set) are also included*

| Loadings (input/hidden) | | | | Loadings (input/hidden) | | |
|---|---|---|---|---|---|---|
| Variable | Variate 1 | Variate 2 | Variate 3 | Variable | Variate 1 | Variate 2 |
| $H_1$ | −0.398 | −**0.633** | −0.028 | $H_1$ | −**0.504** | 0.482 |
| $H_2$ | **0.479** | **0.657** | −0.437 | $H_2$ | **0.978** | −0.055 |
| $H_3$ | 0.144 | −**0.891** | −0.238 | $H_3$ | −0.291 | **0.862** |
| $H_4$ | 0.159 | −**0.647** | −0.632 | $H_4$ | 0.014 | **0.797** |
| $H_5$ | −**0.637** | **0.645** | 0.018 | $H_5$ | −0.074 | −**0.973** |
| L | 0.450 | **0.674** | 0.139 | A | **0.765** | −0.644 |
| P | **0.819** | 0.297 | 0.432 | V | 0.260 | 0.966 |
| Pv | 0.187 | 0.270 | 0.825 | | | |
| S | **0.748** | 0.420 | 0.262 | | | |
| T | 0.264 | 0.478 | 0.151 | | | |
| Tx | **0.608** | 0.280 | 0.217 | | | |
| Canonical correlation | **0.725** | **0.546** | 0.488 | Canonical correlation | **0.987** | **0.984** |
| Percentage | **61.1%** | **23.4%** | 13.8% | Percentage | **66.0%** | **44.0%** |

Bringing both analyses together it is possible to describe qualitatively how the network inputs are propagated through the hidden layer to the network's outputs. By combining the observations obtained from the two previous

CCA's we can approximate the general strategies for input–output (sound features–affective dimensions) mapping:

1. Dynamics (loudness level – L): Higher loudness relates to increased Arousal and decreased Valence.
2. Mean Pitch (power spectrum centroid – P): The highest pitch passages relate to higher Arousal and Valence (quadrants 1, 2 and 4).
3. Pitch Variation (mean STFT flux – Pv): The average spectral variations related positively to Arousal, e.g., large pitch changes are accompanied by increased activation. The pitch changes have both positive and negative effects on Valence, suggesting more complex interactions.
4. Timbre (sharpness – S): Sharpness induced increased Arousal and Valence.
5. Tempo (bpm – T): Fast tempi are related to high Arousal (quadrants 1 and 2), and positive Valence (quadrants 1 and 4). Slow tempi exhibit the opposite pattern.
6. Texture (multiplicity – Tx): Thicker textures have positive relationships with Arousal (quadrants 1 and 2).

### 12.3.1.6  Summary

By considering that music can elicit affective experiences in the listener, we have focused on the sound features as a source of information about this process. We presented and tested a novel methodology to study affective experience with music. By focusing on continuous measurements of emotion, the choice of the modelling technique considered two important aspects of music perception: interactions between different sound features and their temporal behaviour/organisation. We proposed and tested recurrent neural networks as a possible solution due to their adaptiveness to deal with spatio-temporal patterns, using Korhonen's (2004a) experimental data.

Overall, this computational study presented provides a computational framework capable of inferring the affective value of music from structural features of the perceptual (auditory) experience. The neural network model used focused in detecting temporal structure in the perceptual dimensions of dynamics, tempo, texture, timbre and pitch, rather than only on delays between music events and affective responses, as in previous studies (Schubert, 1999a; Korhonen, 2004a). Moreover, it integrates all variables in a single computational model, is able to detect structure in sound – either as interactions among its dimensions or their temporal behaviour – and to predict affective responses from its form. Along with the development of the model, there was a strong emphasis on the generalisation for novel music. A good generalisation performance indicates a potentially valid model and a good platform to analyse the relationships between sound features and affective responses (as encoded by the model). At the analysis level, modelling and experimental results provide evidence suggesting that spatio-temporal patterns of sound resonate with affective features underlying judgements of subjective feelings. A significant part of the listener's affective response is predicted from six psychoacoustic features of sound – tempo, loudness, multiplicity (texture), power spectrum centroid (mean pitch), sharpness (timbre) and mean STFT flux (pitch variation).

## 12.3.2 The role of physiological arousal

Research on emotion has delivered strong evidence that certain patterns of physiological activation are reliable references of the emotional experience. Modern conceptualisations propose that a stimulus (appraised via cortical or subcortical routes) may trigger physiological changes, which in turn facilitate action and expressive behaviour. In this way, together with other components of emotion, physiological activation contributes to the affective feeling state. For instance, physiological Arousal has also been related with psychological representations and determinants of emotion, such as Valence (or hedonic value) and Arousal (Lang *et al.*, 1998). Implicitly individuals may use their body state as a clue to the Valence and intensity of the emotion they feel (Dibben, 2004).

In this section we develop a new simulation experiment which extends the model presented in the previous section to physiological Arousal. Using the same modelling paradigm described in the previous section, we now describe another neural network model that includes not only sound features as inputs, but also physiological variables. We use the data from a new experiment (Coutinho, 2008), which, apart from the self-report data, includes two physiological measures: HR and skin conductance level (both quantities are amongst the most common measures of physiological activation in laboratory-controlled experiments of music perception). Our intention is to test if the subjective feelings of emotion relate not only with the psychoacoustic patterns of music (as shown in the previous experiment), but also with physiological activation. The music pieces used in that experiment are shown in Table 12.4.

In the experimental study it was hypothesised that music can alter both the physiological component and the subjective feeling of emotion in response to music, sometimes in highly synchronised ways. The analysis of the experiment has shown that loudness, tempo, mean pitch and sharpness have a positive relationship with both psychological Arousal and Valence (a stronger effect was observed on the first) in such a way that an increase in those sound features is reflected by an increase in both dimensions of emotion. Regarding the physiological recordings, only the HR showed statistically significant relationships to the Arousal dimension of emotion. Further analysis of the synchronisation between 'peak' changes in the physiological and subjective feeling reports confirmed that result and revealed another: almost half of the strong changes in HR were followed by changes in Arousal and Valence – 43 per cent of the strong changes in HR preceded (up to 4 s) strong changes in the subjective feeling component.

In the analysis of the same experiment it was also shown that there are relevant interactions between the psychological and physiological components of emotion. A higher differentiation on the general levels of Arousal and Valence was obtained when using the physiological dimensions together with psychoacoustic variables, when compared with only psychoacoustic variables.

*Table 12.4    Pieces used in the experimental study (Coutinho, 2008). The stimu-*
*lus materials consisted of nine pieces chosen by two professional*
*musicians (one composer and one performer, other than the author),*
*to illustrate the combination of Arousal and Valence, and to cover*
*the widest area of the 2DES possible (combinations of Arousal and*
*Valence values). The pieces were chosen so as to be from the same*
*musical genre, classical music, a style familiar to participants, and to*
*be diverse within the style chosen in terms of instrumentation and*
*texture*

| Piece ID | Alias | Composer and title | Duration (s) |
|---|---|---|---|
| 1 | Adagio | T. Albinoni – Adagio | 200 |
| 2 | Grieg | E. Grieg – Peer Gynt Suite No.1 (IV. 'In the Hall of the Mountain King') | 135 |
| 3 | Prelude | J. S. Bach – Prelude No. 15 (BWV 860) | 43 |
| 4 | Romance | L. V. Beethoven – Romance No. 2 (Op. 50) | 123 |
| 5 | Nocturne | F. Chopin – Nocturne No. 2 (Op. 9) | 157 |
| 6 | Divertimento | W. A. Mozart – Divertimento (K. 137) (II. 'Allegro di molto') | 155 |
| 7 | La Mer | C. Debussy – La Mer (II. 'Jeux de vagues') | 184 |
| 8 | Liebestraum | F. Liszt – Liebestraum No. 3 (S. 541) | 183 |
| 9 | Chaconne | J. S. Bach – Partita No. 2 (BWV 1004) | 240 |

In 81.5 per cent of the test cases, combinations of the sound features resulted in a successful classification of the general affective value (2DES quadrant) of each segment (the pieces were divided into segments with similar affective values and each variable averaged over each segment). By combining sound features and physiological variables, the rate increased to 92.6 per cent. This improvement suggests that physiological cues combined with sound features, give a better description of the self-report dynamics.

### 12.3.2.1    Psychoacoustic encoding (model input data)
The sound features that encode the music as time-varying variables corresponds to the same set used in the previous section: loudness level (dynamics), power spectrum centroid (mean pitch), mean STFT flux (pitch variation), sharpness (timbre), beats-per-minute (tempo) and multiplicity (texture) (see page 343).

### 12.3.2.2    Subjective feelings (model output data)
Participants reported their emotional state by using the EMuJoy software (Nagel, Kopiez, Grewe, & Altenmüller, 2007), a computer representation of a 2DES. The physiological measures were obtained using a WaveRider bio-feedback system (MindPeak, USA). Leads were attached to the subjects for measuring HR and skin conductance level. The self-report data was later synchronised with physiological data.

### 12.3.2.3   Simulation procedure

The simulation methodology for this model is similar to the one presented in the previous section (we will refer to this model as 'Model 2'). We consider the sound features (loudness level, mean pitch, pitch variation, sharpness, tempo and multiplicity) and physiological cues (HR and SCR) as inputs for the model. The self-report of emotion (Arousal and Valence) act as the outputs. The aim is to predict subjective feelings of emotion reported by participants from the dynamics of psychoacoustic and physiological patterns in response to each piece. In relation to the previous simulation experiment ('Model 1') we want to investigate the specific contribution of the physiological input to subjective feeling response (peripheral feedback).

The 'training set' includes five of the pieces used in the experiment (pieces 1, 4, 5, 6 and 8). The 'test set' includes the remaining four pieces (pieces 2, 3, 7 and 9). The pieces were distributed between both sets in order to cover the widest range of values of the emotional space. The logical basis behind this decision is the fact that, for the model to be able to predict the emotional responses to novel pieces in an ideal condition, it is necessary that it must have been exposed to widest range of values possible.

At each training iteration, the task ($t$) is to predict the next ($t + 1$) values of Arousal and Valence. The 'teaching input' (or target values) are the average A/V pairs obtained experimentally. The range of values for each variable (sound features, self-report and physiological variables) was normalised to the range between 0 and 1 in order to be used with the model. The learning process was again implemented using the standard back-propagation technique (Rumelhart *et al.*, 1986).

For each replication of the simulations, the network weights were initialised with different values randomly distributed between $-0.05$ and $0.05$ (except for the connections from the hidden to the memory layer which are set constant to 1.0). Each trial consisted of 80 000 iterations of the learning algorithm. The training stop point was estimated *a posteriori* by calculating the number of training iterations which minimise the model outputs error for both training and test sets. This is a fundamental step to avoid the over-fitting of the training set. During training, the same learning rate and momentum were used for each of the three connection matrices. The learning rate was set to 0.075 and the momentum to 0.0 for all trials. The *rmse* error will be used to quantify the deviation of the model outputs from the values observed experimentally. The model performance will also be assessed with the linear ($r$) correlation coefficient.

### 12.3.2.4   Results

The initial simulations experiments aimed at testing the model performance with the inclusion of physiological cues (see Table 12.5). In a set of four simulations, we tested the model with only the sound features ($M_2I_1$ – a model with an architecture similar to 'Model 1'), and then with the addition of HR ($M_2I_2$), SCR ($M_2I_3$) and both ($M_2I_4$), as inputs to the model.

Table 12.5    *'Model 2': simulations with different combinations of inputs (average rmse errors over all trials). Loudness level, L; mean pitch, P; pitch variation, Pv; timbre, S; tempo, T; texture, Tx*

| Sim. ID | Inputs | Training set | | Test set | | Average |
|---------|--------|:---:|:---:|:---:|:---:|:---:|
| | | A | V | A | V | |
| $M_2I_1$ | L, T, P, Pv, Tx, S | 0.070 | 0.060 | 0.087 | 0.078 | 0.075 |
| $M_2I_2$ | L, T, P, Pv, Tx, S + HR | 0.077 | **0.057** | **0.082** | 0.078 | **0.073** |
| $M_2I_3$ | L, T, P, Pv, Tx, S + SCR | 0.082 | 0.067 | 0.089 | 0.076 | 0.079 |
| $M_2I_4$ | L, T, P, Pv, Tx, S + HR, SCR | **0.070** | 0.060 | 0.087 | **0.075** | **0.073** |

The best average performances were achieved with models $M_2I_2$ (sound features + extra HR input) and $M_2I_4$ (sound features + extra HR and SCR inputs). Although very similar, the first simulation had a lower error for the test set ($rmse(\text{HR}) = 0.080$ and $rmse(\text{HR,SCR}) = 0.081$ – these values correspond to the mean $rmse$ errors for both outputs of the Test data set). Because the addition of SCR alone does not have a positive impact on the model performance (note that $M_2I_1$ has a better performance than $M_2I_3$), the set of inputs in simulation $M_2I_2$ was selected for the final configuration of Model 2. This means that the HR variations may contain significant information for the self-report of emotion in music, since its inclusion as an input has a positive influence on the model performance. This result is consistent with the preliminary analysis of the experimental data reported by Coutinho (2008), where it was shown that HR significative changes also preceded significative changes in the subjective feeling component (see also page 351).

The following analysis is conducted on the model the input layer of which includes the HR input plus the six key sound features included in 'Model 1': loudness level (L), mean pitch (P), pitch variation (Pv), timbre (S), tempo (T) and texture (Tx). After retesting the hidden and memory layers size with different number of nodes, five units was found again to convey the best performance. Figure 12.3 shows the final model architecture.

Table 12.6 shows the *rmse* error and the linear correlation coefficient ($r$) to describe the deviation and similarity between the model outputs and experimental for the best trial of simulation $M_2I_2$. The values are shown for each piece separately (1–9), together with the averaged values (mean) across all pieces.

The model responded with low *rmse* error for both variables and almost all pieces ($rmse_{mean}(Arousal) = 0.069$, $rmse_{mean}(Valence) = 0.050$). The exceptions are pieces 1, 7 and 8, with an error higher than 0.8 for Arousal predictions. While for piece 1 the model output also does not show a significant correlation with the experimental data, pieces 7 and 8 have a high linear correlation coefficient ($r_7(Arousal) = 0.775$ and $r_8(Arousal) = 0.880$). Regarding the Valence output, all pieces show a low *rmse* error, despite the fact that pieces 7

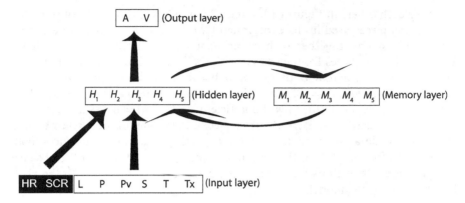

Figure 12.3  *Neural network architecture and units identification for 'Model 2'.*
*Inputs (physiological variables and sound features): heart rate*
*(HR), loudness level (L), mean pitch (P), pitch variation (Pv),*
*timbre (S), tempo (T) and texture (Tx). Outputs: Arousal (A)*
*and Valence (V)*

Table 12.6  *Model 2: rmse errors and r coefficient, per variable, for each music*
*piece for best trial of simulation $M_2I_2$ (\*p < 0.0001)*

| Piece ID | rmse | | r | |
|---|---|---|---|---|
| | A | V | A | V |
| 1 | 0.089 | 0.034 | 0.065 | 0.550* |
| 2 | 0.036 | 0.026 | 0.939* | 0.834* |
| 3 | 0.072 | 0.051 | 0.869* | 0.949* |
| 4 | 0.047 | 0.026 | 0.884* | 0.873* |
| 5 | 0.063 | 0.079 | 0.605* | 0.944* |
| 6 | 0.042 | 0.060 | 0.805* | 0.892* |
| 7 | 0.102 | 0.074 | 0.815* | 0.011 |
| 8 | 0.093 | 0.027 | 0.770* | 0.627* |
| 9 | 0.079 | 0.070 | 0.813* | 0.166 |
| Mean | 0.069 | 0.050 | | |

and 9 do not have a significant linear correlation between model output and
experimental data ($r_7(Valence) = 0.011$ and $r_9(Valence) = 0.166$).

Despite some of the higher *rmse* errors and lack of significant linear correlations for the mentioned pieces and outputs, an inspection of the model
outputs versus the experimental data shows nevertheless that the time series

resemble each other. In Figures 12.4 and 12.5 we show the model's outputs for some of the pieces used in the experiment (two from the Training set and two from the Test set), together with experimental data (target outputs). Figure 12.4 includes the pieces from each set with the highest linear correlation coefficients (averaged across both outputs) between model output and experimental data $(r_4(mean) = 0.879$ and $r_2(mean) = 0.887)$, while Figure 12.5 contains the pieces with the lowest coefficients $(r_1(mean) = 0.308$ and $r_7(mean) = 0.413)$. As it can be seen, for the four pieces, the model outputs resemble the experimental data obtained from human participants (see Coutinho, 2008, pp. 154–156 for the plots of the remaining pieces), suggesting that the absence of significant correlations for some of the pieces may be related with the non-linear nature of the model.

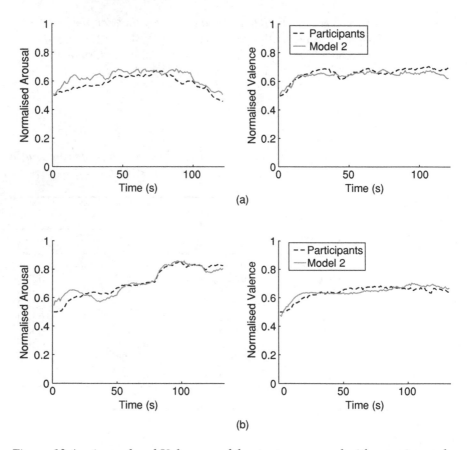

*Figure 12.4    Arousal and Valence model outputs compared with experimental data for two sample pieces (the ones with the highest mean linear correlation coefficient between model outputs and experimental data) from the Training and Test data sets: (a) Piece 4 (Beethoven, Romance No. 2) and (b) Piece 2 (Grieg, Peer Gynt Suite No. 1)*

In the following paragraphs, 'Model 2' is further analysed in order to understand how the sound and physiological inputs relate with the self-report of Arousal and Valence.

### 12.3.2.5 Model analysis

'Model 2' was able to track the general fluctuations in Arousal and Valence for the training data but also to predict human responses to another set of novel music pieces. We recur again to a CCA to reveal some of the strategies the model uses to predict affective responses to music. This method was already used in the analysis of 'Model 1'. Once again the aim is to infer the dynamics of information flow within the model. Table 12.7 shows the CCA results.

*Table 12.7 Canonical correlation analysis (CCA): the canonical correlations (the canonical correlations are interpreted in the same way as the Pearson's linear correlation coefficient) quantify the strength of relationships between the extracted canonical variates, and so the significance of the relationship. To assess the relationship between the original variables (inputs and hidden units activity) and the canonical variables, the canonical loadings (the correlations between the canonical variates and the variables in each set) are also included*

**Loadings (input/hidden)**

| Variable | Variate 1 | Variate 2 |
| --- | --- | --- |
| $H_1$ | 0.046 | **−0.555** |
| $H_2$ | −0.484 | 0.302 |
| $H_3$ | **−0.779** | 0.425 |
| $H_4$ | −0.037 | 0.338 |
| $H_5$ | **−0.899** | −0.347 |
| L | **0.421** | −0.312 |
| P | **0.352** | −0.071 |
| Pv | 0.033 | 0.085 |
| S | **0.430** | −0.284 |
| T | **0.494** | **0.780** |
| Tx | 0.209 | 0.114 |
| HR | **0.892** | 0.416 |
| Canonical correlation | 0.976 | 0.895 |
| Percentage | 76.5% | 15.1% |

Two canonical variables explain 91.6 per cent of the variance in the data. The first pair of variables loads on L, T, P, S, HR (input set), $H_3$ and $H_5$ (hidden layer). The second, loads mainly on T and $H_1$. The first canonical function loads positively on L, T, P, S and HR, and negatively on $H_3$ and $H_5$. $H_3$ has its strongest impact on the Arousal output (with a negative weight,

$w_{H_3-A} = -3.78$ – the weights matrices can be found in Coutinho, 2008), while $H_5$ has a negatively weighted connection to the Valence output ($w_{H_5-V} = -2.27$). In this way those inputs have a general positive effect on both outputs. The second canonical function consists mainly of T and $H_1$. Because the canonical function loads negatively on T and $H_1$, which in turn has a positive weight to the output ($w_{H_1-V} = 2.720$), this dimension conveys a positive effect of T on Valence.

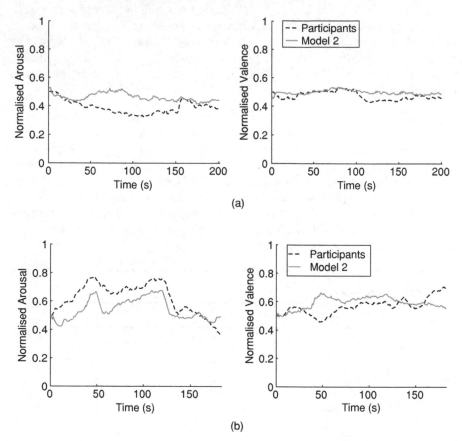

*Figure 12.5    Arousal and Valence model outputs compared with experimental data for two sample pieces (the ones with the lowest mean linear correlation coefficients between model outputs and experimental data) from the Training and Test data sets: (a) Piece 1 (Albinoni, Adagio) and (b) Piece 7 (Debussy, La Mer)*

Pv and Tx are the only variables with no significant linear correlations with the hidden layer. This does not mean that they are not relevant for the model dynamics, but instead of that they may have more complex interactions with the output. The relationships are not necessarily linear and exclusive (as we have observed in the Introduction (Section 12.1) to this chapter, the different sound features are not independent, and show different levels of interaction), making it difficult to analyse the model performance. This fact also suggests that the model representations are distributed, meaning that the hidden units distribute the input signals through the internal representations space. This is a strong argument to suggest that interactions among the sound features are a fundamental aspect of the affective value conveyed.

### 12.3.3   Summary

'Model 2' was essentially an extension of 'Model 1' to include physiological variables as inputs together with the sound features. Using new experimental data, our results have confirmed and reinforced that a subset of six sound (psychoacoustic) features (loudness level, mean pitch, pitch variation, timbre, tempo and texture) contain relevant information about the affective experience with music, and that the inclusion of the HR level added insight to the subjective feeling variations. These results suggest that the HR variations may be used as an insight to the emotion 'felt' while listening to music, as demonstrated by Dibben (2004). A detailed analysis of the model has also shown that the HR has a positive relationship with reports of Arousal and Valence.

Figure 12.6 provides a qualitative representation of the relationships between sound features and affective dimensions (as described in page 350), plus the relationship of the HR level, to Arousal and Valence. Note that this

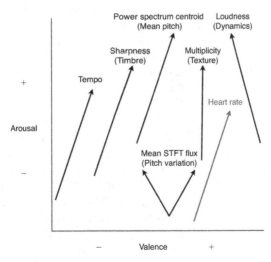

*Figure 12.6    Qualitative representation of individual relationships between music variables, heart rate and emotion appraisals: summary of observations from model analysis. The direction of the arrows indicates an increase in the variable indicated (the arrow sizes and angles formed with both axis are merely qualitative, and cannot be interpreted in mathematical terms)*

representation is merely representative of the main observable fluxes of information in the model, and does not represent the complete interaction between inputs and outputs. As you can see most variables have a positive effect on both Arousal and Valence. The exceptions are loudness, which tends to have a negative effect on Valence when it increases, and the mean STFT flux (pitch variation), that shows more complex relationships with Valence (having both positive and negative effects).

## 12.4    Conclusions and future research

The central focus of this chapter was the description of a computational model capable of predicting human participants' self-report of subjective feelings of emotion while listening to music. It was hypothesised that the affective value of music may be derived from the nature of the stimulus itself: the spatio-temporal patterns of sound (psychoacoustic) features that the brain extracts while listening to music. In this way, the link between music and emotional experiences was considered to be partially derived from the representation of the musical stimulus in terms of its affective value. The belief is that organised sound may contain dynamic features, which may 'mimic' certain features of emotional states. For this reason this investigation was based on continuous representations of both emotion and music.

The complexity of experimental data on music and emotion studies requires adequate methods of analysis, which support the extraction of relevant information from experimental data. We have addressed this question by considering a novel methodology for their study: connectionist models. The ENN, a class of spatio-temporal connectionist models, was suggested as a paradigm capable of analysing the interaction between sound features and the dynamics of emotional ratings. This modelling paradigm supports the investigation of both temporal dimensions (the dynamics of musical sequences) and spatial components (the parallel contribution of various psychoacoustic factors) to predict human participants' emotional ratings of various music pieces.

In a preliminary set of simulations (see Coutinho & Cangelosi, 2009), we identified a core group of variables relevant for this study hypothesis. Those variables are: loudness level (dynamics), power spectrum centroid (mean pitch), mean STFT flux (pitch variation), sharpness (timbre), beats-per-minute (tempo) and multiplicity (texture). Then a series of simulations were conducted to 'tune' and test the model. To train the neural network to respond as closely as possible to the responses of human participants, 486 s of music (three pieces) were used. An additional 632 s of music served as a test data, comprising three pieces unknown to the model. It was shown that the model predictions resembled those obtained from human participants. The generalisation performance validated the model and supported the hypothesis that sound features are good predictors of emotional experiences with music (at least for the affective dimensions considered).

In terms of modelling technique, this model constitutes an advance in several respects. First, it incorporates all music variables together in a single model, which permits to consider interactions among sound features (overcoming some of the drawbacks from previous models, Schubert, 1999a). Second, artificial neural networks, as non-linear models, enlarge the complexity of the relationships between music structure and emotional response observed, since they can operate in higher dimensional spaces (not accessible to linear modelling techniques such as the ones used by Schubert, 1999a and Korhonen, 2004a). Third, the excellent generalisation performance (prediction of emotional responses for novel music stimuli) validated the model and supported the hypothesis that sound features are good predictors of the subjective feeling experience of emotion in music (at least for the affective dimensions considered). Fourth, another advantage, is the ability to analyse the model dynamics, an excellent source of information about the rules underlying input–output transformations. This is a limitation inherent in the previous models we wished to address. It is not only important to create a computational model that represents the studied process, but also to analyse the extent to which the relationships built-in are coherent with empirical research. In this chapter, we found consistent relationships between sound features and emotional response, which support important empirical findings (e.g., Hevner, 1936; Gabrielsson & Juslin, 1996; Scherer & Oshinsky, 1977; Thayer, 1986; Davidson, Scherer, & Goldsmith, 2003; see Schubert, 1999a and Gabrielsson & Lindström, 2001 for a review).

In a new computational investigation, we investigated the peripheral feedback hypothesis, by extending our neural network model to physiological cues. It was shown that the inclusion of the HR level as an input to the model (a source of information to predict the affective response) improves the model performance. The correlation between the model predictions and human participants' responses, was improved by 10 per cent for Arousal and almost 20 per cent for Valence (refer to Coutinho, 2008). This is a supporting argument favouring the peripheral feedback theories (e.g., Dibben, 2004; Philippot *et al.*, 2002), and reinforces the idea that listeners may derive affective meaning from the spatio-temporal patterns of sound. Moreover, the model presented, was tested on two different populations and on different sets of music pieces.

Overall, this chapter presented some evidence supporting the 'emotivist' views on musical emotions.[9] It was shown that a significant part of the listener's affective response can be predicted from the psychoacoustic properties of sound. It was also found that these sound features (to which Meyer referred as 'secondary' or 'statistical' parameters) encode a large part of the information that allows the approximation of human affective responses to music. Contrary

[9] There are two main complementary views on the relationships between music and emotions. 'Cognitivists' claim that music simply expresses emotions that the listener can identify, while 'emotivists' posit that music can elicit affective responses in the listener (Krumhansl, 1997; Kivy, 1990).

to Meyer's (1956) belief, the results presented here suggest that 'primary' parameters (derived from the organisation of secondary parameters into higher order relationships with syntactic structure), do not seem to be a necessary condition for the process of emotion to arise (at least in some of its components). This is also coherent with Peretz *et al.* (1998) study, in which a patient lacking the cognitive capabilities to process the music structure (including Meyer's 'primary' parameters), was able to identify the emotional tone of music. Finally, we provided a new methodology to the field of music and emotion research based on combinations of computational and experimental work, which aid the analysis of emotional responses to music, while offering a platform for the abstract representation of those complex relationships.

### 12.4.1    Future research and applications

The computational model presented was applied to a limited set of music and listeners (western instrumental/art music). We focused on classical music since this musical style is the most often studied in music and emotion research. Another reason is that classical music is widely considered to be a style which 'embodies' emotional meaning. A fundamental aspect that needs to be addressed in future research is certainly the expansion of the musical universe.

Due to the flexibility of connectionist models it is also possible to extend the model presented to incorporate other variables related to perceptual (e.g., psychoacoustic features or higher order parameters such as harmony or mode), physiological (e.g., respiration measures) or even individual characteristics of the listener (e.g., musical training/expertise, personality traits, among others). The modelling process may also be improved by using other types of spatio-temporal connectionist models. Another class of models to consider are the long short-term memory neural networks (Hochreiter & Schmidhuber, 1997). These networks are able to represent more complex temporal relationships of the modelled processes, have a good generalisation performance and the ability to bridge long time lags. This characteristic may be relevant when considering the effects of music on 'mood' states or the accumulated effects of specific variables (sound features or physiological variables).

We are also considering the extension of our model to the analysis of affective cues in speech (Coutinho & Dibben, in press), since behavioural and neurological studies suggest that the affective content of music and speech is dependent on some of the same structural-auditory characteristics and mechanisms (e.g., Deutsch, Henthorn, & Dolson, 2004). Moreover, musical behaviour and vocal communication also share common ancestry (Brown, 2000; Dissanayake, 2000). Our goal is to test some theories which propose that speech prosody and music make use of similar acoustic signals to encode emotion-related information (e.g., Juslin & Laukka, 2003), based on evidence that showing that parameters like pitch height and contour, intensity, tempo, timbre and rhythm are means of conveying emotional meaning in speech as well as in music.

Finally, we also endeavour the application of this and other computational methodologies in healthcare and music therapy scenarios. It is possible to develop models of single and groups of listeners, as an auxiliary tool for the analysis and improvement of sound spaces in healthcare institutions or public spaces. It can also be developed towards other purposes, such as the prediction of affective value of music pieces as a therapeutic tool.

## Acknowledgements

The authors would like to acknowledge the courtesy of Mark Korhonen for sharing his experimental data, and the financial support from the Portuguese Foundation for Science and Technology (FCT).

## References

Balkwill, L., & Thompson, W. (1999). 'A cross-cultural investigation of the perception of emotion in music: psychophysical and cultural cues'. *Music Perception*, **17**(1), 43–64.

Bechara, A., Damasio, H., & Damasio, A. (2003.) 'Role of the amygdala in decision-making'. *Annals of the New York Academy of Sciences*, **985**(1) 356–369.

Berlyne, D. (1974). *Studies in the New Experimental Aesthetics: Steps Toward an Objective Psychology of Aesthetic Appreciation*. London, UK: Halsted Press.

Berntson, G., Shafi, R., Knox, D., & Sarter, M. (2003). 'Blockade of epinephrine priming of the cerebral auditory evoked response by cortical cholinergic deafferentation'. *Neuroscience*, **116**(1), 179–186.

Blood, A., & Zatorre, R. (2001). 'Intensely pleasurable responses to music correlate with activity in brain regions implicated in reward and emotion'. *Proceedings of the National Academy of Sciences*, **98**(20), 11818–11823.

Blood, A., Zatorre, R., Bermudez, P., & Evans, A. (1999). 'Emotional responses to pleasant and unpleasant music correlate with activity in paralimbic brain regions'. *Nature Neuroscience*, **vol. 2,** pp. 382–387.

Bradley, M., & Lang, P. (1994). 'Measuring emotion: the Self-Assessment Manikin and the semantic differential'. *Journal of Behavior Therapy and Experimental Psychiatry*, **25**(1),49–59.

Brown, S. (2000). 'Are Music and Language Homologues'? In N. Wallin, B. Merker, & S. Brown (Eds.), *The Origins of Music* (pp. 271–300). Cambridge USA: MIT Press.

Cabrera, D. (1999). 'Psysound: a computer program for psychoacoustical analysis'. *Proceedings of the Annual Conference of the Acoustical Society of Australia* (pp. 47–54). Melbourne, Australia: Australian Acoustical Society.

Cacioppo, J., Klein, D., Berntson, G., & Hatfield, E. (1993). 'The psycho-physiology of emotion'. In M. Lewis & J. Haviland (Eds.), *Handbook of Emotions* (pp. 119–142). New York and London: The Guildford Press.

Coutinho, E. (2008). *Computational and psycho-physiological investigations of musical emotions.* Unpublished doctoral dissertation, University of Plymouth.

Coutinho, E., & Cangelosi, A. (2009). 'The use of spatio-temporal connectionist models in psychological studies of musical emotions'. *Music Perception.* **27**(1), 1–15.

Coutinho, E., & Dibben, N. (in press). 'Music, Speech and Emotion: psychophysiological and computational investigations'. Submitted for publication.

Damasio, A. (1994). *Descarte's Error: Emotion, Reason and the Human Brain*, New York: Grosset/Putnam Books.

Damasio, A. (2000). *The Feeling of What Happens: Body, Emotion and the Making of Consciousness.* London: Vintage.

Davidson, R., Scherer, K., & Goldsmith, H. (2003). *Handbook of Affective Sciences.* Oxford, New York: Oxford University Press.

Dell, G. (2002). 'A spreading-activation theory of retrieval in sentence production'. *Psycholinguistics: Critical Concepts in Psychology*, **93**(3), 283–321.

Deutsch, D., Henthorn, T., & Dolson, M. (2004). 'Absolute pitch, speech, and tone language: some experiments and a proposed framework'. *Music Perception*, **21**(3), 339–356.

Dibben, N. (2004, Fall). 'The role of peripheral feedback in emotional experience with music'. *Music Perception*, **22**(1), 79–115.

Dissanayake, E. (2000). 'Antecedents of the temporal arts in early mother-infant interaction'. In N.L. Wallin, B. Merker, & S. Brown (Eds.), *The Origins of Music* (pp. 389–410). Cambridge, USA: MIT Press.

Dowling, W., & Harwood D. (1986). *Music Cognition.* San Diego, CA: Academic Press.

Ekman, P., & Davidson, R. (1994). *The Nature of Emotion: Fundamental Questions.* Cambridge, MA: Oxford University Press.

Elman, J. (1990). 'Finding structure in time'. *Cognitive Science*, **vol. 14**, pp. 179–211.

Elman, J. (1991). 'Distributed representations, simple recurrent networks, and grammatical structure'. *Machine Learning*, **7**(2–3), 195–225.

Feldman, L. (1995). 'Valence focus and arousal focus: individual differences in the structure of affective experience'. *Journal of Personality and Social Psychology*, **69**(1), 153–166.

Frijda, N. (1986). *The Emotions.* London, UK: Cambridge University Press.

Fritz, T., Jentschke, S., Gosselin, N., Sammler, D., Peretz, I., & Turner, R. (2009). 'Universal recognition of three basic emotions in music'. *Current Biology*, **19**(7), 1–4.

Gabrielsson, A. (2002). 'Emotion perceived and emotion felt: same or different'? *Musicae Scientiae*, Special Issue 2001–2002, pp. 123–147.

Gabrielsson, A., & Juslin, P. (1996). 'Emotional expression in music performance: between the performer's intention and the listener's experience'. *Psychology of Music*, **24**(1), 68.

Gabrielsson, A., & Lindström, E. (2001). 'The influence of musical structure on emotional expression'. In P. Juslin & J. Sloboda (Eds.), *Music and Emotion: Theory and Research* (pp. 223–248). New York: Oxford University Press.

Giles, C., Lawrence, S., & Tsoi, A. (2001). 'Noisy time series prediction using a recurrent neural network and grammatical inference'. *Machine Learning*, **44**(1/2), 161–183.

Goldstein, A. (1980). 'Thrills in response to music and other stimuli'. *Physiological Psychology*, **8**(1), 126–129.

Greenwald, M., Cook, E., & Lang, P. (1989). 'Affective judgment and psychophysiological response: dimensional covariation in the evaluation of pictorial stimuli'. *Journal of Psychophysiology*, **3**(1), 51–64.

Grewe, O., Nagel, F., Kopiez, R., & Altenmuller, E. (2005). 'How does music arouse 'chills'? Investigating strong emotions, combining psychological, physiological, and psychoacoustical methods'. *Annals of the New York Academy of Sciences*, **1060**(1), 446.

Harrer, G., & Harrer, H. (1977). 'Music and the brain: studies in the neurology of music'. In M. Critchley & R. A. Henson (Eds.), *Music and the Brain. Studies in the Neurology of Music* (pp. 202–216). London, UK: William Heinemann Medical Books.

Hevner, K. (1936, April). 'Experimental studies of the elements of expression in music'. *The American Journal of Psychology*, **48**(2), 246–268.

Hochreiter, S., & Schmidhuber, J. (1997). 'Long short-term memory'. *Neural Computation*, **9**(8), 1735–1780.

Hotelling, H. (1936). 'Relations between two sets of variables'. *Biometrika*, **28**(3), 321–377.

Iwanaga, M., & Tsukamoto, M. (1997). 'Effects of excitative and sedative music on subjective and physiological relaxation'. *Perceptual and Motor Skills*, **85**(1), 287–296.

Jordan, M. (1990). 'Attractor dynamics and parallelism in a connectionist sequential machine'. In J. Diederich (Ed.), *Artificial Neural Networks: Concept Learning* (pp. 112–127). Piscataway, NJ: IEEE Press.

Juslin, P., & Laukka, P. (2003). 'Communication of emotions in vocal expression and music performance: different channels, same code'? *Psychological Bulletin*, **129**(5), 770–814.

Juslin, P., & Laukka, P. (2004). 'Expression, perception, and induction of musical emotions: a review and a questionnaire study of everyday listening'. *Journal of New Music Research*, **33**(3),217–238.

Juslin, P., & Sloboda, J. (2001). *Music and Emotion: Theory and Research*. New York: Oxford University Press.

Kellaris, J., & Kent, R. (1993). 'An exploratory investigation of responses elicited by music varying in tempo, tonality, and texture'. *Journal of Consumer Psychology*, **2**(4), 381–401.

Khalfa, S., Peretz, I., Blondin, J., & Manon, R. (2002). 'Event-related skin conductance responses to musical emotions in humans'. *Neuroscience Letters*, **328**(2), 145–149.

Kivy, P. (1990). *Music Alone: Philosophical Reflections on the Purely Musical Experience*. Ithaca, NY: Cornell University Press.

Koelsch, S., Fritz, T., Cramon, D., Müller, K., & Friederici, A. (2006). 'Investigating emotion with music: an fMRI study'. *Human Brain Mapping*, vol. **27**, pp. 239–250.

Korhonen, M. (2004a). *Modeling continuous emotional appraisals of music using system identification*. Unpublished master's thesis, University of Waterloo.

Korhonen, M. (2004b, August). *Modeling continuous emotional appraisals of music using system identification*. Available from http://www.sauna.org/kiulu/emotion.html [Accessed 8 December 2008].

Korhonen, M., Clausi, D., & Jernigan, M. (2004). 'Modeling continuous emotional appraisals using system identification'. In S. Libscomb, R. Ashley, R. Gjerdingen, & P. Webster, (Eds.), *Proceedings of the 8th international conference on music perception and cognition*. Evaston. Adelaide: Causal Productions.

Kremer, S. (2001). 'Spatiotemporal connectionist networks: a taxonomy and review'. *Neural Computation*, **13**(2), 249–306.

Krumhansl, C.L. (1996). 'A perceptual analysis of Mozart's piano sonata k. 282: segmentation, tension, and musical ideas'. *Music Perception*, **13**(3), 401–432.

Krumhansl, C.L. (1997). 'An exploratory study of musical emotions and psychophysiology'. *Canadian Journal of Experimental Psychology*, **51**(4), 336–353.

Lang, P., Bradley, M., & Cuthbert, B. (1998, September). 'Emotion and motivation: measuring affective perception'. *Journal of Clinical Neurophysiology*, **15**(5), 397–408.

Langer, S. (1942). *Philosophy in a New Key*. Cambridge: Harvard University Press.

Lawrence, S., Giles, C., & Fong, S. (2000, Jan/Feb). 'Natural language grammatical inference with recurrent neural networks'. *IEEE Transactions on Knowledge and Data Engineering*, **12**(1), 126–140.

Lazarus, R. (1991). 'Cognition and motivation in emotion'. *American Psychologist*, **46**(4), 352–367.

Levitin, D. (2006). *This is Your Brain on Music*. New York: Dutton.

Ljung, L. (1986). *System Identification: Theory for the User*. New Jersey: Prentice-Hall.

Madsen, C., Brittin, R., & Capperella-Sheldon, D. (1993). 'An empirical investigation of the aesthetic response to music'. *Journal of Research in Music Education*, vol. **43**, pp. 57–69.

McClelland, J., & Elman, J. (1986). 'The TRACE model of speech perception'. *Cognitive Psychology*, vol. **18**, 1–86.

Meyer, L. (1956). *Emotion and Meaning in Music*. Chicago, IL: The University of Chicago Press.

Mozer, M. (1999). 'Neural network music composition by prediction: exploring the benefits of psychoacoustic constraints and multiscale processing'. In N. Griffith & P. Todd (Eds.), *Musical Networks: Parallel Distributed Perception and Performance* (pp. 227–259). Cambridge, MA: MIT Press.

Nagel, F., Kopiez, R., Grewe, O., & Altenmüller, E. (2007). 'Emujoy: software for continuous measurement of perceived emotions in music'. *Behavior Research Methods*, **39**(2), 283–290.

Nielsen, F. (1987). 'The semiotic web '86: an international year-book'. In T.A. Seboek & J. Umiker-Seboek (Eds.), *The Semiotic Web*, (pp. 491–513). Berlin: Mouton de Gruyter.

Panksepp, J. (1998). *Affective Neuroscience: The Foundations of Human and Animal Emotions*. New York: Oxford University Press.

Panksepp, J., & Bernatzky, G. (2002). 'Emotional sounds and the brain: the neuro-affective foundations of musical appreciation'. *Behavioural Processes*, **60**(2), 133–155.

Parncutt, R. (1989). *Harmony: A Psychoacoustical Approach*. Berlin, Germany: Springer Verlag.

Patel, A., & Balaban, E. (2000). 'Temporal patterns of human cortical activity reflect tone sequence structure'. *Nature*, **vol. 404**, pp. 80–84.

Peretz, I., Gagnon, L., & Bouchard, B. (1998). 'Music and emotion: perceptual determinants, immediacy, and isolation after brain damage'. *Cognition*, **68**(2), 111–141.

Philippot, P., Chapelle, G., & Blairy, S. (2002). 'Respiratory feedback in the generation of emotion'. *Cognition & Emotion*, **16**(5), 605–627.

Resnicow, J., Salovey, P., & Repp, B. (2004). 'Is recognition of emotion in music performance an aspect of emotional intelligence'? *Music Perception*, **22**(1), 145–158.

Rickard, N.S. (2004). 'Intense emotional responses to music: a test of the physiological arousal hypothesis'. *Psychology of Music*, **32**(4), 371–388.

Robazza, C., Macaluso, C., & D'Urso, V. (1994). 'Emotional reactions to music by gender, age, and expertise'. *Perceptual Motor Skills*, **79**(2), 939–944.

Rumelhart, D., Hintont, G., & Williams, R. (1986). 'Learning representations by back-propagating errors'. *Nature*, **323**(6088), 533–536.

Rumelhart, D., & McClelland, J. (1986). 'PDP models and general issues in cognitive science'. In D. Rumelhart & J. McClelland (Eds.), *Parallel Distributed Processing: Explorations in the Microstructure of Cognition*, (vol. 1). Cambridge, MA: MIT Press.

Russell, J. (1980). 'A circumplex model of affect'. *Journal of Personality and Social Psychology*, **39**(6), 1161–1178.

Russell, J. (1989). 'Measures of emotion'. In R. Plutchik & H. Kellerman (Eds.), *Emotion: Theory, Research, and Experience*, (vol. 4). Toronto, Canada: Academic.

Scherer, K. (2004). 'Which emotions can be induced by music? what are the underlying mechanisms? and how can we measure them'? *Journal of New Music Research*, **33**(3), 239–251.

Scherer, K., & Oshinsky, J. (1977). 'Cue utilization in emotion attribution from auditory stimuli'. *Motivation and Emotion*, **1**(4), 331–346.

Scherer, K.R., & Zentner M. (2001). 'Emotional effects of music: Production rules'. In: P.N. Juslin, & J. A. Sloboda (Eds.), *Music and Emotion: Theory and Research*. Oxford, New York: Oxford University Press.

Schubert, E. (1999a). *Measurement and time series analysis of emotion in music*. Unpublished doctoral dissertation, University of New South Wales.

Schubert, E. (1999b). 'Measuring emotion continuously: validity and reliability of the two dimensional emotion space'. *Australian Journal of Psychology*, **51**(3), 154–165.

Schubert, E. (2001). 'Continuous measurement of self-report emotional response to music'. In P. Juslin, & J. Sloboda (Eds.), *Music and Emotion: Theory and Research*, (vol. 393–414). Oxford, UK: Oxford University Press.

Schubert, E. (2004). 'Modeling perceived emotion with continuous musical features'. *Music Perception*, **21**(4), 561–585.

Sloboda, J. (1991). 'Music structure and emotional response: some empirical findings'. *Psychology of Music*, **vol. 19**, pp. 110–120.

Sloboda, J., & Lehmann, A. (2001). 'Tracking performance correlates of changes in perceived intensity of emotion during different interpretations of a Chopin piano prelude'. *Music Perception*, **19**(1), 87–120.

Thayer, J. (1986). *Multiple indicators of affective response to music*. Unpublished doctoral dissertation, New York University, Graduate School of Arts and Science.

Thompson, W.F., & Balkwill, L.-L. (2006). 'Decoding speech prosody in five languages'. *Semiotica*, **158**(1/4), 407–424.

Tzanetakis, G., & Cook, P. (2000). 'Marsyas: a framework for audio analysis'. *Organised Sound*, **4**(03), 169–175.

Witvliet, C., & Vrana, S. (1996). 'The emotional impact of instrumental music on affect ratings, facial EMG, autonomic measures, and the startle reflex: effects of valence and arousal'. *Psychophysiology*, **33** (Supplement 1), 91.

Witvliet, C., Vrana, S., & Webb-Talmadge, N. (1998). 'In the mood: emotion and facial expressions during and after instrumental music and during an emotional inhibition task'. *Psychophysiology*, **35** (Supplement 1), 88.

Wundt, W. (1896). *Grundriss der Psychologie (Outlines of Psychology)*. Leipzig: Engelmann.

Zatorre, R.J. (2005). 'Music, the food of neuroscience?' *Nature*, **vol. 434**, pp. 312–315.

Zwicker, E., & Fastl, H. (1990). *Psychoacoustics*. New York, NY: Springer.

## Chapter 13
# A conceptual model of investor behavior

*Milan Lovric, Uzay Kaymak and Jaap Spronk*

## Abstract

Behavioral finance is a subdiscipline of finance that uses insights from cognitive and social psychology to enrich our knowledge of how investors make their financial decisions. Agent-based artificial financial markets are bottom-up models of financial markets that start from the micro level of individual investor behavior and map it into the macro level of aggregate market phenomena. It has been recognized in the literature, yet not fully explored, that such agent-based models are very suitable tool to generate or test various behavioral hypotheses. To pursue this research idea, first we develop a conceptual model of individual investor that consists of a cognitive model of the investor and a description of the investment environment. In the modeling tradition of cognitive science and intelligent systems, the investor is seen as learning, adapting, and evolving entity that perceives the environment, processes information, acts upon it, and updates its internal states. This conceptual model can be used to build stylized representations of (classes of) individual investors, and further studied within the paradigm of agent-based artificial financial markets.

## 13.1 Introduction

This chapter presents a cognitive conceptual model of the individual investor. By taking a descriptive point of view, we focus our attention on how investors make their investment decisions in a real-world setting, as opposed to rational/optimal behavior proposed by normative financial theories. This conceptual model incorporates results of the research on investment decisions from fields of behavioral finance and cognitive psychology. It is based on a review of existing studies, which themselves were conducted using various research methods. It particularly draws on the heuristics and biases identified in the behavioral finance literature, as well as the dual-process theories of human cognition.

In the rapidly growing field of behavioral finance it becomes more difficult to devise a unifying taxonomy of all the behavioral phenomena, as they arise from various mechanisms, manifest themselves on different levels of behavior and cognition, and have been discovered using various methodologies (e.g., experiments vs. quantitative analysis of market data). However, a conceptual model that presents some of the most important findings from the existing research on investor behavior, and provides a parsimonious description of how individual investors make their investment/trading decisions is relevant from both a theoretical and a practical point of view.

On the basis of a conceptual model, a computational model of the individual investor (or stylized classes of various market participants) can be implemented, and further studied using the paradigm of agent-based artificial financial markets (AFMs). Such agent-based simulations have a potential for studying the market-wise implications of individual investor behavior. They provide answers to questions like how behavioral biases affect the performance of market participants, how they influence market dynamics, and what the effects of learning/debiasing are.

A more detailed version of this conceptual model is available in Reference 1. A similar interdisciplinary framework for software-agent design in economics, based on the integration of experimental economics, computational economics, and neuroeconomics, has been presented in Reference 2.

## 13.1.1  Behavioral finance

A research into individual investors and their behavior has received a lot of consideration during the past, and is increasingly in the focus of interest of many scientists, being not confined only to economists. However, the particular way of looking at individual investor has been subject to a great paradigmatic shift with the inclusion of findings and methodology from psychology into financial studies.[1] Despite many ongoing debates, this has slowly led to the establishment of behavioral economics and behavioral finance as subdisciplines.

In social sciences, particularly economics, the term *Homo Economicus* has been used for a formal representation of an individual, who acts as a utility maximizer, given his preferences and other constraints. An economic man adheres to the axioms of rational choice theory. Even though this hypothetical construct has been useful in formulating economic theories and models, over the past decades psychologists and behavioral scientists have documented robust and systematic violations of principles of Expected Utility Theory, Bayesian Learning, and Rational Expectations – questioning their validity as a descriptive theory of decision making [4]. Furthermore, Herbert Simon, to whom the term *bounded*

---

[1] Statman, however, gives another perspective: 'Some people think that behavioral finance introduced psychology into finance, but psychology was never out of finance. Although models of behavior differ, all behavior is based on psychology' [3]. Another possible view is that economists distanced themselves from the psychological foundations of individual behavior during the development of neo-classical economics.

*rationality* is usually attributed, has emphasized '*the limits upon the ability of human beings to adapt optimally, or even satisfactorily, to complex environments*' [5]. Individual investors – who use heuristics, depend on framing of the problem, and are prone to biases, which in turn may lead to various anomalies at the market level – are subjects of research in the area of behavioral finance.

The efficient markets hypothesis (EMH) posits that market prices fully reflect all available information [6]. Efficient markets do not allow investors to earn above-average returns without accepting above-average risks [7]. To reconcile the theory of efficient markets with behavioral finance, Lo proposes an alternative theory in which both can coexist – the Adaptive Markets Hypothesis [8]. In this evolutionary model individual market participants adapt to changing environment by using heuristics. In other words, Lo provides us with a theoretical framework in which we could easily fit our conceptual model of the individual investor.

A number of surveys and books on behavioral finance and behavioral economics has been published, including some popular books aimed at individual investors [9–11].

## 13.1.2 Agent-based artificial financial markets

Another novel approach to studying financial market dynamics, intricately related to studies on individual investor behavior and its computational implementations, comes from the area of computational finance in the form of agent-based AFMs (or, more specifically, artificial stock markets). Agent-based AFMs can be either mathematical or computational models of financial markets, usually comprised of a number of heterogeneous and boundedly rational agents, which interact through some trading mechanism, while possibly learning and evolving. These models are built for the purpose of studying agents' behavior, the price discovery mechanisms, the influence of market microstructure, and the reproduction of the stylized facts of real-world financial time series (e.g., fat tails of return distributions and volatility clustering). A number of reviews of studies with AFMs are available, e.g., for computational models [12], and for mathematical models [13]. Similar bottom-up approach has been used in the computational study of economies modeled as evolving systems of autonomous interacting agents [14]. Analogous methodology to agent-based modeling comes from the physical sciences in the form of the Microscopic Simulation – a tool for studying complex systems by simulating many interacting microscopic elements [15].

Despite many studies with AFMs, not many attempts have been made to incorporate complex behavioral phenomena into agents' behavior. These attempts may have been hindered by many reasons. Most notably, complex behavior implies highly parameterized models which are difficult to examine and that often lie beyond analytical tractability. However, an agent-based approach inspires us to seek for homeomorphic models [16], which not only reproduce the stylized facts of real-world markets, but also achieve them through processes that are grounded on reasonable (psychologically plausible) assumptions, and resemble actual human behavior and realistic market mechanisms. 'Agent-based models can easily accommodate complex learning behavior, asymmetric information, heterogeneous preferences, and ad hoc heuristics' [17]. It is far from the fact that everything is known about human behavior and cognition pertaining to the investment decisions, but as various fields progress to open up these black boxes, a methodology that can utilize such knowledge may be given more opportunities in the future. The complementarity of behavioral finance research and the agent-based methodology is yet to be explored, as explained by LeBaron [12]. Rare examples of agent-based papers that pursue this idea of explicit accounting for behavioral theories in financial market simulations are References 18,19. Additional motivation for our chapter is found in Reference 20.[2]

## 13.2   Conceptual model

The conceptual model of the individual investor behavior presented in this chapter is built for the purpose of further implementation into an agent-based AFM. Of course, individual behavior could be directly implemented into such a computational model. However, building a separate conceptual model has a number of advantages. A conceptual model is a structured representation of the large diversity, complexity, and interdisciplinarity presented in the field. By building a conceptual model we can become aware of the interrelations between components we may not otherwise see, we can indicate the strength of these relationships, and we can identify those areas that are still largely unknown, thus fostering future research. From a modeler's perspective, a conceptual model gives a structure that is easier to implement. On top of the same conceptual model various implementations can be made, for instance, using mathematical or computational modeling techniques.

In this conceptual model, investment decisions are seen as an iterative process of interactions between the investor, as represented by the cognitive model and the investment environment. This investment process is influenced by a number of interdependent variables and is driven by dual mental systems.

### 13.2.1   Cognitive model of the investor

#### 13.2.1.1   Dual-process systems

Our cognitive model of the investor (Figure 13.1) is based on a two-dimensional framework of neural functioning, as proposed by Reference 21, which

---

[2] 'It is the learning, adapting and evolving abilities that we give to our agents in the agent-based models that can deal with some issues which complement the neoclassical and behavioral approaches. Instead of extraneously imposing a specific kind of behavioral bias, e.g., overconfidence or conservatism, on the agents, we can canvass the emergence and/or the survivorship of this behavioral bias in the highly dynamic and complex environment by computer simulation. Agent-based modeling may lead us to some viewpoints by pushing beyond the restrictions of the analytical approach' [20].

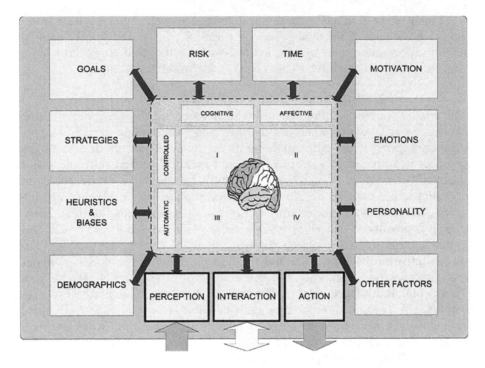

*Figure 13.1   Cognitive model of individual investor*

distinguishes between affective and cognitive, and between controlled and automatic processes.

A distinction between controlled and automatic processes can be found in psychology literature under various names of dual-processing theories (see Reference 21). Controlled processes are serial (step-by-step), evoked deliberately, cause the subjective feeling of effort, and are accessible by introspection. Automatic processes operate in parallel, they are relatively effortless, and are inaccessible to consciousness [21].

A distinction between affect and cognition is also pervasive in contemporary psychology and neuroscience literature [21]. Affective processing is associated with feeling states when affect states reach a certain threshold level. However, most of the affective processing operates unconsciously. The central role of affective processing is in human motivation – affects address '*go-no go*' questions, while cognitive processes address '*true or false*' questions [21].

In this dual-process theory the choice is determined as the result of the interplay between the cognitive and the affective system. This interaction can be collaborative (when both systems work in the same direction) or competing (in which one system wins and overrides the other system). A number of variables can influence the relative strength of these systems, e.g., cognitive load can easily undermine controlled cognitive processes.

Much of the findings in behavioral economics and finance can be interpreted in the light of the dual-process theories. So, for a modeler, it might be just an appropriate level of abstraction, without the need to go into more complex details of cognitive mechanisms and neural functioning.

### 13.2.1.2    Risk attitude

The crucial concept for investments, and decision making in general, is the concept of risk. Unfortunately, there are many definitions of risk with its meaning varying across different domains. In standard decision theory, a risky prospect is expressed as a set of events and event-contingent outcomes, with probabilities assigned to each event. Knight [22] made the distinction between decisions under risk and decisions under uncertainty, with risk being measurable (quantitative) and uncertainty non-measurable (non-quantitative). However, it is possible to conceive decision under risk as a special case of decision under uncertainty, where objective probabilities are known and used in place of subjective probabilities. The most influential theories for decisions under risk and uncertainty are known as Expected Utility Theory [23,24], Prospect Theory [25], Rank-Dependent Utility Theory [26,27], and Cumulative Prospect Theory [28]. Traditional economics and finance have been dominated by these probabilistic models of uncertainty. However, other theories for dealing with uncertainty, ambiguity, or vagueness exist, e.g., Fuzzy Set Theory [29], Dempster–Shafer Theory [30], and Rough Set Theory [31].

A decision maker's attitude toward risk can be characterized as risk aversion, risk seeking (risk tolerance, risk taking, risk loving), or risk neutrality; and can be defined in a classical sense as a preference between a risky prospect and its expected value (the method of revealed preference). In these theoretical considerations risk attitude is usually captured through the curvature of utility function, or alternatively, through non-linear weighting of probabilities. A strong empirical phenomenon that is driving risk aversion to a large extent is known as *loss aversion* [25,32]. '*Losses loom larger than gains*,' and while people are typically risk averse for gains, they are risk seeking in the domain of losses [25]. This highlights reference dependence, i.e., the importance of reference point against which outcomes are coded as losses or gains. Loomes [33] suggests that the current evidence in literature is more in favor of the notion that individuals have only basic and fuzzy preferences, and that each decision problem triggers its own preference elicitation. This is in line with the claim that preferences are constructed (not elicited or revealed) as a response to a judgment or choice task [34].

Slovic [35] argues that although knowledge of the dynamics of risk taking is still limited, there is an evidence of little correlation in risk-taking preferences across different domains and situations. Only those tasks highly similar in structure and payoffs have shown any generality. Also, previous learning experiences in specific risk-taking settings seem more important than general personality characteristics. Furthermore, risk attitude can change depending on the outcomes of previous decisions. Thaler and Johnson [36] found that previous gains increase risk-seeking behavior (*house money effect*), while in the presence of previous losses, those bets that offer a chance to break even seem particularly attractive (*break-even effect*). These are examples of what Thaler refers to as *mental accounting*. It is an important question whether investors have risk attitudes related to the gains and losses defined on the individual stock level (*narrow framing*), on parts of the portfolio, on the overall portfolio, or on their total wealth.

Risk in investments is usually considered as the standard deviation of asset returns (volatility). In addition to volatility, other common measures of risk are downside risk, shortfall probability, and Value-at-Risk. Finance and econometrics literature are the source of other more sophisticated risk-modeling techniques. In an early study of the judgmental processes of institutional investors, Cooley [37] found that most investors perceive variance as a synonym or a large part of investment risk. A substantial number of investors, however, identified an additional dimension of risk in asymmetry (left skewness). Kurtosis, on the other hand, was perceived as risk reducing. Investor perception of risk in security valuation can be biased. Shleifer [38] illustrates this on an example of value and growth stocks, and conjectures that if this biased perception affects the demand for securities, while having nothing to do with the fundamental risk of a portfolio, it would still generate the same returns as we can observe in the data.

Risk attitude is influenced by the competition and collaboration between the cognitive and the affective system. Cognitive system could be assumed to deal with risk in a probabilistic fashion, similar to traditional choice theories. Risk-averse behavior is driven by fear and anxiety responses to risk and the stored pain of experienced losses [21]. Risk-taking behavior is driven by the pleasure of gambling [21]. In the light of the Prospect Theory, Loewenstein and O'Donoghue [39] propose that the affective system contributes to the risk attitude through loss aversion and non-linear (usually S-shaped) probability weighting. However, the deliberative system responds to risk in a way predicted by Expected Utility Theory (or perhaps Expected Value).

In the literature we can find some evidence of interaction between the risk and other components of the model, e.g., risk attitude can be related to risk factors such as gender and age. A meta-analysis study [40] confirmed a significantly higher propensity for risk taking in male participants. In addition, they found age-related shifts in this gender gap, particularly the tendency of the gender gap to decrease with age. Donkers *et al.* [41] show that an individual's risk attitude and probability weighting function is influenced by gender, education level, age, and income. Besides demographic factors, such as gender and age, an important variable influencing risk attitude is the investment horizon. '*Most investment practitioners subscribe to the* time diversification *principle, which states that portfolio risk declines as the investment horizon lengthens. Accordingly, practitioners commonly advise younger clients to allocate a larger proportion of their retirement money to risky assets than older clients do. In contrast, many respected theorists argue that time diversification is a fallacy*'

[42]. In their opinion, the answer lies in the psychology of risk taking, particularly as it relates to time horizon. In favor of the time diversification position, they *'argue that risk perception is not only a function of age (and other cross-sectional idiosyncratic factors) but also of the temporal distance between the initial investment point and the cash-out point typically represented by the individual's retirement'* [42]. Gilovich *et al.* [43] have studied the effect of temporal perspective on subjective confidence, and they found that people tend to lose confidence in their prospects for success as they come closer to the 'moment of truth,' i.e., *'the risk-assessment becomes more conservative with shorter temporal distance.'*

### 13.2.1.3 Time preference

Time preference is in standard economic theory captured with the discount factor of the discounted utility (DU) model [44]. However, a behavioral point of view suggests that modeling time preference with a constant discount rate may not be suitable for descriptive purposes. Hyperbolic discounting has been proposed to capture an empirical observation that between now and a point in the future people discount more than between two other temporally equidistant points in the far future, i.e., the discount rate is declining over time [45]. An opposing finding [46] suggests that an observed pattern of time preference could be also explained by subadditive discounting, or a finding that the amount of *'discounting over a delay is greater when the delay is divided into subintervals than when it is left undivided'* [46].

Time domain enters the conceptual model as the investment horizon, the frequency of update, as well as in planning, forecasting, and discounting. An interplay between cognitive and affective mechanisms might give even more refinement in modeling time preference. Affective system is inherently myopic and impulsive, motivating behaviors that have short-term goals, whereas higher order cognitive functions of the prefrontal cortex can take long-term consequences and planning into account [21,47]. According to Reference 21, factors which strengthen or weaken an affective or cognitive system will influence people to behave more or less impulsively. Any factor that imposes cognitive load on the prefrontal cortex, i.e., the controlled cognitive system, will decrease the influence of this system on behavior. Other factors which can diminish the power of self-control are a previous exercise of self-control, alcohol, stress, and sleep deprivation. Analogously, *'the activation of affective states should accentuate temporal myopia'* [21].

### 13.2.1.4 Personality

Psychological literature on personality has settled around a five-factor model [48]: *Extraversion, Agreeableness, Conscientiousness, Neuroticism,* and *Openness.* Recent studies have examined a possible influence of personality traits on financial decisions, particularly in the context of daily traders. A study among professional traders [49] showed that successful traders tend to be emotionally stable introverts open to new experiences. Contrary to these results, Lo [8]

found the lack of correlation between personality traits and trading performance. So, given the current inconclusive results, the link between personality
traits and investment performance might still be far-fetched. However, the
relationship between personality and risk attitude, time preference, investment
strategies, or susceptibility to particular behavioral biases might be relevant for
practical investment purposes – especially given the availability of various
batteries for testing personality types, and given the stability of personality
traits during a long period of a lifetime. The link between personality traits and
risk propensity has been fairly studied in the literature: sensation-seeking, a sub
scale of the Extraversion dimension, was found to be highly correlated with
most risk-taking domains, while overall risk propensity was higher for subjects
with higher Extraversion and Openness scores and lower for subjects with
higher Neuroticism, Agreeableness, and Conscientiousness scores [8, 50].

### 13.2.1.5 Goals and motivation

Investor behavior, like most human behaviors, can be conceptualized as goal
oriented, which means that investors make decisions to reach their various
financial and non-financial goals. Goals are in the broad sense defined as
mental (internal) representations of desired states [51].

Modern theories of motivation and goal-directed behavior have been
summarized in Reference 52: '*the probability that a given goal state is set,
adopted, and enacted depends on people's ability (a) to mentally access the
representation of the goal; (b) to subjectively assess the expected (or incentive)
value of the goal state; (c) to activate, select, and execute instrumental actions;
(d) to detect, assess, and reduce the discrepancy between the actual and desired
state.*' In this framework, positive affect linked to a goal representation is
capable of directly feeding the motivation system, thus, propelling the goal
pursuit behavior (aimed at attaining the desired state) [52].

While economic agents are classically assumed to be only self-interested,
the list of motivators could be enriched by taking into account also social
preferences, such as fairness, altruism, revenge, status seeking, and survival.

### 13.2.1.6 Strategies

There is an abundant pool of strategies that investors use for the valuation of
assets, for stock picking, and for market timing. To obtain an overview of the
basic groups of strategies, first we looked at a standard investment book. In
Reference 53, investment decision making is described as an iterative process
comprised of several steps. *Investment Policy* refers to determining investor's
objectives and constraints, e.g., financial goals in terms of risk-return trade-off,
the amount of investable wealth, tax status, policy asset mix, investment benchmarks, etc. *Security Analysis* is the analysis of individual securities within previously identified broad asset classes, and can be either *technical* (forecasting
future price movements on the basis of historical data) or *fundamental* (estimating
the intrinsic price as the present value of future cash flows). *Portfolio Construction*
means determining in which assets to invest and what proportion of wealth.

*Portfolio Revision* is a periodic repetition of previous steps, as over time investors, objectives and prices of securities can change. *Portfolio Performance Evaluation* involves measuring the return and the risk of portfolio, as well as benchmarking.

The prescriptive approach of standard investment books is mostly a top-down approach in which an important step is asset allocation, i.e., investment decision on the level of broad asset classes. However, a fully bottom-up approach occurs in practice too. It focuses on specific assets that offer most attractive investment opportunities, without much concern for the resulting asset allocation, and as such may lead to a portfolio that is industry- or country specific, or exposed to one source of uncertainty [54].

Modern Portfolio Theory (MPT) [55] lays the foundations of portfolio allocation from the normative point of view. Behavioral finance takes a descriptive perspective, by studying how individual investors actually allocate their portfolios. In portfolio theory, one of the crucial concepts is diversification, a risk-management technique where various investments are combined to reduce the risk of the portfolio. However, many investors do not (sufficiently) diversify their portfolios. This may be due to beliefs that the risk is defined at the level of an individual asset rather than the portfolio level, and that it can be avoided by hedging techniques, decision delay, or delegation of authority [4]. Benartzi and Thaler [56] studied *naive diversification* strategies in the context of defined contribution saving plans. They found an evidence of $1/n$ *heuristic*, as a special case of diversification heuristic, in which an investor's contributions are spread evenly across available investment possibilities. As the authors convey, such a strategy can be problematic both in terms of ex ante welfare costs, and ex post regret (in case the returns differ from historical norms). However, as shown in Reference 57, naive diversification portfolio strategy is actually a very strong benchmark. By comparing out-of-sample performance of various optimizing mean-variance models, they found that no single model consistently beats the $1/n$ strategy in terms of Sharpe ratio or certainty-equivalent return. Poor performance of these optimal models is due to errors in estimating means and covariances.

Benartzi and Thaler [56] also found a support for *mental accounting* on the *company stock*, by which this stock is treated separately from the rest of equity classes. Another robust finding in portfolio allocation is the *home bias* – despite the advantages of international portfolio diversification, the actual portfolio allocation of many investors is too concentrated in their domestic market [58]. It is argued in Reference 59 that '*familiarity breeds investment*,' and that a persons are more likely to invest in the company that (they think) they know. Instances of this familiarity bias is investing in domestic market, in company stocks, in stocks that are visible in investors' lives, and in stocks that are discussed favorably in the media. Goetzmann and Kumar examined the diversification of investors with respect to demographic variables of age, income, and employment [60]. They found that low income and non-professional categories hold the least diversified portfolios. They also found that young active investors are overfocused and inclined toward concentrated, undiversified portfolios, which might be a manifestation of overconfidence.

Before discussing some of the main findings on how investors further manage their portfolios, it is noteworthy mentioning that both EMH as well as much empirical evidence undermine practical relevance of active portfolio management.[3] While financial literature on active portfolio management offers various techniques for beating the benchmarks, behavioral literature focuses on how individual investors manage (or make changes to) their portfolios. A common tendency to hold losers too long and sell winners to soon has been labeled as the *disposition effect* [62]. They attributed their findings to loss aversion, the issue of self-control, mental accounting, and the desire to avoid regret. Some investors trade too much, which might be a manifestation of their *overconfidence* [63]. For successful investors this overconfidence can be reinforced through *self-attribution bias*, i.e., belief that their trading success should be attributed mostly to their own abilities [63]. Furthermore, while some investors may trade too much and often change their strategies, others may exhibit the tendency of '*doing nothing or maintaining one's current or previous decision*,' which is how [64] defined the *status quo bias*.

### 13.2.1.7 Emotions

Emotions have powerful effects on decisions, and decision outcomes have powerful effects on emotions [65]. Emotions can have both predecision and postdecision effects. Most of the research focused on a unidimensional model in which a predecision emotion can be either positive or negative. However, a more detailed approach is needed, given the variety and domain specificity of emotions [65]. Positive emotions are shown to increase creativity and information integration, and promote variety seeking. They also cause overestimation of the likelihood of favorable events, and underestimation of the likelihood of negative events. Negative emotions promote narrowing of attention and failure to search for alternatives. They promote attribute- versus alternative-based comparisons [65]. One of the most studied emotions that can follow a decision is the feeling of regret. In the short run, people experience more regret for actions rather than inaction, while in the long run they experience more regret for their inactions [66]. Anticipated emotions, such as regret and disappointment, have drawn most attention of economists, whereas immediate emotions (experienced at the moment of decision making) have been mainly studied by psychologists [67]. Loewenstein [67] emphasizes that economists should also pay attention to immediate emotions and a range of visceral factors, which influence our decisions.

One evidence for the importance of emotions in decision making comes from patients with brain lesions in regions related to emotional processing. In an experiment with investment decisions (represented by risky prospects with

---

[3] 'Switching from security to security accomplishes nothing but to increase transaction costs and harm performance. Thus, even if markets are less than fully efficient, indexing is likely to produce higher rates of return than active portfolio management. Both individual and institutional investors will be well served to employ indexing for, at the very least, the core of their equity portfolio' [61].

positive expected value) [68], patients with brain lesions in emotion-related areas of brain made more investments and thus earned more on average. Normal and control patients seem to have been more affected by the outcomes of previous decision – upon winning or losing they adopted a conservative strategy and invested less in subsequent rounds. However, the inability to learn from emotional signals (Somatic Marker Hypothesis, [69]) can also lead to unadvantageous decisions such as excessive gambling. Thus, emotion and cognition both play a crucial role in decision making.

Goldberg and von Nitzsch [10] describe a personal experience of a trader (market participant) who goes through various emotional states during profit-and-loss cycles. The feelings of hope and fear, depending on the success or failure on the market, can be transformed to the states of euphoria or panic. During these transitional states there is a selective perception of information – positive information is perceived and often exaggerated, while negative information is ignored. In final states of euphoria or panic, information has almost no role to play [10].

### 13.2.1.8   Heuristics and biases

Much of the behavioral finance literature deals with individual investor psychology and studies the use of heuristics and various biases in judgment. Organizing and presenting all these heuristics and biases is by no means an easy task, given that they arise from a wide range of cognitive mechanisms. However, a simple but comprehensive framework is important if one aims to use it for descriptive or prescriptive purposes.

A traditional approach describes decisions as choices between risky prospects. A decision maker forms beliefs about probabilities of events and about values (utilities) of outcomes contingent on those events. Finally, preferences are made between risky options. Biases can arise both in the process of forming beliefs and preferences. In the more general sense, a bias can be defined as a departure from normative, optimal, or rational behavior. In their seminal work [70], Tversky and Kahneman investigated heuristics to which people are particularly susceptible when making decisions under uncertainty. These are *representativeness, availability*, and *availability and anchoring*. Heuristics are mechanisms (rules, strategies) for processing information to arrive at a quick (not necessarily optimal) result with little effort [10]. These authors divide heuristics into two major groups: *heuristics for reducing complexity* and *quick judgments*. Heuristics are rules of thumbs, procedures used for processing information and reasoning, often based on trial and error [9]. They are useful as they make cognitively difficult tasks easier. However, they can also lead to systematic biases.

In Reference 71, Rabin discusses various behavioral aspects he finds important from the perspective of economics. Kahneman and Riepe [72] focus on biases in beliefs and preferences of which financial advisors should be particularly aware, and provide recommendations how to avoid them or mitigate their harmful effects. Shefrin [9] distinguishes between *heuristic-driven bias* (representativeness; gambler's fallacy; overconfidence; anchoring and adjustment; ambiguity aversion; emotion and cognition) and *frame dependence* (loss aversion; mental accounting; hedonic editing; cognitive and emotional aspects; self control; regret; money illusion). Hirshleifer [73] proposed a unified explanation for most known judgment and decision-making biases in the context of investor psychology and asset pricing. A survey of behavioral finance literature [11] lists known biases that arise when people form their beliefs and preferences.

In the behavioral finance literature we often come across the idea of potential *debiasing* of investors, according to which some biases may be avoided by careful framing of problems, or by learning effects of repetition. Unfortunately, there is no guaranty that financial decision problems will be presented in such a way which promotes an unbiased decision-making process [73]. It could be argued that individual biases are not that important. As individuals differ, their biases should cancel out in the equilibrium. However, it is well known that some biases can be systematic and persistent [73]. Repeated patterns can even be used as a basis for prediction of behavior of others, which is very important in the context of financial markets [10].[4]

Finally, heuristics and biases should not be treated as synonymous. Heuristics do not always lead to biases, and even if they do (with respect to some normative rational theory), heuristic behavior may actually result in a successful performance compared to an optimal strategy, which was demonstrated in the well-known paper [74]. A financial example of a successful heuristic is naive diversification ($1/n$ heuristic), which is costly ex ante, but ex post serves as a very strong benchmark.

### 13.2.1.9   Perception–interaction–action

Perception, interaction, and action are processes that describe the way in which investor interacts with the investment environment in which one is situated. Processes such as perception and action are commonly included in cognitive models (e.g., Reference 75).

*Perception* is the process of acquiring information from various sources in the investment environment (classical media, such as television, radio, and paper news, as well as various Web applications). Through communication channels, other market participants (peers) can also serve as sources of information. Information can be quantitative or qualitative, and can include financial data, financial news, and socially shared opinions and tips. Oberlechner and Hocking studied information sources, news, and rumors in the foreign exchange market [76]. In this study foreign exchange traders and financial journalists rated the importance of different information sources,

---

[4] The issue whether this could be exploited so as to create profitable opportunities is still rather controversial, and is related to a stance on the market efficiency. 'I am skeptical that any of the "predictable patterns" that have been documented in the literature were ever sufficiently robust so as to have created profitable investment opportunities, and after they have been discovered and publicized, they will certainly not allow investors to earn excess returns' [7].

such as wire services, personal contacts, analysts, daily newspapers, financial television, etc. An interesting finding is that the information speed, expected market impact, and anticipated market surprise were rated as more important than the reliability of the source and the accuracy of information.

Since market participants are exposed to a constant flow of information, processing all this information can be a daunting task, and so it would not be surprising that during this process people apply many heuristics. According to the *representativeness*, heuristic people may overreact to a series of evidence and see patterns where they do not exist. However, people can sometimes underreact to news, i.e., in the light of a new evidence they update their beliefs conservatively, and not in a fully Bayesian manner. *Confirmation bias* may play an important role in how investors acquire and process this information. It suggests that people have a tendency to search for information that supports their current beliefs and decisions, while neglecting information that confronts those beliefs.

Information processing could also be interpreted in the light of the dual-process systems. For instance, information could be used for controlled deliberative reasoning (e.g., whether it fits expectations), but also by an affective system (e.g., emotional response to price changes). Cognitive system might deal with information processing by applying simplifying heuristics, while emotional system could be susceptible to hedonic editing and confirmation bias. These inputs from the environment are necessary to establish a *feedback mechanism* as the basis for learning. The dual-process view taken in this chapter suggests that learning processes are likely '*a splice of cognitive and affective processes*' [21].

*Interaction* is that part of the model which deals with peer influence and social factors. Financial economists have borrowed more from the psychology of the individual than from social psychology [73]. For example, they have examined how information is transmitted by prices, volume, or corporate actions. However, person-to-person and media contagion of ideas and behavior also seems important. Shiller [77] has emphasized the importance of conversation in the contagion of popular ideas about financial markets. In a survey of individual investors, Shiller [78] found that almost all of the investors who recently bought a particular stock had their attention drawn to it through direct interpersonal communication. The influence of conversation on trading may arise from individuals' overconfidence about their ability to distinguish pertinent information from noise or propaganda [73]. Social psychology provides evidence of various social effects that might be important in the context of financial markets as well. *Conformity effect*, or the tendency of people to conform with the judgment and behavior of others, was studied by Asch [79]. Bond and Smith [80] confirmed the conformity effect, showed its historical change, and emphasized its cultural dependence. Other important effects are *fundamental attribution error* and *false consensus effect* (see, Reference 73). *Herding behavior* or mimetic contagion has been proposed as the source of endogenous fluctuations (bubbles and crashes) in financial markets [81]. This is interesting as it suggests that such market fluctuations may arise irrespectively of exceptional news or other exogenous shocks to the market.

*Action* is performed when an investor wants to change the current portfolio. It can be characterized by the security that investors wants to buy or sell, a type of order, the size of the order, the timing, and other parameters characteristic to a particular type of order (e.g., price contingencies). The most common types of orders are *market order* and *limit order*. More about various types of orders and their properties can be found in Reference 82. Our conceptual model can account for various underlying causes for a particular action, such as a trading strategy, peer influence, a habit, or an emotional response to financial news or price changes.

### 13.2.2 Investment environment

The investment environment in which investors operate (Figure 13.2) is based on the description of trading industry in Reference 82. Harris defines traders as people who trade, and who may either arrange trades for themselves, have other people arrange trades for them, or arrange trades for others. Furthermore, trading industry can be grouped into two sides of traders: *Buy Side* – traders who buy exchange services, such as liquidity, or the ability to trade when they want

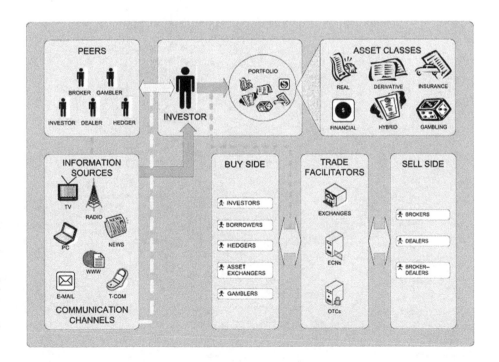

*Figure 13.2 Investment environment*

(e.g., individuals, funds, firms, governments); and *Sell Side* – traders who sell exchange services to the buy side (e.g., dealers, brokers, and broker-dealers) [82].

*Trade facilitators* are institutions that help traders trade [82]. Exchanges provide forums for traders (dealers, brokers, and buy-side traders) to meet and arrange trades. At most exchanges only members can trade, while non-members trade by asking member-brokers to trade for them. Historically, traders used to meet at the trading floor, but now they meet via electronic communication networks (ECNs). At some exchanges traders arrange trades when they want, while other exchanges have order-driven systems that arrange trades by matching buy and sell orders according to a set of rules. Over the counter (OTC) is trading that occurs outside exchanges, arranged by dealers and brokers. Trade facilitators which help traders settling their trades are known as Clearing Agents, Settling Agents, and Depositories and Custodians [82].

Modeling trade facilitators is a challenge itself, particularly if one wants to model financial exchanges as realistically as possible. Most existing agent-based studies, however, do not go into that level of detail, and instead use simple and abstract market mechanisms (Figure 13.3). LeBaron in Reference 12 describes four such mechanisms:

1. The first mechanism is based on a *price adjustment process*, where new price is determined from the past price and the net order scaled by a factor called

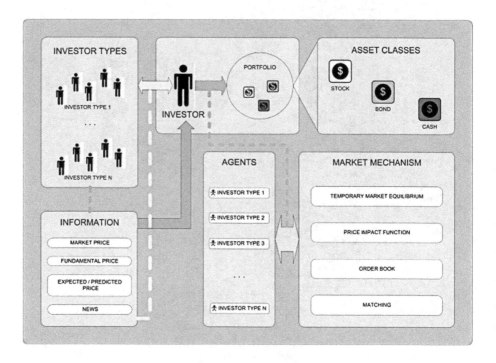

*Figure 13.3    Agent-based implementations*

liquidity. This captures the basic intuition that excess demand raises the price, while excess supply lowers the price.

2. The second mechanism is clearing in a *temporary market equilibrium*, where the price is determined so that the total demand equals the total number of shares in the market.

3. The third mechanism is a more realistic one, where an actual *order book* is simulated, and buy and sell orders are crossed using a certain well-defined procedure.

4. The final mechanism is the one in which *agents (randomly) meet*, and, if it suits them, trade with each other.

   Trading instruments can be organized into classes, such as real assets, financial assets, derivative contracts, insurance contracts, gambling contracts, and hybrid contracts [82]. Most of the (behavioral) finance literature focuses on financial assets, particularly on equities. The same is also true for the agent-based literature, as most AFMs are actually artificial stock markets. Asset classes used in agent-based models are usually risky asset (stocks), risk-less asset (bond), and cash. Many agent-based models study only one risky asset. This does not necessarily represent a too strong restriction, particularly if the focus is on modeling a whole market and its index. However, a more realistic approach should deal with multi-asset environments and the intricacies of correlations among them.

## 13.3   Implementation examples

In this section, we explain, in terms of the proposed conceptual model, three existing agent-based models of financial markets that have accounted for behavioral phenomena:

In Reference 83, the overconfidence bias of investors has been studied using the Levy, Levy, Solomon (LLS) model [15].

In a similar model [84] investor optimism/pessimism has been modeled using a fuzzy connective.

The model of Reference 19 is particularly interesting as it puts emphasis on social theories in the agent-based modeling of financial markets.

### 13.3.1   Study of investor overconfidence

#### 13.3.1.1   Investment environment

In the LLS model, the market consists of two types of investors: Rational Informed Investors (RII) and Efficient Market Believers (EMB). There are two *asset classes* in the market: a risky stock that pays a dividend following a multiplicative random walk and has a finite number of outstanding shares; and a risk-less bond that pays a sure interest and has an infinite supply. At the beginning of the simulation all investors are endowed with the same amount of wealth that is comprised of cash and a number of shares. The trade facilitators and the sell side of financial services are not explicitly modeled in the LLS model. Instead, the pricing mechanism based on the *temporary market equilibrium* [12] is used to determine the price in such a way that the total demand for the risky asset equals the total number of outstanding shares.

### 13.3.1.2    Individual investors

The *goal* of all the investors in the LLS model is the maximization of the expected utility of the next period wealth. The *risk* attitude of the investors is risk aversion, and is captured by the parameter of the utility function. In the LLS model [15], a myopic power utility function with Decreasing Absolute Risk Aversion (DARA) and Constant Relative Risk Aversion (CRRA) properties is used. Due to this myopia property of this utility function, it can be assumed that investors maximize their one-period-ahead expected utility, regardless of their actual investment horizon [15]. An additional *temporal* characteristic of EMB investors is their memory length of past return realizations, which are used in the prediction of future returns. Even though both RII and EMB investors have the same goal of expected utility maximization, their *strategies* are different because of the differences in information that they possess. RII investors know the properties of the dividend process, and can estimate the fundamental price of the risky asset as a discounted stream of future dividends. That fundamental price is used in their prediction of the next period return. EMB investors, however, do not know the dividend process, and must use ex post distribution of returns to estimate ex ante distribution. EMB investors use a rolling window of a fixed size, and in the original model they are called unbiased if, in the absence of any additional information, it is assumed that returns come from a discrete uniform distribution. In Reference 83, we introduce normal EMB investors who predict future returns using a normal distribution based on the mean of the sample of past returns. Overconfidence *bias* is modeled as miscalibration, i.e., underestimated standard deviation of the return distribution. In addition to overconfidence bias, a self-attribution *bias* is modeled to study the emergence of overconfidence. In each period, investors acquire information about the new price and new dividends, so they also receive information about the recent return. This *perception* of the current return is used in a feedback mechanism to evaluate the accuracy of return predictions. A biased self-attribution means that a good prediction is rewarded relatively more than a bad prediction, which results in an increased overconfidence. In the LLS model, investors do not *interact* (exchange information) directly among themselves. If there is some volume (the change in portfolio holdings) of an individual investor, we can say that some trading occurred because the shares have exchanged hands. However, in this model we are not concerned with how that exactly happened. *Actions*, such as market orders, are not modeled explicitly, since this is a model based on the temporary market equilibrium. Both types of investors are expected utility maximizers, which is considered as the cornerstone of rationality. Nonetheless, with EMB investors some noise is added to the optimal proportion to account for *other factors* that could cause such departures from optimal behavior [15].

## 13.3.2    Study of investor sentiment

In Reference 84, based on the same background LLS model [15], we study *emotional* underpinnings of the investor sentiment, namely the investor optimism/pessimism. Optimistic EMB investors predict future returns using a higher value from the past return observations, while pessimistic EMB investors focus on lower values from the rolling window. This is modeled by using a generalized average operator and interpreting its parameter as an index of optimism.

## 13.3.3    Social simulation of stock markets

### 13.3.3.1    Investment environment

*Asset classes* available in the market are one risky stock and cash. The *market mechanism* is based on the order book, using which the limit orders sent by agents are matched and executed. Market price is calculated as the average of the bid and ask prices, weighted by the number of asked and offered shares. Such pricing mechanisms based on order books are more realistic, as they are more similar to those used in actual financial markets. The model of Reference 19 also models *news* arrival process, as a normally distributed noise around the current price.

### 13.3.3.2    Individual investors

The investors are characterized by their level of confidence, which determines how much their private information (price expectation) is weighted compared to the expectations of their neighboring investors. *Strategy* used by investors is based on the comparison between the current market price and the expected market price of the stock. When the expected price is higher than the current price, it is attractive to invest in stock, and when the expected price is lower, it seems attractive to divest. The strategy also determines the proportion of cash to invest or the proportion of stock to divest. The perception of *risk* depends on investors' level of confidence. Investors who have high confidence perceive lower risk. Those investors who have lower levels of confidence perceive high risk and apply *risk-reducing strategies*, which can be a simplifying strategy (*heuristic*) or a clarifying strategy (collecting more *information*). *Time* perspective of investors is myopic. They base their decisions only on the currently available information and expectations for the next period. Agents *perceive* information about the current market price, the news about the stock, and the price expectation of investors in their social network, which are all finally translated into their own market price expectation. The model of Reference 19 is interesting as it pays special attention to the social *interaction* between agents. In different experiments the investors are connected in two different types of social networks, which are used for the dissemination of market price expectations. Investors who exhibit social type of simplifying risk-reducing strategy copy the behavior of other investors in the social network, while those who exhibit social type of clarifying risk-reducing strategy ask other investors for more information. The investors *act* by sending limit orders, which consist of the number of shares that they want to buy or sell, and the limit price, which is set to their expected price.

## 13.4  Conclusion

This chapter presented a descriptive model of individual investor behavior in which investment decisions are seen as an iterative process of interactions between the investor and the investment environment. This investment process is driven by dual mental systems, the interplay of which contributes to boundedly rational behavior where investors use various heuristics and may exhibit behavioral biases. In the modeling tradition of cognitive science and artificial intelligence, the investor is seen as learning, adapting, and evolving entity that perceives the environment, processes information, acts, and updates its internal states. Finally, the investor behavior is influenced by social interactions with other market participants. The aim of this model is descriptive, that is, to be able to capture and describe observed behavioral phenomena. However, these behavioral phenomena stem from a wide variety of cognitive mechanisms and other sources, which makes it difficult to integrate them into a simple and parsimonious framework. Furthermore, in case a normative model exists, a descriptive model is expected to be more general (to capture the behavior of those who depart from the norm, and those who act accordingly). The resulting model is an eclectic model with many interdependent variables, i.e., every variable is influenced by the current states of other variables, by cognitive and affective processes, and by the feedback mechanism. It is rather ambitious task to determine the nature of all these relationships. However, it is possible to indicate some of the relationship by finding evidence in the existing behavioral finance literature.

Traditionally, behavioral finance has relied mostly on experiments. Whereas experiments are good in capturing behavior in controlled environments, they might lack generalizability to real financial markets.[5] Should similar behavior occur in financial markets, how would it affect the performance of individual investors, and how would it aggregate to the market level? Quantitative analysis of financial data may answer some of these questions, but it is not always clear what is the exact behavior of market participants that lead to it. Agent-based AFMs are potentially able to fully bridge this gap between individual investor behavior and aggregate market phenomena, by allowing

---

[5] Given that most biases shown in these experiments come from hypothetical static risky choice problems, a great caution should be exercised when introducing such results into highly dynamic financial models [20]

the modeler to specify the behavior of market participants, to implement various market mechanisms, and to analyze the resulting asset prices. In such a way, AFMs can be used as a tool to generate and/or test various behavioral hypotheses and theories, and to study scenarios for which empirical data does not exist, or is too difficult to obtain. Conceptual model of individual investor behavior presented in this chapter aims to summarize and structure part of the vast knowledge about investor behavior that is present in the (behavioral) finance field. In such a way, it can serve as a pool of ideas for agent-based models of financial markets that aim at a more complex and a more realistic representation of investor behavior.

# References

1. M. Lovric, U. Kaymak, and J. Spronk, 'A conceptual model of investor behavior,' Erasmus Research Institute of Management, ERIM Report Series ERS-2008-030-F&A, May 2008.
2. S.-H. Chen, 'Software-agent design in economics: An interdisciplinary framework,' *IEEE Computational Intelligence Magazine*, **3**(4): 18–30, 2008.
3. M. Statman, 'Behavioral finance: Past battles and future engagements,' *Financial Analysts Journal*, **55**(6): 18–27, 1999.
4. W.F. DeBondt, 'A portrait of the individual investor,' *European Economic Review*, **42**(3–5): 831–844, 1998.
5. H.A. Simon, 'Bounded rationality and organizational learning,' *Organization Science*, **2**(1): 125–134, 1991.
6. E.F. Fama, 'Efficient capital markets: A review of theory and empirical work,' *The Journal of Finance*, **25**:(2): 383–417, 1970.
7. B.G. Malkiel, 'The efficient market hypothesis and its critics,' *The Journal of Economic Perspectives*, **17**(1): 59–82, 2003.
8. A.W. Lo, D.V. Repin, and B.N. Steenbarger, 'Fear and greed in financial markets: A clinical study of day-traders,' *American Economic Review*, **95**(2): 352–359, 2005.
9. H. Shefrin, *Beyond Greed and Fear – Understanding Behavioral Finance and the Psychology of Investing*. Boston: Harvard Business School Press, 2000.
10. J. Goldberg and R. von Nitzsch, *Behavioral Finance*. Chichester: John Wiley and Sons Ltd, 2001.
11. N.C. Barberis and R.H. Thaler, 'A survey of behavioral finance,' in *Handbook of the Economics of Finance*, 1st edn, G.M. Constantinides, M. Harris, and R.M. Stulz (eds.) Amsterdam, North-Holland: Elsevier, 2003, vol. 1, pp. 1053–1128.
12. B. LeBaron, 'Agent-based computational finance,' in *Handbook of Computational Economics*, 1st edn, L. Tesfatsion and K.L. Judd (eds) Amsterdam, North-Holland: Elsevier, 2006, vol. 2, ch. 24, pp. 1187–1233.

13. C.H. Hommes, 'Heterogeneous agent models in economics and finance,' in *Handbook of Computational Economics*, 1st edn, L. Tesfatsion and K.L. Judd (eds.) Amsterdam, North-Holland: Elsevier, 2006, vol. 2, ch. 23, pp. 1109–1186.

14. L. Tesfatsion, 'Agent-based computational economics: A constructive approach to economic theory,' in *Handbook of Computational Economics*, 1st edn, L. Tesfatsion and K.L. Judd (eds.) Amsterdam, North-Holland: Elsevier, 2006, vol. 2, ch. 16, pp. 831–880.

15. M. Levy, H. Levy, and S. Solomon, *Microscopic Simulation of Financial Markets: From Investor Behavior to Market Phenomena*. San Diego: Academic Press, 2000.

16. R. Harré, *The Principles of Scientific Thinking*. London: MacMillan, 1970.

17. N.T. Chan, B. LeBaron, and T. Poggio, 'Agent-based models of financial markets: A comparison with experimental markets,' MIT Artificial Markets Project. MA: MIT, 1999.

18. H. Takahashi and T. Terano, 'Agent-based approach to investor's behavior and asset price fluctuation in financial markets,' *Journal of Artificial Societies and Social Simulation*, 6(3), 2003. Available from http://jasss. soc.surrey.ac.uk/6/3/3.html [Accessed 18 March 2010].

19. A.O.I. Hoffmann, W. Jager, and E.J.H.V. Eije, 'Social simulation of stock markets: Taking it to the next level,' *Journal of Artificial Societies and Social Simulation*, 10(2), 2007. Available from http://jasss.soc.surrey. ac.uk/10/2/7.html [Accessed 18 March 2010]

20. S.-H. Chen and C.-C. Liao, 'Behavioral finance and agent-based computational finance: Toward an integrating framework,' in *Proceedings of 7th Joint Conference on Information Sciences*, 2003, pp. 1235–1238.

21. C. Camerer, G. Loewenstein, and D. Prelec, 'Neuroeconomics: How neuroscience can inform economics,' *Journal of Economic Literature*, 43(1): 9–64, 2005.

22. F.H. Knight, *Risk, Uncertainty, and Profit*. Hart, Schaffner and Marx; Boston: Houghton Mifflin Company, 1921.

23. D. Bernoulli, 'Exposition of a new theory on the measurement of risk,' *Econometrica*, 22(1): 23–36, 1954.

24. J. von Neumann and O. Morgenstern, *Theory of Games and Economic Behavior*. Princeton: Princeton University Press, 1944.

25. D. Kahneman and A. Tversky, 'Prospect theory: An analysis of decision under risk,' *Econometrica*, 47(2): 263–291, 1979.

26. J. Quiggin, 'A theory of anticipated utility,' *Journal of Economic Behavior and Organization*, 3(4): 323–343, 1982.

27. D. Schmeidler, 'Subjective probability and expected utility without additivity,' *Econometrica*, 57(3): 571–587, 1989.

28. A. Tversky and D. Kahneman, 'Advances in prospect theory: Cumulative representation of uncertainty,' *Journal of Risk and Uncertainty*, 1992.

29. L. Zadeh, 'Fuzzy sets,' *Information and Control*, **8**(3): 338–353, 1965.
30. G. Shafer, *A Mathematical Theory of Evidence*. Princeton: Princeton University Press, 1976.
31. Z. Pawlak, *Rough Sets: Theoretical Aspects of Reasoning about Data*. Dordrecht: Springer, Formerly Kluwer Academic Publishers, 1991.
32. H. Markowitz, 'The utility of wealth,' *The Journal of Political Economy*, **60**(2): 151–158, 1952.
33. G. Loomes, 'Some lessons from past experiments and some challenges for the future,' *The Economic Journal*, **109**(453): 35–45, 1999.
34. J.R. Bettman, M.F. Luce, and J.W. Payne, 'Agent-based approach to investor's behavior and asset price fluctuation in financial markets,' *Journal of Consumer Research*, **25**(3): 187–218, 1998.
35. P. Slovic, 'Psychological study of human judgment: Implications for investment decision making,' *The Journal of Psychology and Financial Markets*, **2**(3): 160–172, 1998.
36. R.H. Thaler and E.J. Johnson, 'Gambling with the house money and trying to break even: The effects of prior outcomes on risky choices,' *Management Science*, **36**(6): 643–660, 1990.
37. P.J. Cooley, 'A multidimensional analysis of institutional investor perception of risk,' *The Journal of Finance*, **32**(1): 67–78, 1977.
38. A. Shleifer, *Inefficient Markets: An Introduction to Behavioral Finance*. Oxford: Oxford University Press, 2000.
39. G. Loewenstein and T. O'Donoghue, 'Animal spirits: Affective and deliberative processes in economic behavior,' August 2004, CAE Working Paper 04-14.
40. J.P. Byrnes, D.C. Miller, and W.D. Schafer, 'Gender differences in risk taking: A meta-analysis,' *Psychological Bulletin*, **125**(3): 367–383, 1999.
41. B. Donkers, B. Melenberg, and A.V. Soest, 'Estimating risk attitudes using lotteries: A large sample approach,' *Journal of Risk and Uncertainty*, **22**(2): 165–195, 2001.
42. S. Jaggia and S. Thosar, 'Risk aversion and the investment horizon: A new perspective on the time diversification debate,' *The Journal of Psychology and Financial Markets*, **1**(3 & 4): 211–215, 2000.
43. T. Gilovich, M. Kerr, and V.H. Medvec, 'Effect of temporal perspective on subjective confidence,' *Journal of Personality and Social Psychology*, **64**(4): 552–560, 1993.
44. P.A. Samuelson, 'A note on measurement of utility,' *The Review of Economic Studies*, **4**(2): 155–161, 1937.
45. D. Laibson, 'Golden eggs and hyperbolic discounting,' *The Quarterly Journal of Economics*, **112**(2): 443–477, 1997.
46. D. Read, 'Is time-discounting hyperbolic or subadditive?' *Journal of Risk and Uncertainty*, **23**(1): 5–32, 2001.

47.  H. Shefrin and R. Thaler, 'The behavioral life cycle hypothesis,' *Economic Inquiry*, **26**(4): 609–643, 1988.
48.  J.M. Digman, 'Personality structure: Emergence of the five-factor model,' *Annual Review of Psychology*, vol. 41, pp. 417–440, 1990.
49.  M. Fenton-O'Creevy, N. Nicholson, E. Soane, and P. Willman, *Traders: Risks, Decisions, and Management in Financial Markets*. Oxford: Oxford University Press, 2004.
50.  R.R. McCrae and P.T.J. Costa, 'Toward a new generation of personality theories: Theoretical contexts for the five-factor model' in *The Five-Factor Model of Personality: Theoretical Perspectives*. New York, NY: Guilford, 1996.
51.  J.T. Austin and J.B. Vancouver, 'Goal constructs in psychology: Structure, process, and content,' *The Quarterly Journal of Economics*, **120**(3): 338–375, 1996.
52.  R. Custers and H. Aarts, 'Beyond priming effects: The role of positive affect and discrepancies in implicit processes of motivation and goal pursuit,' *European Review of Social Psychology*, vol. 16, pp. 257–300, 2005.
53.  W.F. Sharpe, G.J. Alexander, and J.V. Bailey, *Investments*. Upper Saddle River: Prentice Hall, 1999.
54.  Z. Bodie, A. Kane, and A.J. Marcus, *Investments*. New York: McGraw-Hill/Irwin, 2006.
55.  H. Markowitz, *Portfolio Selection: Efficient Diversification of Investments*. New Haven: Yale University Press, 1959.
56.  S. Benartzi and R.H. Thaler, 'Naive diversification strategies in defined contribution saving plans,' *The American Economic Review*, **91**(1): 79–98, 2001.
57.  V. DeMiguel, L. Garlappi, and R. Uppal, 'Optimal versus naïve diversification: How inefficient is the 1/n portfolio strategy?' *The Review of Financial Studies*, **22**(5): 1915–1953, 2009.
58.  K. French and J. Poterba, 'International diversification and international equity markets,' *The American Economic Review*, **81**(2): 222–226, 1991.
59.  G. Huberman and W.J. DeMiguel, 'Offering versus choice in 401(k) plans: Equity exposure and number of funds,' *The Journal of Finance*, **61**(2), 2006.
60.  W.N. Goetzmann and A. Kumar, 'Equity portfolio diversification,' December 2001, nBER Working Paper Series.
61.  B.G. Malkiel, 'Reflections on the efficient market hypothesis: 30 years later,' *The Financial Review*, **40**(1): 1–9, 2005.
62.  H. Shefrin and M. Statman, 'The disposition to sell winners too early and ride losers too long: Theory and evidence,' *The Journal of Finance*, **40**(3): 777–790, 1985.
63.  T. Odean, 'Do investors trade too much?' *The American Economic Review*, **89**(5): 1279–1298, 1999.
64.  W. Samuelson and R. Zeckhauser, 'Status quo bias in decision making,' *Journal of Risk and Uncertainty*, **1**(1): 7–59, 1988.

65. B. Mellers, A. Schwartz, and D. Cooke, 'Judgment and decision making,' *Annual Review Psychology*, vol. 49, pp. 447–477, 1998.
66. T. Gilovich and V. Medvec, 'The experience of regret: What, why, and when,' *Psychological Review*, **102**(2): 379–395, 1995.
67. G. Loewenstein, 'Emotions in economic theory and economic behavior,' *The American Economic Review*, **90**(2): 426–432, 2000.
68. B. Shiv, G. Loewenstein, A. Bechara, H. Damasio, and A. Damasio, 'Investment behavior and the negative side of emotion,' *Psychological Science*, **16**(6), 435–439, 2005.
69. A.R. Damasio, B.J. Everitt, and D. Bishop, 'The somatic marker hypothesis and the possible functions of the prefrontal cortex,' *Philosophical Transactions: Biological Sciences*, **351**(1346), 1996.
70. A. Tversky and D. Kahneman, 'Judgment under uncertainty: Heuristics and biases,' *Science*, **185**(4157): 1124–1131, 1974.
71. M. Rabin, 'Psychology and economics,' *Journal of Economic Literature*, **36**(1): 11–46, 1998.
72. D. Kahneman and M.W. Riepe, 'Aspects of investor psychology,' *Journal of Portfolio Management*, **24**(4): 52–65, 1998.
73. D. Hirshleifer, 'Investor psychology and asset pricing,' *Journal of Finance*, **56**(4): 1533–1597, 2001.
74. G. Gigerenzer and D.G. Goldstein, 'Reasoning the fast and frugal way: Models of bounded rationality,' *Psychological Review*, **103**(4): 650–669, 1996.
75. A. Sloman, 'Beyond shallow models of emotion,' *Cognitive Processing*, **2**(1): 177–198, 2001.
76. T. Oberlechner and S. Hocking, 'Information sources, news, and rumors in financial markets: Insights into the foreign exchange market,' *Journal of Economic Psychology*, vol. 25, pp. 407–424, 2004.
77. R.J. Shiller, 'Speculative prices and popular models,' *Journal of Economic Perspectives*, **4**(2): 55–65, 1990.
78. R.J. Shiller and J. Pound, 'Survey evidence on diffusion of interest and information among institutional investors,' *Journal of Economic Behavior and Organization*, **12**(1): 47–66, 1989.
79. S. Asch, 'Studies of independence and conformity: A minority of one against a unanimous majority,' *Psychological Monographs*, vol. 70, pp. 1–70, 1956.
80. R. Bond and P. Smith, 'Culture and conformity: A meta-analysis of studies using Asch's (1952b, 1956) line judgment task,' *Psychological Bulletin*, vol. 119, pp. 111–137, 1989.
81. R. Topol, 'Bubbles and volatility of stock prices: Effect of mimetic contagion,' *The Economic Journal*, **101**(407): 786–800, 1991.
82. L. Harris, *Trading and Exchanges: Market Microstructure for Practitioners*. New York: Oxford University Press, 2003.
83. M. Lovric, U. Kaymak, and J. Spronk, 'Overconfident investors in the LLS agent-based artificial financial market,' in *Proceedings of the 2009*

*IEEE Symposium on Computational Intelligence for Financial Engineering*, ser. 2009 IEEE Symposium Series on Computational Intelligence, Nashville, TN, Mar. 2009, pp. 58–65.

84.  M. Lovric, R.J. Almeida, U. Kaymak, and J. Spronk, 'Modeling investor optimism with fuzzy connectives,' in *Proceedings of the 2009 IFSA World Congress (IFSA 2009)*, Lisbon, Portugal, July 2009, pp. 1803–1808.

*Chapter 14*

# Decision making under risk in swarm intelligence techniques

*Samia Nefti, Ahmed I. Al-Dulaimy, May N. Bunny and John Gray*

## Abstract

The term Swarm intelligence is associated with the decision processes that determine the overall group behavior of a collection of individuals or sets of simple agents cooperating to achieve some purposeful objective or common goal. However, the usual logical decision processes used in the literature to model individual agent behavior are generally found to be inadequate when the phenomena of uncertainty and risk are factored into process; and these models are usually incapable of fully emulating actual human decision-making behaviors under risk and uncertainty.

This chapter proposes a significant modification to agent reasoning processes employed so far in conventional swarm intelligence techniques. We show that by endowing each agent with some descriptive behavior from the field of psychology named Prospect Theory (PT), we can improve considerably the efficiency of global searching procedures. The efficacy of the technique is illustrated by numerical results obtained from applications to two classical problems quoted in the literature.

## 14.1 Introduction

Computer scientists and biologists in the field of artificial life have long considered how to model biological swarm's behavior such as how a flock of birds' maneuvers in the sky, a group of ants' forages for food, or how a school of fish swims, turns, and flees together. There is a need to understand how groups of such social creatures interact, evolve, make decisions, and achieve goals in some optimal way. Recently there has been renewed interest in this kind of swarm behavior since the resulting "Swarm Intelligence"; can be applied with advantage in many field for example, robotics, optimization procedures, and in the study of traffic patterns in transportation systems. The idea of decision-making process under conditions of risk was first introduced by Daniel Bernoulli in 1738, in a paper entitled "Exposition of a New Theory on the Measurement of Risk" (Bernoulli, 1954; Zhang *et al.*, 2007). Therefore, the study of decision-making process is not recent. It has been evolving with contributions from a number of disciplines for over 300 years and provided mathematical foundations to routine applications in many areas such as finance, medicine, and cybernetics. Moreover, Abraham Wald introduced the concepts of loss and risk functions that had a significant impact on the 20th century for decision-making theory (Wald, 1950; Weiss and Wald, 1997; Zhang *et al.*, 2007). Consequently, decision-making theories have been embodied in several concepts and models, which influence nearly all the biological, cognitive, and social sciences (Doyle and Thomason, 1999).

Human decision-making processes are presented as normative (rational decision making) and descriptive (psychological decision making). In rational decision-making models, the decision makers have analyzed a number of possible alternatives (options) from different scenarios before selecting a choice. Then, these scenarios are weighted by probabilities, and the decision makers can determine the expected scenario for each alternative. As a result, the final choice should be the one presenting the best-expected scenario and with the highest probability of outcome. This type of decision has explained how the decision maker uses a set of alternatives to solve problems (Goodwin and Wright, 1998; Hoch *et al.*, 2001; Oliveira, 2007). In view of this process, Russell and Norvig (2002) and Zhang *et al.* (2007) discuss that Expected Utility Theory (EUT) has been generally accepted as a model of rational choice under risk, based on the assumption that people behave as rational agents and wish to follow EUT for the actual human decision-making process. In this respect, the two prominent psychologists Daniel Kahneman and Amos Tversky proposed a descriptive model of the way that humans make decisions, which was later known as Prospect Theory (PT) (Kahneman and Tversky, 1979; Tversky and Kahneman, 1992). Kahneman and Tversky (1979) claim that PT captures common human decision-making attitudes toward risk that cannot be captured by EUT, i.e., risk aversion and risk seeking (Kahneman and Tversky, 1979; Zhang *et al.*, 2007).

Recently, simple decision-making models have been introduced in swarm intelligence techniques such as particle swarm optimization (PSO), the ant colony algorithm, etc. Although the agents follow very basic rules, there is no centralized control structure dictating how individual agents (particles) should behave.

Other work proposed that each particle within the swarm be endowed with some reasoning to achieve the best decision. For instance, Wang *et al.* (2009) have used group decision making in PSO to coordinate the objective behavior of particles to achieve the best outcome, by analyzing a number of possible alternatives from different outcomes. In this algorithm, each agent is considered as an individual decision maker and uses the basic information of the particle to decide the new global best position of the swarm. Even though incorporation of some intelligence in the decision-making process of each agent seems to improve the performance of the global search, the logical decision process used in each agent's behavior is inadequate when the uncertainty and risk are associated with the decision-making process and these models are incapable of fully capturing actual human decision-making behaviors under risk.

This chapter proposes a significant modification to the agent's decision-making process used so far in conventional swarm intelligence techniques. A descriptive model of decision making under conditions of risk and uncertainty, named Prospect Theory (PT), will be endowed in each agent to model a swarm intelligence technique. This chapter will focus mainly on implementing this model in PSO, but it could later be extended to other particle swarm intelligence techniques. Unlike conventional swarm intelligence techniques, in the proposed method, each agent is endowed with nonrational behavior in the sense of PT, where agents are given a score derived from the objective functions and the probability of violating system constraints. This is defined by value and the weight probability functions, respectively. The personal best positions of each agent will then be selected based on the current and the historical best prospect score of each agent. The proposed technique has been tested with two well-known benchmark examples. The outcome and the results show a significant improvement in the global searching ability of the swarm with lower cost and are more reliable compared to other algorithms.

The structure of this chapter is organized as follows: in Section 14.2, the Prospect Theory will be introduced. The value and weighting functions of the prospect theory are given in Section 14.3. Section 14.4 presents a brief introduction of the particle swarm optimization (PSO) and swarm algorithm for decision-making under risk has been presented in Section 14.5. Simulation and comparison results of two benchmark examples are presented in Section 14.6; conclusions are specified in Section 14.7. Finally, limitations and further study are given in Section 14.8.

## 14.2 Prospect Theory (PT)

Prospect Theory (PT) is the leading behavioral model of decision-making under risk, and the major work for which psychologist Daniel Kahneman was awarded the 2002 Nobel Prize in economics. Moreover, this theory captures the tendency for people to strongly prefer avoiding losses over acquiring gains (Kahneman and Tversky, 1979). Heath and Tversky (1991) and Zhang *et al.* (2007) believe that Prospect Theory reflects two common attitudes toward risk or choice in human decision-making (Zhang et al., 2007, p. 2). The first attitude is risk aversion in the case of gains; people overweight gains that are considered certain, relative to gains that are merely probable. The second attitude is risk seeking in the case of losses; people overweight losses that are merely probable, relative to losses that are considered certain.

According to Heath and Tversky (1991), PT explains two main phenomena of choice that the EUT model has failed to take into account. First, the choice that PT identifies runs completely contrary to the assumption of EUT that states "The rational theory of choice assumes description invariance: equivalent formulation of a choice problem should give rise to the same preference order." Furthermore, Kahneman and Tversky (1979) demonstrated with empirical evidence that this is not so. Therefore, they concluded that "variations in the framing of options yield systemically different preferences." Second, the

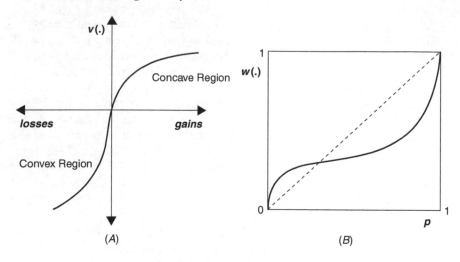

*Figure 14.1    Value and weighting functions for prospect theory. (A) Value function v(.) as a function of gains and losses. (B) Weighting function w(.) for gains as a function of the probability p of a chance event Kahneman and Tversky (1979)*

outcome probabilities of EUT should be linear in shape. As a result, the PT is based on the assumption that people are more focused on changes in their utility states rather than the state itself like EUT (Zhang *et al.*, 2007).

In relation to PT, the decision making under risk can be viewed as a choice between prospects or gambles. The value $V(x, p)$ of a simple prospect that pays $x with probability $p$ (and nothing otherwise) is given by

$$V(x,p) = v(x)wp$$

where $v$ measures the subjective value of the consequence $x$ (Figure 14.1A), and $w$ measures the impact of probability $p$ on the attractiveness of the prospect (Figure 14.1B).

Kahneman and Tversky (1979) and Tversky and Kahneman (1992) clarify the difference between PT and Expected Utility Theory (EUT) in Figure 14.1A and B. First, the utility function of EUT (which represent the utility of obtaining outcome $x$), is replaced by a value function $v(.)$ over gains and losses relative to a reference point (usually the status quo), with $v(0) = 0$. Second, this subjective value function is not weighted by outcome probabilities but rather by a decision weight, $w(.)$ that represents the impact of the relevant probability on the evaluation of the prospect. Decision weights are normalized so that $w(0) = 0$ and $w(1) = 1$. Third, unlike EUT, PT explicitly incorporates principles of editing that allow for different descriptions of the same choice to give rise to different decisions (such as lower outcome as losses and higher outcomes as gains).

## 14.3    The value and weighting functions

Kahneman and Tversky (1979) and Tversky and Kahneman (1992) show that the shapes of the value function $v(.)$ and weighting function $w(.)$ reflect the psychophysics of diminishing sensitivity. The marginal impact of a change in outcome or probability of outcome diminishes with distance from relevant reference points.

For monetary outcomes, the status quo generally serves as the reference point distinguishing losses from gains, so that the function is concave for gains and convex for losses (Figure 14.1A). Concavity for gains contributes to risk aversion for gains with the standard utility function. Convexity for losses, on the other hand, contributes to risk seeking for losses.

### 14.3.1    The value function

The value function of PT is assigned to gains and losses rather than to final assets. This value function is steeper for losses than gains, a property known as loss aversion. Kahneman *et al.* (1997) observe that people typically require more compensation to give up a possession than they would have been willing to pay to obtain it in the first place. Samuelson and Zeckhauser (1988) illustrate that the tendency for the relative disadvantages of alternatives to loom larger than the relative advantages supports a bias toward the status quo.

Kahneman and Tversky (1991) formulate the value function as a power function:

$$v(x) = \begin{cases} x^\alpha & x \geq 0 \\ -\lambda(-x)^\beta & x < 0 \end{cases}$$

where $\alpha, \beta > 0$ measures the curvature of the value function for gains and losses, respectively, and $\lambda$ is the coefficient of loss aversion. Thus, the value function for gains (losses) is increasingly concave (convex) for smaller values of $\alpha(\beta) < 1$, and loss aversion is more pronounced for larger values of $\lambda > 1$. Moreover, Tversky and Kahneman (1992) estimate values of $\alpha = 0.88$, $\beta = 0.88$, and $\lambda = 2.25$ among their sample of college students.

### 14.3.2    Weighting function

Figure 14.1B above shows the weighting function of PT as quoted in Tversky and Kahneman (1992). In this figure, an inverse-S-shaped function that is concave near zero and convex near one (multidimensional character) has been presented. This is because low probabilities are over-weighted (leading to risk seeking for gains and risk aversion for losses) and high probabilities are underweighted (leading to risk aversion for gains and risk seeking for losses). Scholars and researchers such as Camerer and Ho (1994), Wu and Gonzalez (1996, 1998), Prelec (1998), Gonzalez and Wu (1999),

Abdellaoui (2000), and Wakker (2001) argue that the inverse-S-shaped weighting function seems to be consistent with a range of empirical findings.

The weighting function can be parameterized in the following form according to probability weighting function originally proposed by Tversky and Kahneman (1992):

$$w(p) = \frac{p^{\gamma}}{(p^{\gamma} + (1 - p)^{\gamma})^{1/\gamma}}$$

where $p$ is the weighting probability of the distribution of gains or losses and $\gamma > 0$ measures its degree of curvature.

## 14.4  Particle Swarm Optimization (PSO)

Particle swarm optimization (PSO) is a population-based stochastic optimization technique inspired by the social behavior of groups of birds or school of fish. It was first introduced by Kennedy and Eberhart (1995). Eberhart and Kennedy (1995) and Eberhart *et al.* (1996) claim that the PSO is primarily used in solving unconstrained optimization problems. Indeed, several developments evolved to support PSO for use in solving constraint optimization problems.

Eberhart and Kennedy (1995) suggest that in a PSO system, all particles are initialized randomly at the beginning so that they cover the entire search space of the problem, and each particle represents a possible solution to the optimization problem. Through successive generations, each particle evolves, and in the end, the global optimum result is found. In addition, in the case of a standard PSO with *global* neighborhood topology, each particle has its own velocity and the best position is obtained and is known as *personal best* or *PBest* position. The best position among the entire swarm is known as the *global best* or *GBest* position. Furthermore, while in the *local* neighborhood topology each particle has knowledge of the best position obtained within its own neighborhood (known as *local best* or *LBest* position) not of the best global position.

The standard global topology of PSO consists of a swarm of $N$ particles, the position of each particle $i$ at iteration $t$ is written as $x_i(t) = (x_{i1}, x_{i2}, \ldots, x_{iD})$, and the velocity of each particle $v_i(t) = (v_{i1}, v_{i2}, \ldots, v_{iD})$, where $D$ is the dimension of the optimization problem. During iteration, each particle adjusts its position toward its personal best position and the global best position according to the following equation (Shi and Eberhart, 1998):

$$v_{ij}(t+1) = wv_{ij}(t) + c_1 r_1 (PBest_{ij}(t) - x_{ij}(t)) + c_2 r_2 (GBest_{ij}(t) - x_{ij}(t))$$

$$x_{ij}(t+1) = x_{ij}(t) + v_{ij}(t+1)$$

where $c_1$ and $c_2$ are positive constant parameters called the cognitive and social factor, respectively, $r_1$ and $r_2$ are two independently uniformly distributed random variables with the range of [0,1] and $w$ is the inertia weight which controls the impact of previous velocity of particle on its current one. Shi and Eberhart (1998) suggest that $w$ is decreasing linearly with time during the simulation run in order to promote global exploration of the search space and gradually decrease it to get more refined solution. Subscript $j$ indicates problem dimension $j = (1,2, ..., D)$. *PBest* and *GBest* are the personal best and the global best positions respectively, of particle $i$ in dimension $j$.

Parsopoulos and Vrahatis (2002) argue that particle velocities on each dimension should be clamped to the maximum value ($V_{max}$) to control the velocity of the particles. On one hand, if $V_{max}$ is too high, particles might fly past good solutions. On the other hand, if $V_{max}$ is too small, particles may not explore sufficiently beyond locally good regions. Moreover, if a particle happens to fly outside the problem space, its new position is set back to its previous best position (Venayagamoorthy and Doctor, 2004).

## 14.5    A modified particle swarm technique

In the proposed algorithm, we are considering the risk as constraints, and then a decision-making process in the sense of PT will be used to assess and evaluate this risk in the constraints problems. The proposed swarm algorithm consists of the following steps:

*Step 1*: Select swarm parameters; size of swarm $N$, inertia weight $w$, acceleration constant $c_1$ and $c_2$, maximum velocity $V_{max}$, stop criterion, and maximum number of objective function evaluations Max.NFE. Then, select PT parameters such as $\alpha$, $\lambda$, $\beta$ for value function and $\gamma$ for probability weighting function.

*Step 2*: Initialization. For each agent $i$ (where $i$ is index value) in the population:

*Step 2.1*: Initialize the agents randomly within the range ($l_i$, $u_i$), Where $l_i$ and $u_i$ represent the lower and upper bound values of the search space.

*Step 2.2*: Initialize $v_i$ randomly.

*Step 3*: Calculate the objective function value of the problem (fitness) for each agent in the swarm as (Fitness (1), Fitness (2), ..., Fitness (N)).

*Step 4*: Apply the PT to select the *PBest* value of each agent in the swarm.

*Step 4.1*: Calculate the total fitness values of the current swarm (TotFitness):

$$TotFitness = \sum_{i=1}^{N} Fitness(i)$$

*Step 4.2*: Calculate the reference point of prospect value, which is equal to:

$$\text{RefPoint} = \frac{\text{TotFitness}}{N}$$

*Step 4.3*: To find the specific location of each fitness value, either in the gain or loss region of the value function, it is crucial to validate the fitness value calculated in Step 3 based on the following:

- If the fitness value of the specified agent is greater than or equal to the RefPoint, then the values function will be equal to:

$$\text{If Fitness}(i) \geq \text{RefPoint then } v(\text{Fitness}(i))$$

$$= (\text{Fitness}(i) - \text{RefPoint})^{\alpha}$$

- Otherwise, the fitness function will be located in the loss region:

$$v(\text{Fitness}(i)) = -\lambda(\text{Fitness}(i) - \text{RefPoint})^{\beta}$$

*Step 4.4*: The following steps are needed to calculate the probability value for each set of constraints. First, find the constraint values for each agent as $g_1(i)$, $g_2(i)$, $g_3(i)$, ..., $g_n(i)$, where $n$ refers to the number of constraints appearing in the problem. Second, calculate the algebraic sum for all constraints for each particle in the swarm (***TotConst***):

$$\boldsymbol{TotConst} = \sum_{i=1}^{N} g_1(i) + \sum_{i=1}^{N} g_2(i) + \sum_{i=1}^{N} g_3(i) + \cdots + \sum_{i=1}^{N} g_n(i)$$

Third, calculate the probability value for each set of agents (probability $(i)$), which is equal to:

$$\text{Probability } (i) = \frac{(g_1(i) + g_2(i) + g_3(i) + \cdots + g_n(i))}{\boldsymbol{TotConst}}$$

*Step 4.5*: Find the probability weighting function $w(p(i))$ of PT:

$$w(p(i)) = \frac{p(i)^{\gamma}}{(p(i)^{\gamma} + (1 - p(i))^{\gamma})^{1/\gamma}}$$

where $p(i)$ represents the probability calculated in Step 4.4 for each agent.

*Step 4.6*: Find the prospect value by simply multiplying the value function by the weighting function:

$$v(\text{Fitness}(i), p(i)) = v(\text{Fitness}(i))w(p(i))$$

*Step 4.7*: The *PBest* of each agent in the swarm can be found as follows: if the agent's current prospect value is better than the previous stored prospect value, and if the agent is in the feasible space, then set the current positions as the new *PBest*.

*Step 5*: The *GBest* of the current swarm can be obtained as follows: if the fitness value is lower than the best fitness value stored in history, then set current position value as the new *GBest*.

*Step 6*: Perform the flight by calculating the agent velocity and consequently update the agent' positions by:

$$v_{ij}(t+1) = wv_{ij}(t) + c_1 r_1 (PBest_{ij}(t) - x_{ij}(t)) + c_2 r_2 (GBest_{ij}(t) - x_{ij}(t))$$

$$x_{ij}(t+1) = x_{ij}(t) + v_{ij}(t+1)$$

*Step 7*: Repeat Steps 3–6 until the maximum iterations or minimum criteria are achieved.

## 14.6 Simulation and comparison results

To test the performance of the proposed algorithms, two widely used benchmarking engineering design problems have been solved. During the simulations, a swarm size of 20 is used. Acceleration constant factors for $c_1 + c_2 \leq 4$, and $V_{\max}$ is set to a value between 20% and 50% of the size of the search space. A linearly varying inertia weight $w$ over the generations is used, i.e., varying from 1.2 at the beginning of the search to 0.2 during the three quarters of the maximum allowed number of iterations according to Parsopoulos and Vrahatis (2002). The Max.NFE is set to be 2000000 (20 agents × 100000 generations). The test functions selected contain characteristics that are representative of what can be considered "difficult" global search problems for an evolutionary algorithm. All test functions are examined for 10 runs. Tables 14.1 and 14.2 show the parameters used in PT and swarm, respectively.

*Table 14.1   Parameters of the PT*

| Parameters | Value |
|------------|-------|
| $\alpha$ | 0.88 |
| $\lambda$ | 0.88 |
| $\beta$ | 2.25 |
| $\gamma$ | 0.75 |

*Table 14.2    Parameters of the swarm*

| Parameters | Value |
|---|---|
| $N$ | 20 |
| $c_1 + c_2$ | $\leq 4$ |
| $V_{max}$ | 20–50% of the search space |
| $w$ | Varying linearly from 1.2 to 0.2 |
| Max.NFE | 2000000 |

## Example 1

Minimize:

$$f(x) = 3x_1 + 0.000001x_1^3 + 2x_2 + (0.000002/(3)x_2^3)$$

Subject to:

$$g_1(x) = -x_4 + x_3 - 0.55 \leq 0$$

$$g_2(x) = -x_3 + x_4 - 0.55 \leq 0$$

$$h_3(x) = 1000 \sin(-x_3 - 0.25)$$

$$+1000 \sin(-x_4 - 0.25) + 894.8 - x_1 = 0$$

$$h_4(x) = 1000 \sin(x_3 - 0.25)$$

$$+1000 \sin(x_3 - x_4 - 0.25) + 894.8 - x_2 = 0$$

$$h_5(x) = 1000 \sin(x_4 - 0.25)$$

$$+1000 \sin(x_4 - x_3 - 0.25) + 1298.8 = 0$$

Example 1 is a four-dimensional minimization problem (Runarsson and Yao, 2000), subject to two linear inequality and three non-linear equality constraints. The search space is bounded by $0 \leq x_1, x_2 \leq 1200$, and $-0.55 \leq x_3, x_4 \leq 0.55$. The widely known optimum solution is at $[x_1, x_2, x_3, x_4; f] = [679.9453, 1026.067, 0.1188764, -0.3962336; 5126.4981]$. Using the proposed methods the best solution reached by swarm was $[697.82960, 1005.758817582735, 0.105222060067, -0.397798742212; 5078.758276254275]$ and 990340 for NFE. Table 14.3 shows the statistical results obtained by the proposed approach compared with different methods.

*Table 14.3  Comparison of example one*

| Example | Statistical results | Proposed swarm | Runarsson and Yao (2000) | Hu and Eberhart (2002) | Pulido and Coello Coello (2004) | Cagnina *et al.* (2006) |
|---------|---------------------|----------------|--------------------------|------------------------|---------------------------------|-------------------------|
| 1 (Min) | Best | 5078.758276 | 5126.497 | N/S | 5126.640000 | 5126.497 |
|         | Mean | 5105.733330 | 5128.881 | – | 5461.081333 | 5327.9569 |
|         | Worst | 5126.498354 | 5142.472 | – | 6104.750000 | 2300.5443 |
|         | NFE | 990340 | 350000 | – | 340000 | 340000 |

N/S = No solution.

According to the results shown in Table 14.3, the proposed method out-performs clearly the other techniques in term of accuracy. The optimum solution achieved by our algorithm is better than those found by other techniques (Runarsson and Yao, 2000; Pulido and Coello, 2004; Cagnina et al., 2006) so far. We could also notice that the average searching quality (mean) of our proposed method is better than those reported in other techniques. While the worst solutions found by our proposed method are equal or even better than the best solutions found by other techniques.

**Example 2 (pressure vessel design problem)**

The design of the pressure vessel problem is used by Coello Coello (2000) and Worasucheep (2008). This problem is a four-dimensional minimization, subject to four linear and non-linear inequality constraints. The search space is bounded by $1 \times 0.0625 \leq x_1, x_2 \leq 99 \times 0.0625$, and $10.0 \leq x_3, x_4 \leq 200.0$. The well-known optimal solution is at $[x_1, x_2, x_3, x_4; f] = [0.8125, 0.4375, 42.098446, 176.636596; 6059.714335]$. Using the proposed methods, the best solution reached by swarm was $[0.77949012, 0.38532096, 40.38755764, 199.06132290; 5887.649300]$ and 200000 for NFE. Table 14.4 shows the statistical results obtained by the proposed approach compared with different methods.

For the example 2, the results obtained by our proposed algorithm outperform largely the results reported so far in other techniques in term of optimum value, average search quality and the worst solution. The optimum value achieved is much better than the once obtained by other techniques (Ceollo Ceollo, 2000; Hu et al., 2003; Huang et al. 2007; Worasucheep, 2008; Cagnina et al., 2008), the average searching quality of the proposed method is also better than those of other techniques, and finally the worst solution obtained by our proposed method is better than the optimum value reported so far by other techniques; In addition, the new method was able to achieve good results with relatively competitive number of objective functions' evaluations (NFE) compared to other techniques employed.

## 14.7    Conclusions

In this chapter, we have extended a descriptive model of decision making under conditions of risk and uncertainty, named Prospect Theory (PT), to swarm intelligence techniques to model each agent's decision making within the swarm. The proposed model has been used for particle swarm optimization (PSO) but could be adapted to any other swarm intelligence techniques. The proposed algorithm is initialized with a swarm of random feasible and infeasible solutions, and then PT is used to update the personal best position of each agent in the swarm. We have evaluated this technique on two well-known benchmark examples that had previously been implemented using other

*Table 14.4  Comparison of example two*

| Example | Statistical results | Proposed swarm | Coello Coello (2000) | Hu et al. (2003) | Huang et al. (2007) | Worasucheep (2008) | Cagnina et al. (2008) |
|---|---|---|---|---|---|---|---|
| 2 (Min) | Best | 5887.64930 | 6288.74450000 | 6059.131296 | 6059.7340 | 6059.714334 | 6059.714335 |
|  | Mean | 5958.79586 | 6293.84323196 | – | 6085.2303 | 6059.714353 | 6092.049800 |
|  | Worst | 6023.63051 | 6308.14965192 | 6632.5 | 6371.0455 | 6059.714726 | – |
|  | NFE | 200000 | 900000 | 200000 | 204800 | 200000 | 24000 |

techniques found in literature. In all cases, the proposed technique produced better results than those previously reported, as it is able to achieve better performance with relatively small swarms.

The proposed swarm algorithm presents the following advantages from an engineering perspective:

1. Better solutions: The results indicated better solutions than those previously reported in the literature.
2. Broad applicability: This new algorithm can be applied to a broad range of engineering problems with constraints.
3. More robustness: The proposed algorithm is more robust than other techniques due to its ability to handle complex constraints and to find a solution from random combinations of infeasible and feasible solutions.
4. Cost-effectiveness: The proposed algorithm does not need specific knowledge to solve the problem. There are no manipulations or complex calculations required to handle the constraints.
5. Computational requirements: The proposed algorithm was able to achieve the optimal or better solution with relatively competitive number of objective functions' evaluations compared to other techniques employed.

## 14.8   Limitations and further study

Despite all the advantages that this algorithm has, in some more sophisticated risk handling problems, achieving better accuracy requires more fitness function evaluations. Therefore, there is a need for more investigation and research to improve the proposed algorithm in terms of trade-off between accuracy and computational time.

As a part of our further study, we aim at extending the proposed approach to other swarm intelligence techniques such as ant colony, stochastic diffusion search, birds flocking, and fish schooling.

## References

Abdellaoui, M., (2000), 'Parameter-Free Elicitation of Utility and Probability Weighting Functions', *Management Science*, 46(11): 1497–1512.

Bernoulli, D., (1738), 'Specimen Theoriae novae de mensura sortis', *Commentarii Academiae Scientiarum Imperialis Petropolitanae* 5(1738): 175–192 (Translated by L. Sommer, as New Exposition on the Measurement of Risk, *Econometrica*, 22: 23–26).

Bernoulli, D., (1954), 'Exposition of a New Theory on Measurement of Risk', *Econometrica*, 410(1): 23–36.

Cagnina, L. C., Eaquivel, S. C., and Coello Coello, C. A., (2006), 'A Particle Swarm Optimizer for Constrained Numerical Optimization', In *Lectures*

*Notes of Computer Science*, Berlin, Heidelberg: Springer-Verlag, 910–919.

Cagnina, L. C., Eaquivel, S. C., and Coello Coello, C. A., (2008), 'Solving Engineering Optimization Problems with the Simple Constrained Particle Swarm Optimizer', *International Journal of Computing and Informatics*, 32(3): 319–326.

Camerer, C. F. and Ho, T. H., (1994), 'Violations of the between's Axiom and Nonlinearity in Probability', *Journal of Risk and Uncertainty*, 8(2): 167–196.

Coello Coello, C. A., (2000), 'Use of a Self-Adaptive Penalty Approach for Engineering, Optimization Problems', *Computers in Industry*, 41(2): 113–127.

Doyle, J. and Thomason, R. H., (1999), *Background to Qualitative Decision Theory*. AI Magazine, 20, 55–80.

Eberhart, R. C. and Kennedy, J., (1995), 'A New Optimizer Using Particle Swarm Theory', Proceedings of the Sixth International Symposium on Micro Machine and Human Science. Nagoya, Japan, 39–43.

Eberhart, R. C., Simpson, P. K., and Dobbins, R. W., (1996), *Computational Intelligence PC Tools*, Boston, MA: Academic Press Professional.

Gonzalez, R. and Wu, G., (1999), 'On the Shape of the Probability Weighting Function', *Cognitive Psychology*, 38: 129–166.

Goodwin, P. and Wright, G. (1998). *Decision Analysis for Management Judgment*, Chichester, England: John Wiley and Sons Ltd.

Heath, C. and Tversky, A., (1991), 'Preference and Belief: Ambiguity and Competence in Choice Under Uncertainty', *Journal of Risk and Uncertainty*, 4(1): 5–28.

Hoch, S. J., Kunreuther, H. C., and Gunther, R. E., (2001), *Wharton on Making Decisions*, New York: John Wiley and Sons, Inc.

Huang, F., Wang, L., and He, Q., (2007), 'An Effective Co-evolutionary Differential Evolution for Constrained Optimization', *Journal of Applied Mathematics and Computation*, 186(1): 340–356.

Hu, X. and Eberhart, R. C., (2002), 'Solving Constrained Nonlinear Optimization Problems with Particle Swarm Optimization', Proceedings of the Sixth World Multi-conference on Systemics Cybernetics and Informatics, Orlando, FL, USA.

Hu, X., Eberhart, R. C., and Shi, Y., (2003), 'Engineering Optimization with Particle Swarm', Proceedings of the IEEE Swarm Intelligence Symposium, 53–57.

Kahneman, D. and Tversky, A., (1979), 'Prospect Theory: An Analysis of Decision Under Risk', *Econometrica*, 4: 263–291.

Kahneman, D. and Tversky, A., (1991), 'Loss Aversion in Riskless Choice: A Reference-Dependent Model', *The Quarterly Journal of Economics*, 106: 1039–1061.

Kahneman, D., Wakker, P. P., and Sarin, R. V., (1997), 'Back to Bentham?—Explorations of Experienced Utility', *The Quarterly Journal of Economics*, 112: 375–405.

Kennedy, J. and Eberhart, R. C., (1995), 'Particle Swarm Optimization', Proceedings of IEEE International Conference on Neural Networks. Piscataway, NJ, IEEE Service Center, 1942–1948.

Oliveira, A., (2007), 'A Discussion of Rational and Psychological Decision-Making Theories and Models: The Search for a Cultural-Ethical Decision-Making Model', *Electronic Journal of Business Ethics and Organization Studies (EJBO)*, 12(1): 12–17.

Parsopoulos, K. E. and Vrahatis, M. N., (2002), 'Recent Approaches to Global Optimization Problems through Particle Swarm Optimization', *Natural Computing*, 1(2/3): 235–306.

Prelec, D., (1998), 'The Probability Weighting Function', *Econometrica*, 66: 497–527.

Pulido, G. T. and Coello Coello, C. A., (2004), 'A Constraint-Handling Mechanism for Particle Swarm Optimization', Proceedings of the Congress on Evolutionary Computation, 2: 1396–1403.

Runarsson, T. P. and Yao, X., (2000), 'Stochastic Ranking for Constrained Evolutionary Optimization', *IEEE Transactions on Evolutionary Computation*, 4(3): 284–294.

Russell, S. and Norvig, P., (2002), *Artificial Intelligence – A Modern Approach*, Upper Saddle River, NJ: Prentice Hall.

Samuelson, W. and Zeckhauser, R., (1988), 'Status Quo Bias in Decision Making', *Journal of Risk and Uncertainty*, 1: 7–59.

Shi, Y. and Eberhart, R. C., (1998), 'A Modified Particle Swarm Optimizer', Proceedings of the IEEE International Conference on Evolutionary Computation, Piscataway, NJ: IEEE Press, 69–73.

Tversky, A. and Kahneman, D., (1992), 'Advances in Prospect Theory Cumulative Representation of Uncertainty', *Journal of Risk and Uncertainty*, 5(4): 297–323.

Venayagamoorthy, G. K. and Doctor, S., (2004), 'Navigation of Mobile Sensors Using PSO and Embedded PSO in a Fuzzy Logic Controller', Industry Applications Conference, 39th IAS Annual Meeting. Conference Record of the IEEE, 2: 1200–1206.

Wakker, P. P., (2001), 'Testing and Characterizing Properties of Non-additive Measures through Violations of the Sure-thing Principle', *Econometrica*, 69(4): 1039–1059.

Wald, A., (1950), *Statistical Decision Theory*, New York: Mc Graw-Hill.

Wang, L., Cui, Z., and Zeng, J., (2009), 'Particle Swarm Optimization with Group Decision Making', Ninth International Conference on Hybrid Intelligent Systems.

Weiss, L. and Wald, A., (1997), in Johnson, N. L. and Kotz, S. (eds.), *Leading Personalities in Statistical Sciences*, New York: Wiley and Sons, 164–167.

Wu, G. and Gonzalez, R., (1996), 'Curvature of the Probability Weighting Function', *Management Science*, 42: 1676–1690.

Wu, G. and Gonzalez, R., (1998), 'Common Consequence Conditions in Decision Making Under Risk', *Journal of Risk and Uncertainty*, 16(1): 115–139.

Worasucheep, C., (2008), 'Solving Constrained Engineering Optimization Pro-
blem by the Constrained PSO-DD', 5th International Conference on Electrical
Engineering/Electronics, Computer, Telecommunications and Information
Technology, 1: 5–8.
Zhang, Y., Pellon, M., and Coleman, P., (2007), 'Decision Under Risk in
Multi-Agent Systems', IEEE International Conference on System of Systems
Engineering (SoSE 07).

*Chapter 15*
# Towards a cognitive model of interaction with notations

*Maria Kutar and Andrew Basden*

## Abstract

The cognitive dimensions framework provides a mechanism by which we may better understand the factors which influence human interaction with notational systems. Primarily used to inform notational design, or to evaluate programming languages, interaction environments and specification notations, the focus of the framework is in highlighting the things that make using notations for a given task 'hard' or 'easy'. We believe that the dimensions, which have their roots in cognitive psychology, may offer the foundation for a cognitive model of interaction with notations. This work examines the existing body of literature that seeks to understand the cognitive features of human interaction with notations, noting the parallels with the cognitive dimensions framework. We suggest that we may derive a cognitive model of interaction with notations through examination of the dimensions using cognition as a lens. The work provides a contribution to the cognitive dimensions framework itself, highlighting the distinction between cognitive effort and constraints arising at the notational level. It also suggests that interaction with symbols and notations may be seen as a type of embodiment, and we believe that development of a cognitive model of interaction with notations may be of use in better understanding than embodiment.

## 15.1. Introduction

The cognitive dimensions (CDs) framework (Green, 1989) may be used to explore the ways in which humans interact with notational systems. Specifically, the framework enables cognitive understanding of notational systems through a set of dimensions that characterise the aspects of interaction, which can cause difficulties for human users. The framework may be used either to inform notation design, or in evaluation of notations such as programming languages or interaction environments. In this chapter we use the CDs framework as a foundation for the development of a cognitive model of interaction with notations. The model may be of use in developing under-standing of the CDs themselves. The more important contribution in the context of this volume is that it suggests an additional kind of embodiment, and we consider the relationship between this work and that of David Vernon's contribution to the volume in the discussion.

CDs have been demonstrated to have worked in practice for a range of notational systems, and have their foundations in cognitive psychology. We believe that closer examination of the dimensions enables us to discover the aspects of human cognition that inform them. We present the analysis of the individual dimensions with examples and discussion highlighting the underlying cognitive activities involved in interacting with notational systems.

CDs have roots in visual programming language research and have become the de facto standard for usability evaluation of such languages. Blackwell (2006) notes that the framework 'has been adopted as the primary approach to understanding language usability, not only by visual language researchers, but by designers of commercial products such as the Microsoft Visual Studio range'. Further, the framework has been widely applied to other notations, including UML (Kutar *et al.*, 2002), music notation (Blackwell *et al.*, 2000) and theorem provers (Kadoda, 2000).

In a review of the CDs framework, Green *et al.* (2006) recognise that the framework requires some refinement and ask 'Can the dimensions be better defined and where necessary be given an improved specification, to avoid overlaps, or can a better set of dimensions be developed?' Previous work on the CDs framework per se has sought to examine ways in which the framework may be formally underpinned, beyond its original roots in cognitive psychology. Roast and Siddiqi (1996) sought formal definitions for each of the CDs so that the completeness of the suite could be investigated. After a promising start, this exercise was not pursued further, which suggests that attempting precise definitions will not achieve what is sought. Whilst this approach might help judge completeness, it is unlikely to fulfil most of the other requirements mentioned earlier in the text. A 'refactoring' approach is reported by Green *et al.* (2006) which distinguishes between CDs concerned with cognitive prop-erties and those concerned with the structure of information, although this has not yet been developed to cover all of the dimensions. A different approach was taken in (Blackwell and Green, 1999) which examined characteristics of notations that might underlie all CDs. There the analysis considered invest-ment of attention as the underlying theme for the dimensions. This was useful in providing a means by which trade-offs between the individual dimensions might be better analysed and understood. The approach that we take here differs in that we attempt to derive a model which elucidates the cognitive features of human interaction with notations.

In the following sections we first of all provide an overview of the CDs framework ( Section 15.2) and review related work which seeks to understand the cognitive processes involved in interacting with notations (Section 15.3). We then present our analysis of the CDs framework in terms of cognition

(Section 15.4) and present this in terms of the cognitive model of interaction with notations (Section 15.5). Finally, we suggest this chapter can make two contributions, to cognitive models and to a theme that runs through a number of the chapters in this volume, that of embodiment and interaction, and note some directions for further research (Section 15.6).

## 15.2   The CDs framework

The CDs framework is a set of dimensions that are intended as *discussion tools* rather than an analytic method. This makes them stand out against most other approaches to evaluating usability or understandability such as GOMS (Goals, Operators, Methods and Selection rules) (Card *et al.*, 1983) or Task-Action-Grammar (Payne and Green, 1986). Further, they may be applied to static notations as well as in interactive environments, as they go beyond the process of interaction; their focus is the notation, or information artefact. They are equally informative about the cognitive aspects of 'interacting' with tables, graphs or music notation as they are with software environments, word processors or central heating controls. Discussion tools should capture enough of what is important about something to generate informed discussion, providing a basis for informed critique. CDs aim to do just this. The set of 14 orthogonal dimensions is produced in Figure 15.1, together with a short description of each. Detailed discussion of each dimension is provided in Section 15.4.

| Dimension | Description |
|---|---|
| Abstraction | Types/availability of abstraction mechanisms |
| Closeness of mapping | Closeness of representations to domain |
| Consistency | Similar semantics are presented in a similar syntactic style |
| Diffuseness/terseness | Verbosity/succinctness of language |
| Error-proneness | Syntax provokes slips |
| Hard mental operations | High demand on cognitive resources |
| Hidden dependencies | Important links between entities not visible |
| Premature commitment | Constraints on the order of doing things |
| Progressive evaluation | Work-to-date can be checked at any time |
| Provisionality | Degree of commitment to actions or marks |
| Role expressiveness | The purpose of a component is readily inferred |
| Secondary notation | Extra information in means other than formal syntax |
| Viscosity | Resistance to change |
| Visibility and juxtaposibility | Ability to view components easily |

*Figure 15.1   The cognitive dimensions*

In addition to the dimensions themselves, the framework incorporates a number of important concepts. First, the notation must be considered in the context of its environment such that a CDs analysis considers a system comprised of the combination of notation + environment. Second, an individual dimension does not in itself encapsulate some property which is said to be desirable or otherwise. Rather it gives us an understanding of the way in which humans are able to work with artefacts; in some cases this property may constrain, but in others it may have no adverse effects. The activity for which the notation is being used will guide the relative desirability of a given dimension – for example hidden dependencies are unlikely to impact upon a transcription activity but are known to be harmful where modification of the artefact takes place. Finally, although the dimensions are each distinct concepts there are trade-offs between them; they are pairwise independent. Green and Blackwell (1998) describe this as follows:

> *Take a constant mass of gas, with 3 'dimensions': temperature, pressure and volume. If you heat the gas, the temperature rises, and it tries to expand. If the volume is held constant, it can't expand, so the pressure increases. If the pressure is to be kept constant, the volume must increase. So although pressure, temperature, and volume are conceptually independent, for physical objects they are only pairwise independent. Within reason, any combination of pressure and temperature may be obtained, but between them they determine the volume; and likewise for all other combinations. Many of the cognitive dimensions are similarly pairwise independent: any two can be varied independently, as long as some third dimension is allowed to change freely.*

Some of the known trade-offs are illustrated in Figure 15.2.

We believe that the richness of the CDs framework, with its broad applicability to both static and interactive notations, provides a broad basis from which a cognitive model of interaction may be derived. In the following section we consider each dimension in turn, first of all illustrating each with a

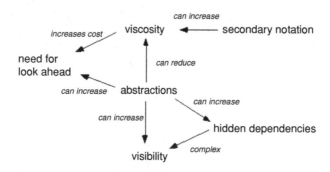

*Figure 15.2    Some trade-offs among cognitive dimensions (reproduced with permission from T.R.G. Green)*

paradigmatic example, and then deriving for each the aspect of cognition that it encapsulates.

## 15.3.   Cognition and representations

There is a large body of research that seeks to understand the cognitive aspects of interaction with representations. The majority of this focuses on graphical or diagrammatic representations, although some, such as Larkin and Simon (1987), contrast diagrammatic and text-based representations. In this section we review such work and examine how it relates to the CDs framework.

### *15.3.1   Representations*

The concept of the *representational effect* is presented in Zhang and Norman (1994), which states that differing isomorphic representations of a common formal structure may result in differing cognitive behaviours. This work focuses on distributed (i.e. some internal and some external) representations that are used for a particular task and emphasises the importance of external representations in problem solving. The CDs framework enables us to recognise the representational effect through analysis of isomorphic representations (see e.g. Kutar, 2001; Kutar *et al.*, 2001). Further, the CDs framework enables examination of the cognitive processes involved in interacting with notations, taking in the impact of the notation on both internal and external representations. For example the dimension of closeness of mapping infers the use of internal representations, whilst others such as visibility focus on external representations.

Larkin and Simon (1987) explore the concepts of information and computational equivalency in representations. Two representations which contain all of the same information (and indeed which may be constructed from the information contained in the other) may be said to be informationally equivalent. However, differences in the way in which that information is presented may influence the ability to draw inferences from the representation. They use the term 'computational equivalence' to describe representations which have both information equivalence and from which inferences may be drawn with similar speed and ease. The CDs framework would have applicability here in enabling greater understanding of whether (and indeed why) representations might be considered to be computationally equivalent. In Larkin and Simon's analysis, deriving meaning from a representation involves the activities of search, recognition and inference. For informationally equivalent representations there may be considerable difference in the ability to search for and recognise information if the representation is diagrammatic rather than sentinel (text or sentence based), and they demonstrate that diagrammatic representations offer considerable support for the recognition activity over sentinel representations. Consequently, sentinel and diagrammatic representations are unlikely to be computationally equivalent as the greater support provided for recognition (and to a lesser extent, search and inference) will enable inference to be drawn from the representation more quickly and more 'easily'.

Other work examines specific features of representations which influence user's ability to draw inferences. One theme is that of *specificity*, the ability of a representation to constrain ways of conceptualising the problem. This is important in supporting comprehension of representations because 'the more limited the expressive power of a system that can fit the data, the more the notation is contributing to an understanding of the phenomena described'. (Stenning and Oberlander, 1995). The relationship of specificity to the CDs framework is complex. It has been proposed as a candidate for an additional dimension (Blackwell *et al.*, 2001). However, it is closely related to the dimension of abstraction, with the underlying premise of Stenning and Oberlander's theory being that graphical representations aid the processing of information because they constrain the use of abstractions. We believe that the abstraction dimension is sufficient to include the notion of specificity, and that the parallels which may be drawn here indicate the importance of the dimension.

## 15.3.2   Visual representations

There is said to be a widespread assumption that graphical representations offer improved readability over textual or linguistic representations, and it is often heard that 'a picture is worth a thousand words'. This assumption is challenged by Petre (1995) who points out that this is only effective in the context of visual programming if the picture is worth the same thousand words to each of those who must interpret it. In particular, Petre suggests that much of the additional cognitive support which may be provided by graphical representations arises from enhanced capacity for the use of secondary notation rather than in the use of graphics per se.

Lohse *et al.* (1994) view visual representations as data structures for expressing knowledge. They identify 11 categories of visual representation; graphs, tables, graphical tables, time charts, networks, structure diagrams, process diagrams, maps, cartograms, icons and pictures. This classification has an interesting relationship to the CDs framework. On the one hand it is apparently broad in nature, including textual visual representations (e.g. tables, time charts) as well as more diagrammatic and pictorial representations. However, on the other hand it appears to have no place for information structures that are textual but with no 'visual' emphasis, such as program code. Thus visual languages would be included, but the remainder would not. The focus of CDs can be distinguished from the narrower field of visual representations, as it is based on the features of the notation being used in a particular environment (a 'system' in CDs terms). Thus notations may be used in a variety of environments to produce different types of visual or textual representation.

A later research by Lohse *et al.* (1995) focuses on representations used in systems development. The aim of the work is to examine how people use information presented in system representation diagrams together with the cognitive support provided by such representations. The authors propose a design process which is intended to enable system representations to be developed which maximise comprehension and communication within development teams, and which is based on their model of visual image processing. Each of these steps can be related to one of the cognitive dimensions in the CDs framework. We present the steps of their design process below together with the related CD to demonstrate that CDs elucidate these steps. We argue that the CDs framework goes beyond Lohse's process because it both encompasses additional cognitive features and has broader application, as it may support both evaluation and design for both visual and textual notations (Figure 15.3).

| Lohse design process steps | Relevant cognitive dimension |
|---|---|
| Identify the requisite information components | Abstraction, role-expressiveness |
| Choose the current local view that expresses the chosen information components | |
| | These steps relate to sub-devices which could be separately evaluated but visibility and consistency relevant here |
| Select a system diagramming tool that represents the chosen information components in a local view | |
| Identify the category of information for each component in the display | Abstraction, role expressiveness, closeness of mapping |
| Identify the most effective visual primitive (colour, texture, shape, value...) for each information component | Secondary notation |
| Identify the desired relationships among information components and organise the requisite information symbolically encoded in the visual primitive into a unified presentation | Hidden dependencies |
| Evaluate the display for unintended ambiguities | Error-proneness, progressive evaluation |

*Figure 15.3   Lohse's design process steps and cognitive dimensions*

It can be seen from this that there are CDs which reflect each of the cognitive aspects of each of the steps proposed by Lohse *et al.* Further, the remaining dimensions enable additional understanding of the cognitive aspects of system development representations which influence comprehension and communication (see e.g. Britton and Jones, 1999).

Other examinations of graphical representations have sought to identify the characteristics which influence ability to carry out cognitive tasks. Two of those identified by Scaife and Rogers (1996) have a particular bearing on the work presented here. These are computational offloading and re-representation. Computational offloading is the extent to which a representation of a problem can reduce the amount of cognitive effort needed to solve the problem by providing the means for direct perceptual recognition of important elements in it. This has relevance to a number of the CDs where a notation's properties directly relate to the potential for computational offloading. Re-representation refers to the way in which different representations which have the same abstract structure influence problem solving. Zhang and Norman (1994) describe the difference between carrying out the same multiplication using Roman or Arabic numerals. The same formal structure is represented by LXVIII × X and 68 × 10 but the former is much more difficult to manipulate to find the solution, for someone who is used to working with Arabic numerals.

## 15.3.3   Categories of users

There is a wealth of research that examines the differences in interaction with various representations for different categories of users. For example it is widely accepted that experts and novices use different cognitive processes to interpret and develop representations. Winn (2004) states that experts are more likely to solve problems through the recognition and interpretation of patterns, than breaking information down into its constituent parts. Petre (1995) identifies that CDs are relevant to all users. Individual dimensions may have more influence on a particular activity for novices than experts, but this is something which can be considered during analysis of the notation and is drawn out via CDs profiles. Whilst there is little work that examines CDs for specific categories of users (although see e.g. Britton and Jones (1999) for application of CDs in specification notations used by novices), this is as a result of under exploitation of the framework rather than restrictions inherent in it.

## 15.3.4   Cognitive models for information artefacts

We are not aware of any existing cognitive models for interaction with information artefacts but a number of cognitive models exist in related fields. Lohse (1993) presents a cognitive model of graph perception, which focuses on elements such as the sequence of eye fixations, short-term memory capacity and duration limits and the degree of difficulty to acquire information in each glance. This clearly addresses a narrower field than the model proposed in this research, as it is concerned only with extracting information from graphs, although there are clear parallels with CDs and the model we propose.

## 15.4   From dimensions to cognition

In this section we examine each of the dimensions in turn and consider the underlying cognitive activities, which influence human interaction with notational systems.

### 15.4.1   Abstraction

An abstraction groups together related elements, enabling them to be considered as a single entity. This may have the effect of changing the conceptual structure of the artefact. Whilst this can have cognitive benefits, and can improve comprehensibility, it also introduces cognitive load. It is widely recognised that thinking in abstract terms is a high level educational skill, and notations which contain large number of abstractions require users to quickly grasp many high-level concepts. The CDs framework introduces the concept of an abstraction barrier, determined by the number of abstractions that a user must learn, which may be further increased if the notation allows for user-defined abstractions, as is common in programming languages. Systems (i.e. the combination of a notation in a specific environment) may be abstraction hungry (requiring the user to define many abstractions), abstraction tolerant (permitting but not requiring abstractions) or abstraction hating (user unable to define abstractions, often coupled with few built-in abstractions). For example the Z specification notation, which requires the user to define abstractions, is abstraction hungry, and has a high abstraction barrier. Styles in word processors, which may be user defined but are not required, may be used to illustrate abstraction tolerance, whilst spreadsheets are abstraction hating. There is a known trade-off here with the dimension of viscosity; enabling components to be treated as a group reduces viscosity (imagine transcribing or resizing a complex picture drawn in Word without first grouping the elements together).

In cognitive terms we can identify that *an abstraction is an effective substitution between reality and the human mind*; it enables us to relate various entities together, but comes with a cost in terms of the cognitive effort required.

### 15.4.2   Closeness of mapping

This dimension is concerned with the closeness of a representation to the domain which it denotes. An empirical study by Hundhausen *et al.* (2003), which explored the translation of alternative descriptions of algorithms into programming languages, supports this with participants performing more quickly where there was a closer match between the description type (pseudo-code) and the programming language (Java) than in the alternative case (visual examples to Java). There is an obvious trade-off here with error-proneness (too much similarity may make it difficult to distinguish elements, inviting confusion and error). In cognitive terms it appears that humans find it 'easier' to work with representations, which they are able to more closely match to the subject of the representation; similarities are useful.

*Cognitive support is offered where it is possible to draw similarities between a representation and the domain.*

### 15.4.3  Consistency

Consistency of syntax and semantics is known to improve learnability. It is recognised as an important property in human–computer interaction (Dix *et al.*, 2004) although it is suggested that it is of importance to learning rather than to usability and there is evidence that it is of less importance than mnemonic labelling (Cramer, 1990) (which is not a structural property and therefore outside of the scope of the CDs framework). Consistency is important in cognitive terms because it supports inductive reasoning rather than immediate knowledge. For example we may identify a previously unseen desktop icon through generalisation to the notion of icons together with recognition (in some cases) of the icon's image.

*We are able to work well with notations which allow us to carry meaning across components of that notation.*

### 15.4.4  Diffuseness/terseness

The diffuseness, or verbosity of a notation has obvious implications for our ability to work with it. Flow charts which cannot fit onto a single page require much additional cognitive processing to make the necessary links between components, and at a simpler level we are unable to hold large volumes of material in working memory. Where there are many tokens (words, symbols) on a page which must be read, it is difficult to identify the significance of each. There is a relationship here with abstractions, as enabling tokens to be treated as a group rather than individually reduce the number of items which must be held in memory. It should be noted that excessive terseness may also cause difficulties, particularly if it is combined with abstractions; here the issue is that cognitive load is increased through the need to decipher individual tokens to understand the representation; the Z specification notation falls into this category.

*Humans like the number of tokens in a group that has meaning as a whole to be small.*

### 15.4.5  Error-proneness

Error-prone notations invite mistakes – i.e. there is an interplay with the elements of human cognition where mistakes are likely. For example Fortran uses the identifiers I and O which are easily confused with one and zero. There is little protection from this such as check digits.

## 15.4.6 Hard mental operations

Hard mental operations are ones which arise at the notational rather than the semantic level. It is recognised that both conditionals and negatives exist at the notational level through evidence that the cognitive problems that they present can be reduced if they are represented in alternative formats such as decision tables (Curtis *et al.*, 1989; Wright and Reid, 1973). Combining several of the features which give rise to hard mental operations will vastly increase the difficulty, as is the case with, for example multiple negatives, complex Booleans and self-embedded sentences.

*Notational features that require significant cognitive processing increase difficulty.*

## 15.4.7 Hidden dependencies

Hidden dependencies are an important feature not because of the existence of the dependency but because of the fact that it is hidden. They are commonly found in spreadsheets where a formula shows which other cells it takes its value from, but not which other cells take their value from it, and any changes to the formula may have knock-on effects which cannot easily be determined. There is evidence that this is responsible for a high error frequency in spreadsheets (Panko and Halverson, 1994). This relates to visibility but is concerned with relationships rather than the gestalt. It codifies what the human needs to understand of issues to operate well.

*It is important that we can see relationships and that dependency is visible.*

## 15.4.8 Premature commitment

Premature commitment requires decisions to be made before all of the necessary information is available. Sketching a map onto paper requires that we have a clear mental picture of the overall size of the map to be drawn before we start; commonly we will find that we have actually started to draw the first feature too close to one side of the paper so that the whole map appears too close to the edge, or cannot be fitted onto the page at all. Many notations require that we make decisions about what should be included at too early a stage, not necessarily a problem if changes can be easily made, but frequently imposing such a high cost that the original decision is retained even when it is unsuitable.

*People do not like making decisions before all of the information is available.*

## 15.4.9 Progressive evaluation

Progressive evaluation enables easy assessment of where you are up to. Notations that allow progressive evaluation help people to work incrementally/in chunks (i.e. they support decomposition) and to regularly check their performance at any time. It is recognised that novices benefit from being able to evaluate problem-solving progress at regular intervals, and there is evidence that experts use a similar strategy. For example experts have been observed to run their programs more frequently than novices when debugging (Gugerty and Olson, 1986).

*Being able to evaluate progress supports notational activity.*

### 15.4.10    Provisionality

This describes the degree of commitment to actions or marks – in simple terms whether 'undo' is supported. Writing on paper with pencil provides a high level of provisionality as the marks may easily be removed. By contrast, using ink has low provisionality. Provisionality is an important dimension for exploratory activities where it enables possible solutions to be attempted, with the ability to back track if required. It may be less influential for activities such as transcription.

*Being able to make tentative marks may support design activity.*

### 15.4.11    Role expressiveness

This refers to the ease with which the meaning of components may be inferred. Figure 15.4 below shows an excerpt from a webmail view.

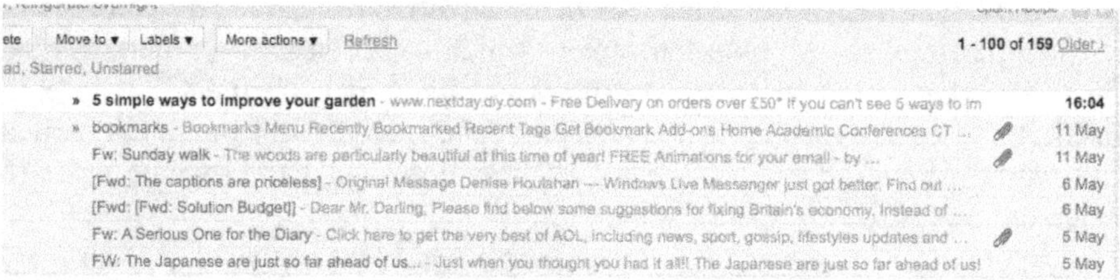

*Figure 15.4    Excerpt from webmail view*

The column on the far right-hand side shows temporal information. In the first row we can infer that the combination 'digit digit colon digit digit', with values of $< 12$ for the first pair of digits and $< 60$ for the second, is indicative of a time (this builds on cultural knowledge). It is also possible to relate this to other components in the view; the use of rows and columns indicates a table. Combining this knowledge allows us to infer that the information is sorted by linear time in the far right column. Thus we could say that this component of the representation has high role expressiveness as the purpose of the components may be inferred. It is important to note that role expressiveness enables us to draw on existing knowledge to add meaning to a representation.

*We depend upon existing, often cultural knowledge to determine the purpose of an entity.*

### 15.4.12    Secondary notation

This dimension refers to the extent to which it is possible to add additional information outside of the formal syntax of a notation. This may include the ability to annotate a representation, or to enrich an object according to the user needs. Redundant recoding describes the use of indentation in computer programmes, or the grouping of related objects on a physical interface such as a car radio. In both of these cases, the user is supported in their understanding of the representation by extra-syntactic clues. A further variation of this dimension is described as 'escape from formalism'. Here the additional information actually adds to the information available. Payne (1993) reported that entries into paper calendar or diary systems are often enriched through the use of colour, underlining, etc. – not a formal part of the notational system but one which is added by the user to add meaning to the representation. There is a trade-off here with viscosity; representations which include secondary notation may become difficult to change as the additional meaning may become irrelevant when a change is made, or it cannot be easily transferred.

*The ability to introduce additional meaning to a notation may support its use.*

## 15.4.13   Viscosity

Viscosity describes the resistance to change that the notation imposes. This is particularly important for design-led activities and can be seen to relate to provisionality. In a viscous representation a user must execute many internal tasks to achieve a single external or plan level task. Green (1990) distinguishes between 'knock-on' and 'repetition' viscosity. Knock-on viscosity means that to achieve goal A, the user must carry out tasks B, C and D which may not be directly goal related. For repetition viscosity the user must carry out many (similar) activities to reach goal A – think for example of changing the heading font in a long document without the use of styles. Both types of viscosity are significant because they require extra work by the user to reach the goal, although the work may be repetitive rather than cognitively challenging. Viscousness may be reduced through the introduction of additional abstractions (e.g. the header style) although this entails a trade-off with the introduction of additional cognitive load.

*If more actions or steps are required to reach a particular goal then difficulty is increased.* Note that difficulty here is not simply in the sense of increased cognitive load but also in terms of a barrier to carrying out the action (i.e. in the case of repetition viscosity).

## 15.4.14   Visibility/juxtaposibility

The final dimension refers to the ability to view components easily and is related to the notion of juxtaposability, the ability to view two or more components at the same time, for example by being able to view two portions of a program at the same time, or by being able to place windows side by side in a display. Spreadsheets display asymmetric visibility; in the data view it is possible to see the formula for only a single cell. Whilst the formulae for all cells may easily be viewed by changing the display mode for the sheet, this has the effect of removing the ability to view the value for each cell. If components may not be viewed easily, or together, the user must rely on the use of working memory (which must be frequently refreshed) or introduce some external memory such as a hard copy of the component, which may be viewed simultaneously.

*It is important to the human cognitive system to have items in view.* (Visibility is concerned with the gestalt or whole.)

## 15.5    Towards a cognitive model

In Figure 15.5 we collect together the underlying mechanism for each of the dimensions.

| Dimension | Cognitive observation |
| --- | --- |
| Abstraction | An abstraction is an effective substitution between reality and the human mind |
| Closeness of mapping | Cognitive support is offered where it is possible to draw similarities between a representation and the domain |
| Consistency | We are able to work well with notations which allow us to carry meaning across components of that notation |
| Diffuseness/terseness | Humans like the number of tokens in a group that has meaning as a whole to be small |
| Error-proneness | – |
| Hard mental operations | Notational features that require significant cognitive processing increase difficulty |
| Hidden dependencies | It is important that we can see relationships and that dependency is visible |
| Premature commitment | People do not like making decisions before all of the information is available |
| Progressive evaluation | Being able to evaluate progress supports notational activity |
| Provisionality | Being able to make tentative marks may support design activity |
| Role expressiveness | We depend upon existing, often cultural, knowledge to determine the purpose of an entity |
| Secondary notation | The ability to introduce additional meaning to a notation may support its use |
| Viscosity | If more actions or steps are required to reach a particular goal then difficulty is increased |
| Visibility and juxtaposibility | It is important to the human cognitive system to have items in view |

*Figure 15.5    Cognitive observations of the dimensions*

It can be seen from this that the cognitive observations that we have made fall into two categories: those which are directly related to the amount of cognitive processing required to carry out a particular activity, and notational features which in some way constrain the activities which a user must carry out. It is important to note that in the context of CDs a system is defined as 'notation + environment'. However, we argue that our analysis has encompassed a variety of activities and environments and so we feel that it goes beyond a particular notation and environment, and considers systems widely.

## 15.6   Discussion

The CDs framework of Green (1989) is an empirical distillation of everyday experience of what is found to be important in making human interaction with symbolic notations easy and effective. It is reasonable to assume (though it would be difficult to prove or disprove) that the easiness and effectiveness of such interactions arises from the structure of the human cognitive system. So this chapter has suggested that the CDs framework may be used to understand human interaction with the world of notations.

This suggestion makes two contributions in the context of the other chapters in this volume. One is that it points to a cognitive model that might be constructed, and the second suggests a strategy to testing the model. In moving towards constructing a model, it becomes clear that at least two things are important in the interaction with notations: the cognitive effort involved and the constraints imposed. How each suggests a cognitive model calls for further exploration. Further, this indicates that the suggestion of 'refactoring' the dimensions (Green *et al.*, 2006) may be a fruitful avenue of research in developing a better understanding of the CDs framework itself.

The other contribution is that it might point to a different kind of 'embodiment'. Embodiment is a major theme that runs through this work, but it is a rather ill-defined concept, especially in relation to that of interaction. David Vernon's chapter in this work suggests there are five types of embodiment, each of which involves a different kind of interaction with the environment, and thus with a different 'world':

- structural interactions are the most general, merely involving a world of some kind;
- historical interactions involve history, and hence presuppose a world that persists;
- physical interactions involve physical force, with a physical world;
- organismoid interactions involve sensory-motor actions like those found in known biological species;
- organismic, which refers to interactions that maintain a living organism's integrity and an equilibrium not dictated by the environment, with a world that involves other living things.

While these categories might be appropriate for the kinds of applications Vernon discusses, none of them helps us understand interaction with a world of notations. Yet human beings, and possibly future types of robots, interact with, and are embodied in, a very diverse world, which includes symbols and notations among other things. The suggestion in this chapter is that this is an important type of interaction and 'embodiment', and we may thus suggest a sixth type to add to the above list: *notational interaction, with a world of symbolic notations: notational embodiment.* (There might be yet other types, and the final chapter in this volume by Basden and Kutar attempts to begin discussion of this.)

Notational interaction can be seen as 'enactive' in Vernon's terms. It involves embodiment with notations. It involves experience, in that interaction with the notations presupposes understanding them, and ability to understand is built by experience. Interaction with notations cannot be reduced to any of the other types, and hence may be seen as 'emergent'. The agent that interacts with notations has some freedom in doing so. As the agent interacts with notations, it will invariably be making sense of them, and also of what the notations signify. The meaning of the notations or symbols is co-determined by the agents and the extant symbols themselves. Making sense of notations or utterances involves not just semantics but also pragmatics. In the context of natural language processing and data and knowledge engineering, in which text is the primary type of notation, Basden and Klein (2008) argue that the sense of the notations cannot be confined to 'text-meaning' but always involves 'life-meaning', which is their meaning in the everyday life of those who interact with them. Life meaning extends and might even subvert the text meaning at three levels: those of illocutionary intention, conversation and diversity of types of meaning. Not only is the agent embedded within a world of notations, but the notations are also embedded in the lifeworld of agents. This approach helps to break down what is sometimes an artificial separation between subject and object.

So we offer two things to the discourse in this work, first the idea of a world of notations with which an agent might interact in a way that is irreducibly distinct from physico-organic modes, and second that Green's (1989) CDs framework provides a useful start to understanding the nature of the processes by which notational embodiment is possible.

## References

Basden, A. and H. K. Klein (2008). 'New research directions for data and knowledge engineering: A philosophy of language approach'. *Data & Knowledge Engineering* **67**(2): 260–285.

Blackwell, A. F. (2006). 'Ten years of cognitive dimensions in visual languages and computing'. *Journal of Visual Languages and Computing* **17**(4): 285–287.

Blackwell, A. F., C. Britton, A. Cox, T. R. G. Green, C. Gurr, G. Kadoda, M. S. Kutar, M. Loomes, C. L. Nehaniv and M. Petre (2001). 'Cognitive Dimensions of Notations: Design Tools for Cognitive Technology'. Cognitive Technology 2001 (LNAI 2117), Warwick, UK, Springer-Verlag, pp. 325–341.

Blackwell, A. F. and T. R. G. Green (1999). 'Investment of Attention as an Analytic Approach to Cognitive Dimensions'. Collected Papers of the 11th Annual Workshop of the Psychology of Programming Interest Group (PPIG-11), Leeds, UK, pp. 24–35.

Blackwell, A. F., T. R. G. Green and D. J. E. Nunn (2000). 'Cognitive Dimensions and Musical Notation Systems'. *ICMC 2000 Workshop on Notation and Music Information Retrieval in the Computer Age*, Berlin.

Britton, C. and S. Jones (1999). 'The untrained eye: How languages for software specification support understanding in untrained users'. *Human-Computer Interaction* **14**(1): 191–244.

Card, S. K., T. P. Moran and A. Newell (1983). *The Psychology of Human-Computer Interaction*. Hillsdale, NJ, Lawrence Erlbaum Associates.

Cramer, M. (1990). 'Structure and mnemonics in computer and command languages'. *International Journal of Man Machine Studies* **32**(6): 707–722.

Curtis, B., S. Sheppard, E. Kruesi-Bailey, J. Bailey and D. Boehm-Davis (1989). 'Experimental evaluation of software documentation formats'. *Journal of Systems and Software* **9**(2): 167–207.

Dix, A., J. Finlay, G. Abowd and R. Beale (2004). *Human-Computer Interaction*. Englewood Cliffs, NJ, Prentice Hall.

Green, T. R. G. (1989). 'Cognitive dimensions of notations'. In A. Sutcliffe and L. Macaulay (eds.), *People and Computers V*. Cambridge, Cambridge University Press, pp. 443–460.

Green, T. R. G. (1990). 'The cognitive dimension of viscosity: A sticky problem for HCI'. *Human-Computer Interaction — INTERACT'90*. In D. G. D. Diaper, G. Cockton and B. Shackel (eds.). Amsterdam, Elsevier.

Green, T. R.G. and A. F. Blackwell (1998). 'A tutorial on cognitive dimensions'. Available from URL *http://www.cl.cam.ac.uk/%7Eafb21/Cognitive-Dimensions/CDtutorial.pdf*.

Green, T. R. G., A. E. Blandford, L. Church, C. R. Roast and S. Clarke (2006). 'Aims, achievements, agenda – where CDs stand now'. *Journal of Visual Languages and Computing* **17**(4): 328–365.

Gugerty, L. and G. M. Olson (1986). 'Comprehension differences in debugging by skilled and novice programmers'. In E. Soloway and S. Iyengar (eds.), *Empirical Studies of Programmers*. Norwood, NJ, Ablex.

Hundhausen, C., R. Vatrapu and J. Wingstrom (2003). 'End-User Programming as Translation: An Experimental Framework and Study'. *Proceedings of 2003 IEEE Symposium on Human-Centric Computing*. Piscataway, NJ, IEEE, pp. 47–49.

Kadoda, G. (2000). 'A cognitive dimensions view of the differences between designers and users of theorem proving assistants'. In A. F. Blackwell and

E. Bilotta (eds.), *Proceedings of the Twelfth Annual Meeting of the Psychology of Programming Interest Group*. Corigliano Calabro, Cosenza, Italy, pp. 123–136.

Kutar, M. S. (2001). 'Specification of temporal properties of interactive systems'. Department of Computer Science, University of Hertfordshire. PhD thesis.

Kutar, M., C. Britton and T. Barker (2002). 'A Comparison of Empirical Study and Cognitive Dimensions Analysis in the Evaluation of UML Diagrams'. In J. Kuljis, L. Baldwin and R. Scoble (eds.), *Proceedings of the Fourteenth Annual Meeting of the Psychology of Programming Interest Group*. Uxbridge, UK, pp. 1–14.

Kutar, M. S., C. L. Nehaniv, C. Britton and S. Jones (2001). 'The Cognitive Dimensions of an Artifact Vis-a-Vis Individual Human Users: Studies with Notations for the Temporal Specification of Interactive Systems.' Cognitive Technology 2001 (LNAI 2117), Warwick, UK, Springer-Verlag, pp. 342–355.

Larkin, J. H. and H. A. Simon (1987). 'Why a diagram is (sometimes) worth ten thousand words'. *Cognitive Science* **11**(1): 65–100.

Lohse, G. L. (1993). 'A cognitive model for understanding graphical perception'. *Human-Computer Interaction* **8**(4): 353.

Lohse, G. L., K. Biolsi, N. Walker and H. H. Rueter (1994). 'A classification of visual representations'. *Communications of the ACM* **37**(12): 36–49.

Lohse, G. L., D. Min and J. R. Olson (1995). 'Cognitive evaluation of system representation diagrams'. *Information & Management* **29**(2): 79–94.

Panko, R. R. and R. P. Halverson, Jr. (1994). 'Individual and Group Spreadsheet Design: Patterns of Errors'. *Proceedings of 27th Hawaii International Conference on System Sciences*, Hawaii.

Payne, S. J. (1993). 'Understanding calendar use'. *Human-Computer Interaction* **8**(2): 83–100.

Payne, S. J. and T. R. G. Green (1986). 'Task-action grammars: A model of the mental representation of task languages'. *Human-Computer Interaction* **2**(2): 93–133.

Petre, M. (1995). 'Why looking isn't always seeing: Readership skills and graphical programming'. *Communications of the ACM* **38**(6): 33–44.

Roast, C. R. and J. I. Siddiqi (1996). 'The Formal Examination of Cognitive Dimensions'. *HCI'96 Industry Day & Adjunct Proceedings*, pp. 150–156.

Scaife, M. and Y. Rogers (1996). 'External cognition: How do graphical representations work'? *International Journal of Human-Computer Studies* **45**(2): 185–213.

Stenning, K. and J. Oberlander (1995). 'A cognitive theory of graphical and linguistic reasoning: Logic and implementation'. *Cognitive Science* **19**(1): 97–140.

Winn, W. (2004). 'Cognitive perspectives in psychology'. In D. H. Jonassen (ed.), *Handbook of Research on Educational Communications and Technology*. Mahwah, NJ, Lawrence Erlbaum Associates.

Wright, P. and F. Reid (1973). 'Written information: Some alternatives to prose for expressing the outcome of complex contingencies'. *Journal of Applied Psychology* **57**(2): 160–166.

Zhang, J. and D. A. Norman (1994). 'Representations in distributed cognitive tasks'. *Cognitive Science* **18**(1): 87–122.

*Chapter 16*

# How to choose adjudication function in distributive cognition?

*Mourad Oussalah and Lorenzo Strigini*

## Abstract

Modular redundancy, in which a function is performed by multiple, redundant agents – machines or humans – requires an 'adjudication' stage to produce a single 'opinion' out of the multiple ones produced. Commonly used adjudication functions are, e.g. majority voting and averaging of the multiple results, but many alternatives are possible and in actual use. The adjudication problem has also emerged as a key part in distributed cognition systems where the decision making process is inferred from a set of multiple agents.

The choice of an adjudication function is of paramount importance for the overall dependability of a redundant system. This problem has been addressed separately by scholars in different disciplines, e.g. in the normative study of group decision making and in computer design. Each discipline has typically addressed a specific set of scenarios, with its own assumptions, which we believe would be of interest in other disciplines as well. All can benefit from taking into account results developed in other disciplines, and seeing their separate sets of assumptions and results in a common framework.

This chapter explains the adjudication problem, investigates the different criteria that help the designer in choosing an appropriate adjudication function and proves some interesting properties that characterize an optimal adjudication function, given a specific utility function and failure hypothesis for the system in which it is to be used.

## 16.1 Introduction

The need to obtain a single result out of multiple attempts at solving the same problem, some of which may give wrong solutions, arises in circumstances as diverse as medical decisions, where several doctors give opinions and one course of action must be chosen. As well in automatic control of aircraft, multiple computers propose commands for an aircraft's control surfaces. Similarly, in distributive cognition, in which the (cognitive) task is distributive across multiple individuals (agents), it is acknowledged that the issue of group's performance depends crucially on its organizational structure whose a key part is the group's 'aggregation procedure' as defined in social choice theory (List, 2005). The 'aggregation procedure', defined as a mechanism by which the group can generate collective judgments on the basis of the group members' individual judgments (List, 2006), is very much related to the adjudication function that performs such aggregation. Formally, an aggregation procedure is a function which assigns a corresponding set of collective judgments to each combination of individual judgments across the group members. A simple example is 'majority voting', whereby a group judges a given proposition to be true whenever a majority of group members judges it to be true. A panel of experts discussing their argumentation to reach a common agreement regarding the trust of some hypothesis is another example (Oussalah, 2004). Strictly speaking, the group members are not only restricted to human (experts) as it may sound like, but any software agent that provides a framework for representing the underlying entities, which motivates the cognitive aspect, is a potential candidate.

In these and many other cases, we are aware of the possibility of errors, and we adopt redundancy to protect against it. In engineering, we build replicas of the system (all with the same function but perhaps using different mechanisms) so that if one of the replicas fails, the outputs of the others can be used to ensure adequate service. For human expert opinions, we do use multiple experts so that larger the number of experts supporting a given opinion, higher our confidence on that particular opinion, and so on. How best to use this redundancy has thus been studied by multiple research communities, e.g. computer engineering, control engineering, psychology, theory of decision-making, etc. See Cooke (1991), Genest and Zidek (1986) and McConway (1981) for a probabilistic investigation of the problem of pooling expert opinions.

There is much in common between the scenarios studied in these various application areas. Most can be represented as in Figure 16.1, which also introduces the terminology we are going to use throughout this chapter. This is perhaps the simplest, practically very important, scheme of application of redundancy. The multiple 'processors' – computers, human experts or whatever – are asked to answer one question at a time. In response to each question, each 'processor' separately produces its own answer – 'opinion' – in the figure. An 'adjudicator' processes the multiple 'opinions' and produces the whole system's 'adjudged output' for that particular input (question). So, the adjudicator inputs are constituted of different opinions, whereas the input of the redundant system corresponds to an invocation.[1] That is, each 'invocation' of the system – submitting one input – will result in one invocation for each of the processors and one of the adjudicator.

---

[1] If no further indication is given, then following the scheme in Figure 16.1, an input would correspond to an invocation.

**Modular Redundant System**

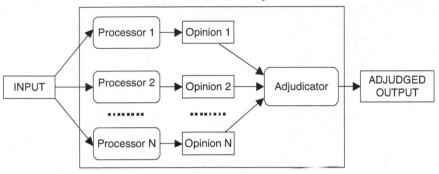

*Figure 16.1    General scheme of a multiple-redundant system, and the role of adjudication. Rounded box stands for a physical/software system while an angular box stands for an input/output of system (subsystem)*

Now considering the meaning of these labels in different disciplines, it is worth mentioning that in engineering arena the input is some mechanically or electrically encoded information. Whereas in psychological arena, the input would be a set of cues or sources of information that the human processors use to make a judgment. The multiple processors in a redundant computer system are usually referred to as channels, lanes, versions or replicas. In the psychological arena, we would rather speak of multiple 'judges' or 'experts'.

The problem itself (which starts from the input and gets the adjudged output) has been called 'voting' or 'adjudication' in control and computer engineering, 'sensor fusion' in other areas of engineering, 'expert pooling' or 'combination of expert opinions' in psychology and economy, etc.

What we call 'adjudicator' (after Anderson, 1986) (see also, Anderson and Knight, 1983) is usually called 'decision maker' in the psychological arena. Instead of 'adjudication', psychologists tend to speak of combination or aggregation of opinions or estimates; the term voting is also used. In the engineering of multisensor systems, the term 'integration' is often used, sometimes, without clear distinction between the adjudicator and the adjudication procedure, as it sounds from the debate on fusion versus integration of sensorial information. For the final result of the process, which we call the 'adjudged output' or 'adjudged result', psychologists would use words, such as 'aggregated opinion', 'combined estimate', 'integrated judgment', etc.

We are interested in the process the adjudicator uses to arrive at the adjudged result. For instance, if we are asking a simple question with a 'yes/no' answer, and we expect the processors to be more likely to give a correct answer, then it would seem reasonable for the adjudicator to follow the majority opinion. But there are many alternative possibilities, particularly when some of the processors are more reliable than others, and, therefore, choice is not always obvious.

From surveying the literature in these different fields, it appears that each research community has focused on some of the possible scenarios of use of redundancy and adjudication, and neglected others, which nonetheless are also of practical importance. However, it is often the case that appropriate solutions for these other possible scenarios can be found in the results obtained by another community.

In this chapter, we give a few examples of the possible variations of the adjudication problem and its solutions. Then, we give a coherent, general description of the requirements for an adjudicator and show how this implies general mathematical properties. In this sense, we show a general solution (extending the previous generalization in Di Giandomenico and Strigini's work, 1990) of which many previously published results are but special cases.

This study is part of the interdisciplinary project DIRC (Interdisciplinary research collaboration on the dependability of computer-based systems) which focuses on systems that include both people and machines. There are various reasons for DIRC interest in this topic:

- Designers and assessors of these systems must obviously consider the effects of both human and mechanical parts of the system, and thus study the adjudication problems for each of them, making sure that their assumptions are consistent and realistic.
- It is not unusual for the same redundant systems to include both humans and computers in the role of the 'processors' in Figure 16.1 (or in more complex configurations). Any such redundant system must, of course, be designed and assessed as a whole. As an example, the recent mid-air collision on July 1, 2002 over Southern Germany, between a Bashkirian Airlines Tupolev 154 and a DHL Boeing 757-200, involved a pilot required to 'adjudicate' between contrasting 'opinions' given by an automated collision-avoidance system and a ground-based air traffic controller. Incidentally, given the short time available for the pilot's decision, there is little doubt that such decisions must be mostly the result of pre-learned procedures, rather than of re-examining the merits of the two opinions in detail. One must make sure that one does not just inappropriately extend the analysis methods and assumptions typical of, e.g. computer design to humans, and vice versa.
- As pointed out earlier, a cross-fertilization between the various sub-disciplines, which have studied the adjudication problem would be useful for all, and the fact that people from multiple disciplines collaborate in DIRC gives an opportunity for pursuing this potential synergy.

There are useful surveys of adjudication functions in different contexts: e.g. Parhami (1994) in the computing and sensor fusion areas, Clemen (1987, 1989) and French (1985) in the area of psychology of decision-making and others in other fields.

There have also been many attempts to compare alternative adjudication functions. But many are limited to just a few alternatives, and thus fail to show that the same comparison criteria would allow one actually to choose among the whole set available; others are purely experimental, and ignore the fact that certain aspects of the issue can be solved universally by proof. The reader may consult, for instance, references Blough and Sullivian (1990, 1994), Kanekawa *et al.* (1989), Leung (1995), Lorczak *et al.* (1989), McAllister *et al.* (1990), Shabgahi *et al.* (2000), Vouk *et al.* (1990), Scott *et al.* (1987), McAllister & Scott (1991), Nicolas & Goyal, (1990) and Belli and Jedrzejowicz (1990).

Strictly speaking, from the aforementioned overall picture of the adjudication problem, there are some more general presentations of the issue, which allow one to derive the adjudicator design directly from one's utility function for the output of the redundant system, and one's assumptions about the processors' (stochastic) failure behavior (e.g. Di Giandomenico and Strigini, 1990). Apart from their value in clarifying the general issues, these methods allow one to design an adjudication function in some practical cases in which none of the simple adjudication functions (mean, median, etc.) would be a good solution. In this chapter, we will present a further generalization (to a wider class of cost functions, and removing the assumption of the previous paper concerning a uniform prior distribution of the correct answers).

## 16.2 Examples of adjudication functions and difficulty of comparison

Many types of adjudication functions have been investigated in the literature. A basic distinction is between those that simply choose one of the opinions and those which may produce an adjudged output different from all the opinions (e.g. the average of the opinions; or 'Help, I cannot compute an answer that I can trust!'). There are good surveys provided in other studies. Some typical examples are:

- *Majority adjudicator*[2]: chooses an adjudged result which is equal to at least $[(N + 1)/2]$ out of the $N$ opinions. $N$ is often odd to make ties less likely.
- *Consensus adjudicator*: as above, but, if there is no $[(N + 1)/2]$ majority, the opinion which is shared by the largest subset of processors is used as the adjudged output.
- *Median adjudicator*: selects the mid value of the opinions. In other words, if the opinions are $X_1, X_2, \ldots, X_N$, this adjudicator will order them as $X_{\sigma(1)} \leq X_{\sigma(2)} \leq \cdots \leq X_{\sigma(N)}$, where $\sigma$ stands for some permutation of set of indices $\{1, 2, \ldots, N\}$, and take the median: $X = X_{\sigma(N+1)/2)}$.
- *Mean adjudicator*: usually calculates the arithmetic mean of opinions $X_i$ ($i = 1$ to $N$), i.e. $X = 1/N\Sigma_i X_i$. Other types of mean can also be applied, such as geometric mean, harmonic mean, etc.

---

[2] Note the names we give here are taken from the literature, but are not standard across the literature and are often misleading. For example, there are many functions that correspond to the common English meaning of 'majority voting' (depending, e.g. on how ties are resolved), but some authors restrict this name to one of these functions, some to another one. Also note that we will often use the words 'voter' and 'adjudicator' interchangeably.

- *Weighted average adjudicator*: uses a weighted sum of the opinions, with positive weights $w_i$ adding to 1. So, $X = \Sigma_i w_i X_i$. This can also be extended to other forms of mean.

An adjudicator may be designed to use additional inputs besides the opinions of the processors, e.g. in a computer system, it may receive information about whether each processor has passed self-tests, and thus how likely it is to be functioning properly.

Consider, for instance, a redundant system in which three processors (e.g. thermometers or people) have to give the exact temperature in a room. Their outputs are likely to be different. Since we know that they are affected by small random errors, we would probably think it is reasonable to take an average of the three.

But suppose our system has instead to decide whether a two-color traffic light should show red or green, and we have again three processors. If their opinions are different, averaging them makes no sense (literally: no sensible meaning of 'an average between green and red' exists for this system). On the other hand, taking a majority opinion seems (and actually is, in most realistic scenarios) a good idea, provided each processor is right more often than it is wrong.

If we make these scenarios slightly more complex, the answers become less obvious.

If our traffic light has multiple possible light configurations (e.g. usual British ones: red, green, yellow, red-and-yellow, flashing yellow), there might be cases in which the three processors give three different opinions. Should the adjudicator choose with the majority, or choose a 'safe' option, such as red, or flashing yellow?

Every thermometer suffers from a systematic bias in addition to random noise. If the designer of the adjudicators could measure these biases once and for all for each thermometer, he could choose a better, but less obvious, adjudication function than the mean. It is also possible for a thermometer to fail and remain stuck at an end of its scale. Thus, the designer may decide that the adjudicator will discard outliers before averaging, and so on.

For all these scenarios, there are standard solutions adopted by the pertinent technical community. There are also theoretical proofs about which adjudicator should be chosen under specific assumptions on the scenario of use. For example, in the field of pooling of expert opinions, the mean adjudicator has been proven to be optimal under specific assumptions, e.g. McConway (1981) proved the uniqueness of linear pooling under assumptions of zero preservation and marginalization. But it is clear that these are not universally valid. For example, suppose the question is whether a certain prediction is true, and two experts provide respectively probabilities of 0.05 and 0.95. This probably implies that they use different assumptions, and choosing the mean of the two answers, 0.5, would be inconsistent with both experts' opinions, and less likely to be correct than either one.

There have also been attempts to cover various scenarios under a common theoretical umbrella (e.g. Di Giandomenico and Strigini, 1990). However, choices made by practitioners do not usually refer to general theory, and are often motivated by considerations of convenience from limited viewpoints.

For instance, in redundant computing systems for automated control, the median adjudicator is often applied to numerical 'opinions'. The rationale is often given as follows. The correct answer for each specific input value is defined as a certain number plus or minus a certain acceptable approximation error. Indeed, different correct processors often produce different answers, because they read slightly different readings of variables (temperatures, speeds, etc.) or due to the limitations of computer arithmetic ('floating point' representation of real numbers), or due to using different numerical algorithms to produce approximate answers. If at most a minority of the processors produce a wrong opinion, then the median of the opinions is guaranteed to be within the range of correct answers (one can easily convince oneself of this theorem by sketching the possible scenarios on a piece of paper).

This explanation shows that the median adjudicator will cope well with an interesting set of scenarios. But are we guaranteed that it is the best we can do? The answer is not obvious.

So, even in the simple scenario described earlier in this section, the criteria for choosing among alternative adjudication functions may not be obvious.

Basically, we are interested in the design of the adjudicator; namely, we have to decide which procedure the adjudicator in a given system must apply in response to future questions (equivalently, which mathematical function it must implement that maps a set of opinions into an adjudged output). More formally, what is the best adjudication function, considering the following.

- With reference to Figure 16.1, the specification of what the processors are required to do, as a function of the input, is the same as that of what the whole system is required to do.[3] This specification indicates, for the specific system and for each possible input value, a *correct answer* (or, more generally, a set of answers that are correct). That is, at each invocation, some possible answers are right and others are wrong, and some opinions may be wrong as well. The adjudicator's purpose is, thereby, to produce a correct answer nonetheless. We know in advance that it will sometimes fail, and thus the system will produce a wrong answer. Yet, we want this to be a less serious risk if we use the whole redundant system than if we use just one of the processors.
- The designer of the adjudication may know much about the nature of the questions that the processors have to answer. For instance, 'by how much should the rudder of a ship be turned, given an intended and a current direction of motion – the inputs', or 'produce a weather forecast for tomorrow, given as input a set of measurements of the state of the

---

[3] This is only roughly speaking. Particularly, there may be differences like, e.g.: the processors are allowed to say 'I don't know' or 'Don't trust my answer', but the system (and thus the adjudicator) is not, or vice versa, or both are or are not allowed.

atmosphere'. But the designer cannot know about the specific question and set of opinions on any future invocation (i.e. the designer cannot simply pre-calculate the answer to the $n$th future question).

- The adjudicator's inputs will give hints about which answer is probably correct, at any future invocation, but no certainty. Patterns of erroneous responses may be misleading – it is even possible, though improbable, that all opinions are identical and wrong. The adjudication function must be a reasonable compromise, which will decide 'well enough' 'often enough', though at times, inevitably, it will go wrong.

In summary,

1. Different systems will require different adjudicators; no mathematical function of the opinions, that will be the best adjudicator function for all systems, exists.
2. The information that – if we have it – should restrict our choice of adjudicators for a given system includes:
   - which kind of data the redundant system has to produce (e.g. 'green or red' vs. a temperature reading);
   - how likely the various possible values of a correct answer are, in the environment of use (e.g. if a system of multiple thermometers is being used for outdoor temperatures in a tropical area, we may want a differently calibrated adjudicator from that for Antarctica);
   - how likely the processors' opinions are to be wrong, or – better – the probabilities of the various wrong opinions. We typically obtain this from knowledge of the failure modes of the processors;
   - the costs of wrong adjudged outputs (in the environment of use of the redundant system), and whether some wrong outputs are better than others (especially whether in the environment of use there is some 'safe' output; e.g. a red traffic light, which should have been green is less likely to cause accidents than a green light which should have been red).

An important aspect of the requirements for an adjudicator is thus its generality. For a start, designing an adjudicator differs from the common human task of deciding about a specific set of opinions: we need to design *one* adjudicator to deal with *any* set of opinions that may arise in the system of which the adjudicator is part. Furthermore, we often try to design an adjudicator that will work with a whole class of systems.

How effective an adjudicator we can build depends on the information available about the system of which it is part of in at least two ways:

1. *Design-time information*: information that is available to the designer about the system's intended function, environment of use and physical make up. For example, if I know that the three processors in Figure 16.1 are thermometers, used in an environment which guarantees that their readings should only differ by 5° at most, I can use this knowledge in my adjudication function: any discrepancy of more than 5° indicates that one of the

thermometers is faulty which may enable the adjudicator to identify it and discard its opinion. In this example, the system-related information used by the designer concerns the *relationship* between two or more opinions. Information will also be available which can be used in conjunction with each opinion *individually*, allowing designers to build many checks and corrections into the system. These may best be seen, for the purpose of understanding adjudicator design, as adjuncts to the functions of each processor individually. The augmented processor thus created presents to the adjudicator a more refined or a more trustworthy opinion. For instance, a designer can use the information that all Celsius temperatures must be greater than $-273°$, and make a processor effectively declare 'I have no useful opinion' when all opinions violate this constraint. The designer may have more specific information, e.g. know the range of possible temperatures for a specific system, or know that this individual thermometer has a systematic error of $+0.15°$. Such reasonableness checks can be made more effective in some cases, sometimes leading to almost certain detection of erroneous opinions; e.g. for a processor whose task is to compute the square root of a number, one can design a checker that squares the result and compares it with the input.

2. *Run-time information*: information that the adjudicator is designed to receive as input during its functioning. For example, the adjudicator can be fed, besides the multiple opinions, by the results of reasonableness or correctness checks as outlined above. In a computer system, the designer can feed to the adjudicator the results of periodic diagnostic checks on the individual processors, indicating whether they seem to be functioning properly. With a group of human experts, the adjudicator can use knowledge about their relative abilities. As a substitute for diagnostic checks, some designs exploit the history of past adjudication: the more often a processor has disagreed with the adjudged result, the less reliable it will be considered.

With both forms of additional information, using more information means specializing the adjudicator design for a narrower class of uses. Most of the literature, which proposes simple adjudication functions, like majority voting or averaging, assumes that the adjudicator design needs to be extremely general. We will see here that some general results can be stated that apply to all adjudication designs, irrespective of exactly what information is given as input to the adjudicator.

## 16.3  Framework for comparison of different adjudicators

### 16.3.1  *Does the adjudicator match the functional system specifications?*

As shown in our earlier examples, there are cases in which the type of data formulates the opinions and the required output of the system simply precludes the use of some types of adjudicators.

Each adjudication function can only be applied to certain types of opinions (inputs to the adjudicator). For example:

1. The median adjudicator requires the inputs (set of opinions) to be defined in an ordered space. That is, the set of opinions must be endowed with a total ordering relation: for any two possible opinions, it must be possible to tell us whether they are equal, or which one is higher in the order. Otherwise, one cannot decide that a given opinion is greater or smaller than another one.
2. The consensus and majority adjudicators only require that for any two opinions one can tell whether they are equal or not. There are also variations of these adjudicators that work on the basis of a non-transitive 'equivalence' relationship (as, e.g. for numbers, the equivalence relationship defined as '$x$' is equivalent to $y$ if '$|x-y| < \varepsilon$', where $\varepsilon$ stands for some real valued threshold value). See (4) below.
3. The mean adjudicator requires the space of inputs to be a vector space, such that the mean of two opinions can be computed.
4. Other adjudicators require opinions to be defined on a 'metric' space, i.e. one which is endowed with a distance structure. These adjudicators compute the distance among different opinions and compare these values to some real valued threshold to decide whether they are 'close enough to be in agreement'.

So, opinions that can be any real number allow all these kinds of adjudications. Opinions that are points or vectors in a multi-dimensional Cartesian space allow the mean but not the median adjudication (unless the application of the system implies a reasonable ordering relationship over this space), and so on. Despite its importance, this first-cut method for selecting an adjudicator is not sufficient to decide between the different adjudicators that fit a given system.

### 16.3.2 How to choose between adjudicators that match the functional system specifications?

Several criteria have been explored in the literature. We describe the most common ones briefly, discard some of them and then focus on the last one.

#### 16.3.2.1 Deterministic comparison of how the adjudicators would behave on various combinations of correct and wrong opinions

Many authors or designers start with an adjudication function that looks plausibly good (e.g. majority), then wonder whether under some specific combinations of opinions, or conjectured processor faults (in the sense of 'causes of wrong opinions') a slightly different adjudication function would be better. For example, one might say 'suppose that, in a system with four processors, I have a majority adjudicator, M, requiring a majority of 3. If there is no majority, the adjudicator just produces an alarm signal. But imagine now a situation – call it Scenario A – in which two processors fail and give two different wrong opinions. So, the four opinions are divided into three "parties" of 2, 1 and 1 respectively. This clearly indicates that the two "minorities of one" must be wrong. So, let's change the adjudicator M into a better one, M', so that if the opinions are split this way, a majority of 2 is sufficient'.

Unfortunately, if we thus choose adjudicator M', which will choose well in Scenario A, we guarantee that the adjudicator will choose wrongly in some other situation. Indeed, consider a Scenario B, in which three processors fail, giving only two different wrong opinions. Two wrong opinions will agree and will be chosen by M' as the adjudged output. I can choose between M and M' only by considering whether, e.g. the added value of M' producing a correct system output in Scenario A compensates for the risk of it producing wrong decisions in Scenario B.

So, these purely deterministic arguments are generally insufficient for design decisions.

### 16.3.2.2 Complexity of the voters or other implementation-related costs and constraints

Ample literature exists in comparing adjudication functions on the basis of, e.g. the number of elementary operations (e.g. comparisons) the adjudicator performs to do its job. And, hence, how long it takes (for sequential execution by software or people), how much hardware it requires (for a hardware implementation), how much electrical power it will consume (which could be important in some applications), etc.

The usefulness of this kind of assessment thus changes with changing technologies. For example, how many diodes are needed to implement a majority voter on 4-bit numbers is no longer interesting for most designers. Vice versa, whether a human adjudicator can be expected to compute the required adjudication function mentally is not an important consideration in many cases.

Another important implementation-related aspect is how reliable the adjudicator's implementation can be made, and at what cost. For example, any adjudicator can be implemented itself in modular-redundant fashion, and some can be made self-checking in various ways. Most people can count votes for a majority vote, but far fewer can compute a mean of real numbers reliably. If we build an adjudicator in multiple redundant copies in multiple computers, the cost of communications between all these copies, and the probability of the messages being corrupted by noise, become a factor.

So, all these comparison criteria are meaningful, and the specific trade-offs that are appropriate will depend on details of implementation constraints, which we will not discuss here.

However, in many cases there are multiple adjudication functions that can be implemented without noticeable differences in cost, achievable reliability and such criteria. Then, the only question the designer needs to address is our original question: what is the best adjudication function, regardless of how exactly it will be implemented?

### 16.3.2.3   Algebraical properties of the adjudication function

Several authors have shown that the choice of a particular adjudication function can be justified as a straightforward result of requiring that the adjudged output satisfies some 'desirable' properties. Examples of such properties are for instance

- invariance to opinion permutation (the result of voting among the opinions of Dick, Tom and Harry should not change if Dick and Harry swap their opinions);
- modularity properties, like associativity, to allow an iterative process where the result of the adjudicator for $n-1$ opinions can be combined with the $n$th opinion to yield the result for $n$ opinions;
- 'boundary properties' concerning the behavior of the adjudicators on boundary cases, and so on. Intriguingly, sometimes imposing a set of such properties happens to lead to a unique adjudication function. For example, each one among the mean, majority and median adjudication can be fully characterized as the only possible choice if one requires a certain simple set of such algebraical properties.

The reader may consult the Aczel's book (Aczel, 1966) that contains several uniqueness theorems related to this topic. Clearly, despite its mathematical interest, this approach does not address the question of which of these properties are relevant to the designer. In particular, these properties are defined on the basis of the relationships between the opinions and the adjudged output, without reference to the required correct result. The more difficult problems are thus left open. Referring, e.g. to the 'invariance to input permutation' property, the designer is left to answer questions like 'Should I require this property, given that I trust Dick a lot more than Harry to give a sensible opinion?', or 'Why exactly, and under which conditions, would this invariance property help me to get a sensible adjudged output more often?'

### 16.3.2.4   Probabilistic/statistical objective measures

The purpose of an adjudicator is to form, together with the processors, a redundant system which performs its specified function more dependably than anyone of the processors would do on its own.

So, a reasonable criterion for deciding between two adjudicators A and B could be: 'Will my redundant system with adjudicator A deliver a correct result more often than with adjudicator B?'

This is not the only reasonable criterion, though. In some cases, we might want to use the different criterion 'Will my redundant system with adjudicator A deliver a dangerously wrong result less often than with adjudicator B?', and so on. In short, for every system there will be an (cost or utility) objective function, which we would want to improve. Referring to such objective functions (e.g. associating different costs to a traffic light's 'red in all directions' failures – which will cause traffic jams – and 'green in all directions' failures – which may cause accidents) allows me to calculate their expected values, for given assumptions on the probabilities of different wrong opinions, for each proposed adjudication function, and thus choose between alternatives. Furthermore, in many practical cases we can design an adjudicator which optimizes a given cost function, directly on the basis of the cost function itself and of my knowledge about the system, rather than laboriously starting from a set of plausible adjudication functions (e.g. from the literature), and comparing them until the best of the bunch (not necessarily the best possible) is left.

So, adjudicator design is reduced to a special application of statistical decision theory, in which the designer has to choose, instead of the answer to a specific question (invocation of the system), a function that chooses such answers for every possible questions, using the limited information given by the opinions, and chosen among all the functions defined over a certain domain[4] so as to optimize the appropriate cost function for the system in question.

Other objective functions might be, e.g. the mean square error in the adjudged result, or other statistics of this error.

We can take either a statistical, frequentist view or a probabilistic view of these measures, the latter being able to support design decisions for single decisions and even for systems that will operate only once (say, ammunition fuses, or launch of ICBMs – inter-continental ballistic missiles – under 'double-key' systems).

This last criterion for choosing adjudicators will be substantially developed in the next sections as it sounds the best that supports wholly rational design choices. Yet we should not ignore the possible problems in applying it. For instance, which objective function is adequate will not always be obvious.

## 16.4 Choice of adjudications based on objective functions and optimality

The aim in this section is to look at the conditions under which one adjudication function is deemed to be optimal. As mentioned in the previous section, we

---

[4] In many cases, the designer does not really consider the domain of all possible combinations of possible opinions, but a suitably abstracted representation of it. For example, a majority voter can be seen as taking as its input the information about which opinions are equal and which differ. This greatly reduces the size of the domain, which may make the adjudicator easier to understand and/or to implement. For example, in a three-processor system in which the opinions and the adjudged result must be 20-bit numbers, the number of possible combinations of opinions is $(2^{20})^3 = 2^{60}$. The adjudication function must give an adjudged output for each of these, i.e. it can be seen as a list of $2^{60}$ items, each one of which can take one among $2^{20}$ values. The total number of different possible lists, i.e. of possible adjudicator functions, is thus $2^{20}$ to the power of $2^{60}$, i.e. more than 1 million to the 4th power.

To design, e.g. a majority voter we can take a compressed view of the possible opinions, considering which opinions are equal to each other. The input to the adjudicator becomes three pieces of information, each with two possible values ('equal' or 'different'), i.e. 3 bits. There are at most eight possible combinations of values, and at most three adjudged outputs to choose from (the three opinions) ('at most' because some combinations are impossible). So, we have to choose among fewer than $3^8$ possible adjudication functions.

shall investigate those objective functions that involve probabilistic reasoning, which may account for the reliability of the different processors and designer's prior knowledge.

### 16.4.1    Historical motivations and relationship to previous work

Traditionally, two aspects of adjudication function contrast. The former views the adjudication function as an aggregation focusing on how individual opinions are aggregated. This has especially been studied in both computer systems and social choice theory. The second stream views the adjudication as a concise form of communication about an individual's preference. Especially, this communication effect is more apparent when the adjudication function involves more than one round before yielding the adjudged output.

Historically, literature on adjudication in the computing area was originally dominated by proposals of ad hoc designs. For example, majority voting was easy to implement in digital logic and could be shown to yield a correct adjudged output given few enough processor failures. New designs were mostly justified in terms of our criteria A and B in the previous section: by noting a scenario in which the new design would yield a correct result while the old one would not, or by considerations about the ease of building the new design and making it reliable enough. A criterion that was often used in specifying requirements for redundant systems was how many processor failures they should tolerate, and this translated directly into requirements for adjudicators.

As we noted before in Section 16.3.2, these criteria are often unsatisfactory. Whether it is more important for the adjudicator to react correctly to one or another combination of wrong opinions really depends on how likely each combination is to occur, and whether the cost of wrong adjudged outputs differs between those circumstances. As probabilistic reliability models became more commonly used, designers would then be more likely to compare systems on the basis of their predicted reliability, which of course would depend on the adjudicator design. By comparing reliability predictions for two hypothetical system designs, using different adjudicators, one could judge which adjudicator design would be better for a given system. However, the number of feasible adjudicator designs to choose from was large, and became larger as adjudicators began to be implemented in software and even complex algorithms became cheaper to implement. In 1990, the two papers (Blough and Sullivian, 1990; Di Giandomenico and Strigini, 1990) proposed an optimization-based approach and the observations are mentioned below.

1. The adjudication function need not be a simple function like averaging or taking a majority. It may be any mathematical function that maps the set of inputs to the adjudicators ('syndrome' in the language of Di Giandomenico and Strigini, 1990) into an adjudged output. Instead of describing the function by an algebraic expression, it could be described as a precomputed table with a line for every possible syndrome. During operation, the adjudicator would look up in the table the syndrome presented to it,

and finds the pre-computed adjudged output. With a software implementation, a look-up table is generally adopted.

2. The designers can thus ignore the pre-defined lists of adjudication functions in the literature, and concentrate on building the best possible adjudicator, given: (i) which inputs will be available to be fed to it in operation, and (ii) the information that the designers have about the system in which the adjudicator will be used.

Both papers then presented an optimal adjudicator for the case in which the adjudged result is required to be one of the opinions, and the requirement is that the adjudged result be as likely correct as possible. This adjudicator is built by looking at each possible syndrome in turn. Then one considers the probabilities of all the scenarios (combinations of component failures) that would produce that syndrome, and which may imply different correct results, and on this basis choosing that opinion that is most likely to be correct, *conditional* on the observed syndrome.

Di Giandomenico and Strigini (1990) also: (i) considered the case in which the inputs to the adjudicator include results of reasonableness tests and (ii) generalized the optimality criterion to account for safety applications, in which sometimes it is better to choose a safe output rather than that opinion that is most likely to be correct. For this purpose, Blough and Sullivian (1990) introduced the notion of 'worth', which might be expressed as a function of safe state in this case.

One may question whether an optimal decision is what is required. If, e.g. it turned out that using an ordinary majority adjudicator produces a system that is only slightly less dependable than with an optimal adjudicator, would we not be better off using the former? Then, why go to the expense of devising and calculating the more complex, optimal adjudication function? This is a valid objection when considering adjudicators that use as little information as majority voters, at least for cases in which there is an actual cost difference between the two designs. For example, if the adjudicator function is performed by a person, counting votes may be faster and less error-prone (and feel less silly) than looking up the answer in a printed table. For a software implementation, on the other hand, table look-up is often very cheap to execute and is a standard function for which reliable implementations are available. However, the reason for proposing the optimization approach is that a designer may be in a more complex situation, for which simple reasoning like that which leads to choosing majority voting is impossible. For instance, the processors' probabilities of error may be different, but the system may also provide to the adjudicator additional information, like results of reasonableness tests on the opinions, of diagnostic checks on the processors, and so on. Then, the well-known results on the properties of textbook adjudicators, some of which are pointed out in Section 16.3, do not apply. So, the complexity of deciding whether a simple adjudicator that seems satisfactory is actually almost as good as an optimal one may approach or equal the complexity of simply designing the optimal one.

We continue here from these two papers, and generalize their results slightly.

- Both papers assumed a cost function, which was only a function of whether the adjudged result is correct: we consider cost functions that depend on the magnitude of the error in the adjudged result.
- Both papers considered designers with knowledge about the probabilities of errors by the processors, but not of the relative frequencies of different inputs. So, they assumed 'prior indifference' among all possible values of the required output. Here, we generalize to the case in which the designer does know the probabilities of the different possible values of the required output.

Section 16.4.2 provides a more formal description of our probabilistic model.

## 16.4.2    *Probabilistic basis and assumptions*

It is worth recalling that the adjudged output is always a function of the different opinions. Since these opinions are unknown for the adjudication function beforehand and may vary arbitrarily within the range allowed by the processors, it makes sense to consider the adjudged output $\theta$ as some realization of a random variable $\hat{\Theta}$. On the other hand, the adjudged output, even if being correctly processed by the adjudication function, may not be correct. To see it, it is enough to notice, for instance, that a set of experts may all agree on a wrong opinion. The correctness of the opinions is rather a matter of the extent to which the system fulfils the system requirements as already mentioned in Section 16.2, which obviously always occurs in case of ideal system but may fail in practice. Incorrectness of the real system to satisfy the system requirement is described through system failures. As a matter of fact, since the correct answer, say, $\theta$ may as well vary arbitrarily, it makes sense to assume it as a realization of another random variable, say, $\Theta$. From this perspective the adjudged output is deemed acceptable if it agrees with the correct answer. Graphically, one may illustrate this phenomenon in Figure 16.2. In the latter, rounded box stands for a physical/software system while an angular box stands for an (numerical) input/output of system (subsystem). Dashed line means that there is no physical relationship between both entities.

Let a syndrome $r$ be a vector of all collected opinions supplied by the different processors for a given invocation, i.e. $r = [r_1\ r_2 \cdots r_N]$. The following probabilities are of interest

- $p(\theta)$ is the prior density probability that summarizes the user prior knowledge about the distribution of the correct answers. Some of the outputs are, for instance, more likely to occur than others. In the absence of such knowledge an obvious solution is to assign a uniform probability where all alternatives are equally likely.

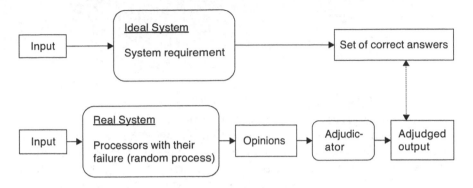

*Figure 16.2    Origin of randomness in adjudication function*

- $p(r|\theta)$ is the likelihood density corresponding to the conditional probability of the occurrence of each syndrome given that the correct answer takes a precise value $\theta$. Typically, this probability describes the failure assumptions of the different processors.

In other words, the distribution $p(\theta)$ characterizes our confidence that the (possible) outcome $\theta$ is a correct answer. For instance, if we knew that any faulty processor would deliver by default an opinion between 0 and 0.4, then it is unlikely that such range of values will be assigned relatively high probability value $p(\theta)$.

While the distribution $p(r|\theta)$ accounts for the reliability attached to the different processors, as well as for the dependency of the information supplied by these processors. Indeed, since the syndrome $r$ is in vector form, this leads to a joint probability that fully characterizes the dependency among its different components. Further, the elicitation of the probability $p(r|\theta)$ often requires its decomposition with respect to either the different opinions (components of $r$), or correctness of the processors. See, for instance Haimes *et al.* (1994) and Trivedi (1982). For the latter case, one may have the following decomposition, which results from some assumptions about system failures that will be discussed in detail later in the text:

$$p(r|\theta) = \sum_{h=0}^{\kappa(\theta,r)} P\{r \text{ occurs}|(\theta \text{ is correct}) \cap (h \text{ processes are non-faulty})\}$$

$$P\{h \text{ process are non-faulty} \mid \theta \text{ is correct}\}$$

where $\kappa(\theta, r)$ stands for the number of processors that produced $\theta$ as their result.

The previous two probabilities allow, through Bayes theorem, determining the posterior probability $p(\theta|r) = p(r|\theta)p(\theta)/p(r)$, with $p(r) = \perp p(r|\theta)p(\theta)d\theta$, or, in finite case,

$$p(r) = \sum_{v \in \Omega} p(r|v)p(v)$$

## 16.4.3   Adjudication and objective functions

The optimization process constructs an estimation $\hat{\theta}$ of the random variable $\Theta$ of the unknown true value of the adjudged output based on the available set of opinions, the prior knowledge about the distribution of the true value and the failure assumptions of the processors as summarized through the above conditional probabilities. For this purpose, a rationale is to minimize the so-called Bayes risk, or in other terms, the average cost. The latter is mainly dependent on the chosen cost function.

More formally, if $c(\theta(r), \theta)$ stands for the cost function, then the optimization looks at minimizing the average quantity (expected value)

$$J = E[c(\hat{\theta}(r), \theta)]$$

i.e.

$$J = \int_{\theta} \int_{r} c(\hat{\theta}(r), \theta)p(\theta, r)d\theta dr = \int_{r} p(r) \left[ \int_{\theta} c(\hat{\theta}(r), \theta)p(\theta|r)d\theta \right] dy,$$

where the integration is performed over all the state of true values of adjudged output and the set of all possible syndromes.

Now since $p(r) \geq 0$, and because one requires an optimal adjudication function for each chosen syndrome, minimizing $J$ with respect to $\theta$ boils down to minimizing

$$J' = \int_{\theta \in \Omega} c(\theta(r), \theta)p(\theta|r)d\theta$$

Let us consider some typical examples of cost functions, namely: squared error, absolute error and hit-and-miss. Below are results of the above optimization problem. The proofs of these results are reported in the appendix of this chapter.

- Squared error: $c(\hat{\theta}(r), \theta) = (\hat{\theta}(r) - \theta)^2$

  Then minimizing

  $$J = \int_{\theta \in \Omega} (\hat{\theta}(r) - \theta)^2 p(\theta|r)d\theta$$

leads to the solution

$$\hat{\theta}_{mn} = E[\theta \,|\, r] = \int \theta p(\theta \,|\, r) d\theta$$

In other words, the mean adjudicator[5] is the result of minimizing a Bayes risk when the cost function is constituted of squared error.

- *Absolute error:* $c(\hat{\theta}(r), \theta) = |\hat{\theta}(r) - \theta|$
  Similarly, minimizing $J = \int_{\theta \in \Omega} |\hat{\theta}(r) - \theta| p(\theta \,|\, r) d\theta$ leads to

$$\hat{\theta} = \text{median } p(\theta \,|\, r)$$

We notice that the median[6] of a probability distribution is given as

$$\hat{\theta}_{md} = \inf \left[ \theta, F(\theta \,|\, r) \geq \frac{1}{2} \right]$$

where $F(\theta|r)$ is the cumulative distribution associated with the density distribution $p(\theta|r)$. Equivalently, this corresponds to the solution in $\hat{\theta}$ to

$$\int_{-\infty}^{\hat{\theta}} p(\theta \,|\, r) d\theta = \frac{1}{2}$$

The median is then determined as through, possibly a non-linear system resolution involving integration of the conditional posterior distribution.

- *Hit-and-miss error:* $c(\hat{\theta}(r), \theta) = I(|\hat{\theta}(r) - \theta| \varepsilon)$,
  where $\varepsilon$ takes a sufficiently small value, possibly zero value, and aims at capturing the similar elements as discussed in Section 16.3.1. $I(.)$ stands for the characteristic function of the entity under bracket, i.e. takes a value one if the entity under bracket holds and zero otherwise. Then the minimization of $J = \int_{\theta \in \Omega} I(|\hat{\theta}(r) - \theta| > \varepsilon) p(\theta \,|\, r) d\theta$ is equivalent to maximize

$$J' = \int_{\hat{\theta}(r) - \varepsilon}^{\theta(r) + \varepsilon} p(\theta \,|\, r) d\theta$$

- So, as $\varepsilon$ tends toward zero, the optimal estimator coincides with the maximum a posteriori estimator, i.e. $\hat{\theta}_c = \max_\theta p(\theta \,|\, r)$.

---

[5] We should stress that the definition of the mean adjudicator here is somehow different from that described in Section 16.2 of this chapter, since here it relies on the expectation of a probability distribution, while it coincides with the standard arithmetic mean operation in most adjudication literature.
[6] Similarly to mean adjudicator, this definition of the median is different from its standard definition, as it becomes here a feature of the underlying probability distribution.

The latter expression is somewhat similar to the consensus or majority adjudicator in the sense that it aims at finding the alternative $\theta$ that has the greatest frequency of occurrence in the sense of the posterior probability. Such alternative might be the majority if such frequency is greater than half of the total number of opinions. Otherwise, we end up with the consensus voter.

On the other hand, the precedent is very similar to what Di Giandomenico and Strigini (1990) called optimal adjudication function, which, for each syndrome, has the highest probability of selecting the right value (correct value) provided the occurrence of that syndrome. Leung (1995) had as well expressed the same concern, under the name of maximum likelihood voting. Likewise, the investigation of Blough and Sullivan (1990) lies in the same direction.

### 16.4.4   Discussions and exemplification

The aforementioned probabilistic approach has certainly the clear advantages of using the prior knowledge available to the user and makes use of the likelihood function, which summarizes how likely each possible candidate might be the true adjudged output once the set of opinions are available.

It should be stressed that the mean, median, consensus and majority adjudication functions are somehow different from the standard ones in the sense that they are defined here as features of the posterior probability $p(\theta|r)$, while the standards ones are only special cases of this more general class. So, putting the arithmetic mean, for instance, as a substitute to $\hat{\theta}_{mn}$ will not lead to an optimal value for the underlying objective function.

Strictly speaking, when we are concerned only with obtaining correct answers, the optimal strategy sounds the one that provides a value, which is most likely to be correct given the particular results generated by the processors. That is, the optimal adjudication function would give $\hat{\theta}$ so that the probability $P(\theta|r)$ is maximal. While the previous section shows this strategy is indeed optimum only under the constraint of requiring the correct value to be the most frequent one among the set of results provided by the processors.

On the other hand, as discussed by Blough and Sullivan (1990), there are several situations where the maximum likelihood approach is not the most reasonable one. Indeed, this approach is based on the idea that the correct values (ideal values) are the most desired ones. However, in safety-critical applications, for instance, safe values, which are not necessary the correct ones, should be considered the most interesting values for the designer. These safe values can be produced if the probability of dangerous events becomes too high even if it is not the most likely event. Hence, even if from a purely probabilistic viewpoint, it is likely that the majority of sensors are correct, it might be better to choose the safe alternative. This motivated the authors to propose the concept of 'worth', which can be calibrated according to the context. From this viewpoint, it is easy to see such proposal as similar in spirit to our optimality based classification. Namely, the maximum likelihood approach leads to an optimal solution when the concept 'worth' is defined as the 'hit-and-miss' principle. While the use of mean and median-like adjudicators corresponds to the application of other concepts of 'worth', optimal mean squared error and optimal mean absolute error respectively.

The proposed optimization-based framework relies on the assumption that all the data are available in numerical setting. This excludes from our analysis those data which are only given for instance in ordered space or non-metrical space.

The modeling aspect in the form of conditional probabilities may hide several other important features, like dependence among failures and forgotten parameters. The dependency as explained by Littlewood and Miller (1989) (also in Ashton, 1986; Deb, 1988; Avizienis, 1985; Eckhardt *et al.* 1991; Kelly *et al.* 1988; Eckhardt & Lee, 1985) is of paramount importance in software engineering. On the other hand, in several practical situations, the set of syndromes and information about the correct answers may not be enough to fully characterize the adjudication function. For instance, information, regarding whether the adjudicator has received all opinions, or because of time delay, only some of them have been received, is missing. Consequently, a more realistic model would involve additional parameters in condition part of $p(r|\theta)$. In this course, the distribution $p(r|\theta)$ is less informative in terms of dependency of information supplied by different processors.

Further, the modeling framework provides useful insights on the quality of the estimation. This includes, for instance, the variance of that estimator on the basis of which two adjudication functions may be compared and evaluated. That is, one may compute, for instance, the variance of the random variable $\theta - \hat{\theta}$ based on the knowledge of the prior probability $p(\theta)$. On the other hand, the problem of eliciting the probability $p(r|\theta)$, which summarizes the failure assumption still is an open issue, and, does basically, requires further assumptions as it is given in detail in the examples below.

Now let us provide some examples of eliciting the aforementioned probabilities, when the reliability of different processors is known.

## Example 1

Consider a case of five binary processors; i.e., the output is either 0 or 1 (confirm or deny a target). Suppose that the processors have reliability 0.7, 0.8, 0.94, 0.85, 0.97.

Assumptions

1. The processors are mutually independent.
2. For a given input, and a fixed output cardinality $R$, only one of the possible $R$ outputs is correct, and $R-1$ are incorrect.
3. A failed processor gives one of the $R-1$ incorrect outputs with equal probability.

Assumption 1 ensures there is no correlation among the various opinions supplied by the various sources. A typical implication of this assumption is that if the sources stand for experts, then the latter do not communicate to each other when they deliver their opinions, and also, possibly, the experts come from different background. Assumption 2 expresses that multiple correct answers are not possible. Assumption 3 translates the non-preference among the set of incorrect outputs.

So, if $p_i$ stands for the reliability of the processor $i$, and let us take for instance a syndrome $r = (1, 0, 0, 1, 0)$. Then,

$P(\text{correct output is } 0|r)$

$= P\{(\text{correct output is } 0) \cap r\}/P(r)$

$= \dfrac{1}{P(r)} \cdot P\{(\text{processor 1 is correct}) \cap (\text{processor 2 is correct})$

$\cap (\text{processor 3 is faulty}) \cap (\text{processor 4 is faulty}) \cap (\text{processor 5 is correct})\}$

$= \dfrac{1}{P(r)} \cdot (1 - p_1)p_2p_3(1 - p_4)p_5 = \dfrac{1}{P(r)} \cdot 0.033$

While $P(r)$ can be computed as
$P(r) = P(r \cap \text{correct output is } 0) + P(r \cap \text{correct output is } 1)$

Similarly, we have: $P(r \cap \text{correct output is } 1) = p_1(1 - p_2)(1 - p_3)p_4(1 - p_5) = 0.0001$

So, $P(\text{correct output is } 0|r) = 0.033/(0.033 + 0.0001) \approx 0.99$.

In the same way, one finds

$$P(\text{correct output is } 1|r) = \frac{P\{(\text{correct output is } 1) \cap r\}}{P(r)} = \frac{0.0001}{0.033 + 0.0001}$$

$$\approx 0.01$$

Likewise, other probabilities, for other syndromes or outputs, can be elicited in quite similar fashion. Consequently, the output '0', which corresponds to the result of the majority voter, is the most likely result.

On the other hand, choosing another syndrome, say $r = (0, 0, 1, 0, 1)$ would lead to a different result. Indeed, this leads to

$$P(\text{correct output is } 0|r) = p_1 \cdot p_2(1 - p_3)p_4(1 - p_5)\left(\frac{1}{P(r)}\right) \approx \frac{0.0008}{P(r)}$$

And,

$$P(\text{correct output is } 1|r) = (1 - p_1) \cdot (1 - p_2)p_3(1 - p_4)p_5\left(\frac{1}{P(r)}\right) \approx \frac{0.008}{P(r)}$$

Consequently, in this case the output '1' is the more likely to occur, and even ten times more likely than the output of the majority voter, i.e. '0'!!

This means that reasoning only in terms of majority, while further information regarding the reliability of single processors is provided, is not enough to ensure a satisfactory output in terms of its agreement with the correct answer as it is clearly indicated from the preceding.

Commenting other types of adjudication functions, the consensus voter, which maximizes the probability $P$(correct output is $j|r$), sounds the most favored one. This soundness follows from the binary nature of the outputs. The mean voter would lead an output: $1 \cdot p$ (correct output is $1 \ |r) + 0 \cdot p$ (correct output is $0 \ | \ r) \approx 0.43$, which, after forcing that result to fit the binary assumption, would give an output 0 as well. While the median adjudicator would lead to an alarm signal since the equation $\int_0^{\hat{\theta}} p(\theta \ | \ r) d\theta = 1/2$ has no solution in $\hat{\theta}$.

The above results joint, to some extent, our discussion carried out in Section 16.4.1, as requirement the inputs (to the adjudicator) and the output be binary would exclude some possible scenarios of adjudication functions such as mean and possibly median adjudicator.

Moreover, intuitively, in cases of small output space cardinality, which, of course, include the binary case, the mean and median voters are less convenient as they always induce an approximation of the result to fit the constraint of the finite output space. And even the result may have no solution, like if the result lies exactly in the boundary of two successive output values. This observation has been well examined by McAllister *et al.* (1990), who proved the superiority of the consensus voter in such situation (small output space cardinality).

## Example 2

In contrast to Example 1, we suppose further that the prior probability of each value of the output space to be the correct value is given, and, the faulty processor can produce any value with some probability (not necessary uniform as it was the case in Example 1). More formally, assume now that the following information is taken for granted:

1. The set $\Omega$ of true values, that may be produced by processors, is finite
2. A processor $i$ has a reliability $p_i = p$
3. A value $\theta \in \Omega$ has a prior probability $q(\theta)$ of being the true (correct) value
4. A failed processor will produce the output $\epsilon \ \theta$ with probability $r(\theta)$
5. The processors fail independently.

Further if $R$ stands for the set of results for all processors (syndrome), let $c(R, \theta)$ be the set of processors that produced $\theta$ as their results in $R$.

Then, following Blough and Sullivan's development (1990, 1994), the determination of the conditional probability can be achieved as

$$P\{\theta \text{ is correct} \ | \ R \text{ occur}\} = \frac{P\{R \text{ occurs} \ | \ \theta \text{ is correct}\} \cdot P\{\theta \text{ is correct}\}}{P\{R \text{ occurs}\}}$$

where $P\{\theta \text{ is correct}\} = q(\theta)$, and

$P\{R \text{ occurs} \mid \theta \text{ is correct}\}$

$$= \sum_{h=0}^{|c(R,\theta)|} [P\{R \text{ occurs} \mid (\theta \text{ is correct}) \cap (h \text{ processors are non-faulty})\}$$

$\quad \cdot P\{h \text{ processors are non-faulty} \mid \theta \text{ is correct}\}].$

$$= \prod_{v \in \Omega, v \neq \theta} [r(v)]^{|c(R,v)|} \cdot \sum_{h=0}^{|c(R,\theta)|} (|c(R,\theta)|) \cdot p^h (1-p)^{n-h} \cdot [r(\theta)]^{|c(\theta,R)|-h}$$

While $P(R \text{ occurs})$ is computed as $P(R \text{ occurs}) = \sum_{v \in \Omega} P(R \text{ occurs} \mid v \text{ is correct})$ $P(v \text{ is correct})$.

Likewise Example 1, the discussion concerning the use of mean and median adjudicators for this case are still valid.

The above two examples prove that even if a particular adjudicator function sounds rational in the light of the type of available knowledge, the computation of adjudged output may differ, according to the available set of hypotheses governing the failure of the processors.

## 16.5   Conclusion

In our days, the application of redundant systems becomes increasing, as a result of either the need for compensating the possible failure of one component by another one, and so increases the reliability of the overall system, or to explore the full advantages of using diversely conceived components, though highly reliable, to protect the system from common mode failures. As a result of this development the need for an adjudicator that allows the system to combine the results issued from the different components to get a more meaningful knowledge becomes crucial. Further, in complex human–machine organizations, the adjudication function may have significant role in distributing tasks between different parts of computer systems and human. The recent mid-air collision between a Bashkirian airline and DHL Boeing 757-200, due to con-flicting commands of ground-based air traffic systems and the own pilot's opinion, highlights the importance of the adjudication function that allows us to handle the different opinions. Nevertheless, if under simplified scenarios, the use of majority, mean or median-like adjudicators can be justified, their extension to more complex situations in which the different processors may have different failure assumptions is not so obvious. Choosing a majority adjudicator, for instance, in cases where any processor may fail might lead to situation in which the majority output is a wrong answer. Further, the designer of the adjudicator needs to reason not for a particular scenario of the inputs but for all possible cases. The difficulty is even more stressed since it is not easy to find meaningful criteria that would be universally accepted. This motivates our present analysis, which does not restrict to particular scenarios of inputs. From this perspective, it should be noted that there is no universal adjudicator, which is unanimously better than all other adjudicators that could be designed in all situations. Nevertheless, there are several hints that could help the designer to draw an optimal adjudicator under a large class of scenarios. This includes, the nature of data that the redundant system has to produce, how likely the different possible values might be correct, how likely the various processors may fail and possibly the cost of wrong adjudged output.

This chapter advocates an optimization-based framework, which extends the previous work of Di Giandomenico and Strigini (1990). Particularly, by choosing appropriate cost functions, which describe the relationship between the adjudged output and the correct answer, one leads to classes of adjudicator functions, which are function of the posterior probability $p(\theta|r)$, i.e. probability that the correct output takes a precise value $\theta$ conditional on the occurrence of a syndrome $r$. Formally, three classes of adjudicator function have been found when choosing the cost function as squared error, absolute error and miss-and-hit error respectively. These classes generalize the mean, median and consensus standard adjudicator functions.

However, the result pointed out in this chapter require the data be given in a metric space, which may restrict its application in situations when only ordered space for instance is available. Further, to distinguish between different kinds of failures, say, safe failure and catastrophic failure, additional investigations are required.

## Appendix – Proof of claims made in Section 16.4.3

The aim is to minimize the average cost $c(\hat{\theta}(r), \theta)$ for all possible values of correct answers $\theta$ and all syndromes $r$. That is, minimize $J = E[c(\hat{\theta}(r), \theta)]$. As seen in Section 16.5.2, this comes down to minimizing the quantity $J = \int_{\theta \in \Theta} c(\hat{\theta}(r), \theta) p(\theta \mid r) d\theta$ for each syndrome $r$.

Now let us consider the result of this optimization for special cases of cost function, consisting of squared error, absolute error and hit-and-miss error.

*Squared error*

$$c(\hat{\theta}(r), \theta) = (\hat{\theta}(r) - \theta)^2$$

Minimizing

$$J = \int_{\theta \in \Omega} (\hat{\theta}(r) - \theta)^2 p(\theta \mid r) d\theta$$

leads to

$$\frac{\partial J}{\partial \hat{\theta}(r)} = 2 \int\limits_{\theta \in \Theta} (\hat{\theta}(r) - \theta) p(\theta \mid r) d\theta$$

So, setting $(\partial J/(\partial \hat{\theta}(r))) = 0$ implies

$$\int\limits_{\theta \in \Theta} \hat{\theta}(r) \cdot p(\theta \mid r) d\theta = \int\limits_{\theta \in \Theta} \theta \cdot p(\theta \mid r) d\theta$$

i.e.

$$\hat{\theta}(r) \cdot \int\limits_{\theta \in \Theta} p(\theta \mid r) d\theta = \int\limits_{\theta \in \Theta} \theta \cdot p(\theta \mid r) d\theta$$

Since $\int_{\theta \in \Theta} p(\theta \mid r) d\theta = 1$, it follows

$$\hat{\theta}(r) = \int\limits_{\theta \in \Theta} \theta p(\theta \mid r) d\theta = E[\theta \mid r]$$

which recovers

$$\hat{\theta}_{mn} = E[\theta \mid r) = \int \theta p(\theta \mid r) d\theta$$

*Absolute error*

$$c(\hat{\theta}(r), r) = |\hat{\theta}(r) - \theta| \tag{5.1}$$

Now to minimize $J = \int_{\theta \in \Theta} |\hat{\theta}(r) - \theta| p(\theta \mid r) d\theta$, we should split up the integration in the following way:

$$\int\limits_{\theta \in \Theta} |\hat{\theta}(r) - \theta| p(\theta \mid r) d\theta \, J = \int\limits_{-\infty}^{\hat{\theta}(r)} (\hat{\theta}(r) - \theta) p(\theta \mid r) d\theta + \int\limits_{\hat{\theta}(r)}^{+\infty} (\theta - \hat{\theta}(r)) p(\theta \mid r) d\theta$$

Now to compute the derivative of $J$ with respect to $\hat{\theta}(r)$, we shall use Leibnitz's rule: $\dfrac{d}{da} \displaystyle\int_{-\infty}^{a} f(y) dy = f(a)$ and its extension

$$\frac{d}{da} \int\limits_{h(a)}^{g(a)} f(a, x) dx = \int\limits_{h(a)}^{g(a)} \frac{\partial f(a, x)}{\partial a} dx + f(a, g(a)) \frac{dg(a)}{da} - f(a, h(a)) \frac{dh(a)}{da}$$

So,

$$\frac{d}{d\hat{\theta}(r)} \int_{-\infty}^{\hat{\theta}(r)} (\hat{\theta}(r) - \theta)p(\theta\,|\,r)d\theta = \int_{-\infty}^{\hat{\theta}(r)} p(\theta\,|\,r)d\theta + (\hat{\theta}(r) - \hat{\theta}(r))p(\hat{\theta}(r)\,|\,r)$$

$$= \int_{-\infty}^{\hat{\theta}(r)} p(\theta\,|\,r)d\theta$$

Similarly, one obtains

$$\frac{d}{d\hat{\theta}(r)} \int_{\hat{\theta}(r)}^{\infty} (\theta - \hat{\theta}(r))p(\theta\,|\,r)d\theta = - \int_{\hat{\theta}(r)}^{\infty} p(\theta\,|\,r)d\theta$$

Now setting $((\partial J)/(\partial\hat{\theta}(r))) = 0$ yields

$$\int_{-\infty}^{\hat{\theta}(r)} p(\theta\,|\,r)p(\theta\,|\,r)d\theta - \int_{\hat{\theta}(r)}^{\infty} p(\theta\,|\,r)d\theta = 0$$

which is equivalent to $P(\theta \leq \hat{\theta}(r)) = 1/2$. The latter is nothing else than the median value provided by distribution $p(\theta|r)$.

### Hit-and-miss error

$$c(\hat{\theta}(r), \theta) = I(\,|\,\hat{\theta}(r) - \theta\,| > \varepsilon)$$

i.e.

$$c(\hat{\theta}(r), \theta) = \begin{cases} 1 & \text{if } |\,\hat{\theta}(r) - \theta\,| > \varepsilon \\ 0 & \text{otherwise} \end{cases}$$

Here $J$ can be rewritten as

$$J = \int_{\theta} I(\,|\,\hat{\theta}(r) - \theta\,| > \varepsilon)p(\theta\,|\,r)d\theta$$

$$= \int_{\theta} [1 - I(\,|\,\hat{\theta}(r) - \theta\,| < \varepsilon)]p(\theta\,|\,r)d\theta$$

$$= 1 - \int_{\hat{\theta}(r)-\varepsilon}^{\hat{\theta}(r)+\varepsilon} p(\theta\,|\,r)d\theta, \quad \text{since } \int_{\theta} p(\theta\,|\,r)d\theta = 1$$

Consequently, minimizing $J$ is equivalent to maximizing the quantity

$$J' = \int_{\hat{\theta}(r)-\varepsilon}^{\hat{\theta}(r)+\varepsilon} p(\theta \mid r)d\theta$$

This entails in case $\varepsilon \to 0$, $\hat{\theta}(r) = \max_{\theta} p(\theta \mid r)$.

Indeed, by applying Leibnitz's rule again to expression $J'$, it turns out that the derivative, in case of $\varepsilon \to 0$, vanishes. Consequently $J$ has a constant behavior. It follows the optimal $\hat{\theta}$ is reached for maximal value of $J$, which is nothing else than the supremum of $p(\theta \mid r)$ over $\theta$.

## Acknowledgments

This work is part of Inter-disciplinary project DIRC funded by EPSRC, which is gratefully acknowledged.

## References

Aczel J. 1966. *Lectures on Functional Equations and Their Applications.* Academic Press, New York.

Anderson T. 1986. 'A structured decision mechanism for diverse software', *5th Symposium on Reliability in Distributed Software and Database Systems.* Los Angeles, pp. 125–129.

Anderson T. and Knight J. C. 1983. 'A framework for software fault tolerance in real time systems'. *IEEE Transactions on Software Engineering*, **9**(5), 355–364.

Ashton R. H. 1986. 'Combining the judgement of experts: How many and which ones?' *Organizational Behavior and Human Decision Process*, vol. 38, pp. 405–414.

Avizienis A. 1985. 'The N-version approach to fault-tolerant software'. *IEEE Transactions on Software Engineering*, **SE-11**(12), 1491–1501.

Belli F. and Jedrzejowicz P. 1990. 'Fault-tolerant programs and their relia-bility'. *IEEE Transactions on Reliability*, **29**(2), 184–192.

Blough D. M. and Sullivian G. F. 1990. 'A comparison of voting strategies for fault tolerant distributed systems'. *Proceedings of 9th Symposium on Reliable Distributed Systems*, pp. 136–145.

Blough D. M. and Sullivian G. F. 1994. 'Voting using predispositions'. *IEEE Transactions on Reliability*, **43**(4), 604–616.

Clemen R. T. 1987. 'Combining overlapping information'. *Management Science*, vol. 33, pp. 373–380.

Clemen R. T. 1989. 'Combining forecasts: A review and annotated biblio-graphy'. *International Journal of Forecasting*, vol. 5, pp. 559–583.

Cooke R. M. 1991. *Experts in Uncertainty: Opinion and Subjective Probability in Science.* Oxford University Press, New York.

Deb A. K. 1988. 'Stochastic modelling for execution time and reliability of fault-tolerant programs using recovery block and N-version schemes', Ph.D. Thesis, Syracuse University.

Di Giandomenico F. and Strigini L. 1990. 'Adjudicators for diverse redundant components'. *Proceedings of 9th Symposium on Reliable Distributed Systems*, Huntsville, AL, pp. 136–145.

Eckhardt D. E., Caglayan A. K., Kelly J. P. J., Knight J. C., Lee L. D., McAllister D. F. and Vouk M. A. 1991. 'An experimental evaluation of software redundancy as a strategy for improving reliability'. *IEEE Transactions on Software Engineering*, 17(7), 692–702.

Eckhardt D. E. and Lee L. D. 1985. 'A theoretical basis for the analysis of multi-version software subject to coincident errors'. *IEEE Transactions on Software Engineering*, SE-11(12), 1511–1517.

French S. 1985. 'Group consensus probability distributions: A critical survey'. In: *Bayesian Statistics 2*, J. M. Bernardo, M. H. DeGroot and D.V. Lindley (eds.), Elsevier Science and Valencia University Press, Amsterdam, North-Holland, pp. 183–202.

Genest C. and Zidek J. V. 1986. 'Combining probability distributions: A critique and an annotated bibliography (avec discussion)'. *Statistical Science*, vol. 1, pp. 114–148.

Haimes Y. Y., Barry T. and Lambert J. H. 1994. 'When and how can you specify a probability distribution when you don't know much?' *Risk Analysis*, vol. 14, 661–706.

Kanekawa K., Maejima H., Kato H. and Ihara H. 1989. 'Dependable on board computer systems with new method: Stepwise negotiated voting'. *IEEE 19th International Symposium on Fault Tolerant Computing Systems*, Chicago, pp. 13–19.

Kelly J., Eckhardt D., Caglayan A., Knight J., McAllister D. and Vouk M. 1988. 'A large scale second generation experiment in multi-version software: Description and early results'. *Proceedings of FTCS*, 18, pp. 9–14.

Leung W. 1995. 'Maximum likelihood voting for fault tolerant software with finite output space'. *IEEE Transactions on Reliability*, 44(3), 419–427.

List C. 2005. 'Group knowledge and group rationality: A judgment aggregation perspective'. *Episteme: A Journal of Social Epistemology*, 2(1), 25–38.

List C. 2006. 'The discursive dilemma and public reason'. *Ethics*, 116(2), 362–402.

Littlewood B. and Miller D. R. 1989. 'Conceptual modelling of coincident failures in multiversion software'. *IEEE Transactions on Software Engineering*, 15(12), 1596–1614.

Lorczak P. R., Caglayan A. K. and Eckhardt D. E. 1989. 'A theoretical investigation of generalized voters'. *Proceedings of 19th International Symposium on Fault Tolerant Computing Systems*, Chicago, pp. 444–451.

McAllister D. F. and Scott R. 1991. 'Cost modelling of fault tolerant software'. *Information and Software Technology*, 33(8), 594–603.

McAllister D. F., Sun C. E. and Vouk M. F. 1990. 'Reliability of voting in fault-tolerant software systems for small output spaces'. *IEEE Transactions on Reliability*, **39**(5), 524–534.

McConway K. J. 1981. 'Marginalization and linear opinion pool'. *Journal of American Statistical Association*, vol. 76, pp. 410–414.

Nicola V. F. and Goyal A. 1990. 'Modelling of correlated failures and community error recovery in multi-version software'. *IEEE Transactions on Software Engineering*, **16**(3), 350–359.

Oussalah M. 2004. 'Mechanism of trust in panel system'. In: *Intelligent Sensory Information: Methodology and Application*, D. Ruan and X. Zeng (eds.), Book Series on 'Studies in Fuzziness and Soft Computing' Springer Verlag, Berlin, Heidelberg, New York, pp. 115–137.

Parhami B. 1994. 'Voting algorithms'. *IEEE Transactions on Reliability*, **43**(4), 617–629.

Scott R. K., Gault J. W. and McAllister D. F. 1987. 'Fault-tolerant software reliability modelling'. *IEEE Transactions on Software Engineering*, vol. SE-13, pp. 582–592.

Shabgahi G. L., Bennett S. and Bass J. M. 2000. 'Empirical evaluation of voting algorithms used in fault tolerant control systems'. *Proceedings of International Conference of Parallel and Distributed Computing and Systems*, Las Vegas, pp. 340–345.

Trivedi K. S. 1982. *Probability and Statistics with Reliability, Queuing, and Computer Science Applications*. Prentice-Hall, New Jersey.

Vouk M. A., Caglayan A., Eckhardt D. E., Kelly J. Knight J., McAllister D. and Walker L. 1990. 'Analysis of faults detected in a large-scale multi-version software development experiment'. *Proceedings of DASC '90*, pp. 378–385.

*Chapter 17*
# Diversity in cognitive models
*Andrew Basden and Maria Kutar*

## Abstract

This final chapter looks back and forward. It looks back over the other chapters, to trace a 'thread' that runs through the work as a whole, which can suggest new and interesting avenues for future work. A very obvious theme argued in several chapters is that cognition is developed by interaction with the environment, especially of an embodied kind. Rather taken for granted, but no less important, is the diversity of these (embodied) interactions between the agent and its domain of application. It is diversity as such which this chapter traces as the thread that runs through the entire work, to point a way forward.

Diversity may be understood by reference to Dooyeweerd's philosophical exploration of distinct spheres of meaning and law which, as aspects, enable us to differentiate types of interaction and embodiment. Plotting the chapters by these aspects reveals interesting patterns, which allow a number of suggestions to be made for future research, including focussing on under-represented aspects, providing agents with good quality meta-rules that express the distinct rationality of each aspect, and contributing to meeting a number of challenges. It concludes by suggesting that, in addition to contributions this work makes to the field of cognitive systems and the paradigm of embodied interaction, it can also contribute an understanding of diversity itself.

## 17.1 Introduction

The chapters in this book exhibit a very interesting diversity. How can the reader benefit from this diversity? The reader could treat it as a tasty buffet, taking away what they like, and leaving the rest for others, but that does not do justice to the coherence of this book. The title of this work is *Advances in Cognitive Systems*. As well as each chapter making a separate little advance, we want the whole work to be an advance and we want to understand what that advance is and how all its chapters contribute to that advance.

In this chapter the thread will be traced that runs through the others and stitches them together, which makes what is, perhaps, the greatest contribution of this work. The chapters fall into the general area of cognition – but that is not the thread we seek; cognition is the fabric as a whole. Some of the reported research uses neural nets, others use semantic nets – but the thread of neural versus semantic has rather worn thin from exhaustive debate over the last 30 years. The thread this chapter traces is, rather, that of diversity itself.

Rolf Pfeifer, who contributes to this volume, has published Figure 17.1 in Pfeifer and Bongard (2007; used with permission) (it is also in Pfeifer & Gomez' chapter in this volume).

This contrasts what the authors call traditional and 'modern, embodied' ways of trying to understand (artificial) intelligence. In the traditional approach the focus is on the brain and central processing, while in the modern approach, it is on engagement and interaction with the environment. Cognition, they suggest, emerges from this interaction. Cognition is therefore not the instantiation of some detached reasoning or predesigned representational model on the world, but is itself engaged with that world. Its very processes and model are modified by it.

Not only does the diagram give a message of interaction, but it also speaks another message very loudly: the things that we interact with are highly diverse in a meaningful way. We see: food, painting, clothing, telephoning, muscular exercise, reading, typing, sport and leisure. Presumably whoever drew this diagram did not intend to exhaust all the types of interaction that might occur, but could, if they wished, have also included, for example reprimanding or rewarding children, caring for others, religious worship, tending household crops or chickens and many more, and most likely would have done so had they been from a majority-world rural culture. (These are the types of inter-action that the first author found to be important during a working visit to rural Uganda.) Issues like this, which are important in such cultures, will be added to those found in Figure 17.1 when it is useful to emphasise a diversity that extends beyond that of Western culture.

Three threads run through the chapters in this book. One is, as its title suggests, cognition. As already mentioned, however, this cannot be a thread of major interest here because the debate between neural and semantic net approaches to modelling been well rehearsed over the last 30 years. Moreover, to see cognition as the main thread would be taking the traditional attitude expressed in the left-hand side of Figure 17.1 and would ignore the diverse interactions depicted in the right-hand side.

The next thread is that stressed by Pfeifer and Bongard (2007), that cog-nition itself is constructed by the interaction with the world. To make that point is important because it is still not universally accepted, and how it is to be actualised in information technology is still being explored. Many of the chapters contribute to that exploration.

A final chapter of a collection like this, however, should not just review the others but should, if possible, suggest new and interesting ways forward that

*Figure 17.1   Two ways of approaching artificial intelligence*

are, perhaps unknowingly, already implied in the other chapters. This chapter does both by reference to the third thread, which is meaningful diversity. It finds that the book as a whole, and all chapters individually, contribute something to understanding the diversity of ways in which we interact meaningfully with the world.

If cognition is indeed built up by interaction with the world and if these interactions are highly diverse, as the diagram shows, then it will become necessary to understand the nature of this diversity. This is because, if the types of interaction that contribute to cognition are diverse, then the processes and rules by which cognition is built up might also be diverse – at least it is reasonable to explore whether this is so. We are not concerned with surface diversity, such as the difference between two sports like football and rugby, but with diversity of a more fundamental kind, the difference between, for example playing sport, feeding oneself, socialising, and reprimanding or rewarding children. So what this chapter focusses on is the thread of 'meaningful diversity'.

The diversity that characterises the chapters in this collection lies not so much in the cognitive modelling techniques or technologies employed, as in the ways in which interaction with the world is meaningful. That is, the diversity of interest here is the diversity of applications of cognitive modelling techniques. This diversity is summarised in Table 17.1, where chapters are identified by a number plus initial of name of first author (the number is an arbitrary identification number, which is not the chapter number).

This chapter will open with a general discussion of meaningful diversity and introduce a way of getting to grips with it, drawn from the philosophy of the mid-twentieth century Dutch thinker, Herman Dooyeweerd (1894–1977).

*Table 17.1   Overview of chapters, their applications and issues*

| Chapter | Author(s) | Applications | Issues |
|---------|-----------|--------------|--------|
| 1-M | Montebelli *et al.* | Understanding bodily anticipation | How the biological affects the behavioural |
| 2-M | Montesano *et al.* | Robots imitating humans | Affordance |
| 3-D | Durán | Modelling behaviour of infants | Wrong choices |
| 4-V | Vernon *et al.* | – | Cognitivism v. Emergentism |
| 5-C | Coutinho and Cangelosi | Effect of music on emotions | Modelling how music affects emotions |
| 6-N | Nefti *et al.* | Numerical optimization | Psychology can improve handling of risks and constraints |
| 7-R | Ruini *et al.* | Organism motivations | Strategic v. tactical |
| 8-G | Gemrot *et al.* | Creator of bots for games | Realistic behaviours in games |
| 9-S | Simko and Cummins | Understanding speech production | How organic structure affects this |
| 10-A | Alissandrakis *et al.* | Robots imitating humans | Choosing what to imitate from gestures |
| 11-K | Kutar and Basden | Interacting with symbolic systems | User satisfaction, effectiveness |
| 12-G | Grinsberg *et al.* | Prisoner dilemma games | How cooperation affects success |
| 13-O | Oussalah and Strigini | Several | Adjudication between multiple agents |
| 14-T | Tsagarakis *et al.* | Robot actuators | How to achieve human organic safety |
| 15-L | Lovric *et al.* | Understanding how investors behave | Modelling creative financial decisions |
| 16-P | Pfeifer and Gómez | Sensori-motor activity of robots | How morphology can assist |

This is applied to the chapters of this collection, allowing some patterns to emerge that reveal the thread of diversity. Finally, suggestions of strategic direction for future 'advances in cognitive systems' will be briefly outlined.

## 17.2   Meaningful diversity

One major integrating thread running through this work is meaningful diversity. The word 'meaningful' is used advisedly, and to refer to two things. One is that the diversity is meaningful to those interested in cognitive modelling, in that diversity has an impact on the way we model things and on the way cognition operates and develops. Modelling is not just of a passive world, but is engaged with that world, insofar as different types of things and activities being modelled are reflected not only by differences in the content of the model itself but also in the types of modelling formalisms and technologies used. This type of meaningfulness is returned to later.

The other way the diversity is meaningful is that we live in a meaningful world. The diversity of the world itself is not mere accidental variety that makes it impossible to obtain coherence, but works in such a way that the diverse ways in which things are meaningful all work with each other, support each other and refer to each other. All the activities shown in the right-hand side of Figure 17.1 contribute to the life and very being of that person; all the chapters in this work contribute to its meaning as a whole.

This challenges us to understand diversity. We want to be able to orient ourselves within it, not just so that we know where we are at the moment, but so that we know in which direction we should head and know what we might expect on the way that we have not yet encountered. In relation to this work, these mean that we want to know

- whether and how all the chapters relate to each other,
- how we might usefully continue research in the future, of the type encountered in this book, and
- how to cope with the next application of cognitive modelling that we come across: to what extent should we try to squeeze it into the moulds offered by the chapters of this book?

The importance of the third point becomes clear when considering two examples. Music is important in Uganda. If we wished to model the effect of music on emotions there we could apply the methods of Coutinho and Cangelosi in this volume almost directly even if the music were different.

But suppose we wished to model the rewarding and reprimanding of children in a family setting, so that in future they will learn to behave in a socially responsible way. No chapter in this collection seems to offer much help. Grinberg, Hristova and Lalev's chapter makes a useful contribution to understanding the link between attitude and gamesmanship. It might seem relevant if we were to replace prisoners by children because it speaks of cooperation and defection, but it makes a number of assumptions that reduce its relevance. It could only relate to groups, not single children; it discusses the interaction among the children whereas reward and reprimand is an interaction between children and parents; the chapter is about immediately measurable gain whereas social responsibility is a long-term character trait; use of the chapter would presuppose that children (as prisoners) are pitted against their parents (as prison warders) rather than being in a relationship of care, nurture and respect. Although that chapter might be used to model the situation in dysfunctional families where these conditions apply, doing so would make a very doubtful contribution, not only because this would be a minority of cases, leaving the majority of cases unmodelled, but also because it still overlooks the kernel issue in reward and reprimand: justice, fairness and appropriateness.

The model might in fact make a harmful contribution to the field by perpetrating a replacement of this kernel issue with those of gamesmanship and attitude.

Thus, if we are to benefit from the diversity in this collection, we need to understand the diversity of such fundamental, kernel issues and know how to differentiate the more fundamental diversity from that exhibited by differences between types of music or sport.

## 17.2.1   Towards a coherent view of diversity

An understanding of diversity cannot come from any one science, not even psychology, because each science has its main focus on one aspect of reality. Clouser (2005) argues that when a science tries to explain other aspects in its own terms the result is often a reduction that makes important aspects invisible (such as attempting to explain life in terms of physics alone). A clear example of focus on a single aspect may be found in Coutinho and Cangelosi's chapter on the impact of musical emotions. Music, as an everyday experience, has aesthetic, historical, social and emotional aspects, among others. The chapter looks at music through the eyes of psychology, studying the emotional aspect in detail; it does not claim to address the aesthetic issue of what makes great music, so it does not fall into the trap of reductionism. With Clouser, Dooyeweerd (1955/ 1984) and Hart (1984) argue that science is fundamentally unable to span across the diversity of aspects, but that it is the role of philosophy to attempt this. Therefore, it is to philosophy that we look to understand the nature of diversity. Three approaches will be examined before a proposal is made.

The philosopher Heidegger (1927/1962) explored the nature of existential interaction with the world. An entity gains its 'being' by virtue of its being situated in a world. From Descartes, Western thinking had assumed the notion of subject and object to be one of an active, thinking agent separate from world of objects. This caused many problems, not least of which was the elevation of the aspect of thinking above all others, as depicted in the left-hand diagram above. Heidegger tried to dissolve the distinction between subject and object and investigated the nature of our interaction with, or our being in, the world. Unfortunately, his extensive studies focussed only on the nature of this being-in, and provides very little help in understanding the nature of its diversity.

Evolution theory concerns itself with diversity, insofar as it offers an explanation for how diversity comes about, usually over a long period. Evolutionary biological science has been especially powerful in explaining the origin of types of living thing (species). Evolutionary philosophy has been harnessed to explain the origin of diverse aspects of human functioning, such as language, social activity, ethics and religion. Unfortunately, it is not as much help here as we might have expected, because evolution theory explains only the process of emergence of diversity, and not what diversity we face today. Indeed, it must presuppose the latter in that, before it can begin to construct an explanation of how, for example language or ethics emerged, we must present it with a prior understanding of what language and ethics mean. All we need here is this understanding of what diversity there is, which is prior to evolution theory.

An approach that claims to provide such prior understanding of what diversity there is, is to look at ontologies of the world offered by various thinkers. For example, Bunge (1979) distinguished five 'systems genera' in the world: physical, chemical, biological, technical and social. As soon as we try to apply this, its limitations become clear. We are interested in diversity of interactions; these must first be rendered as systems (example from Figure 17.1: the system might be person, frying pan, egg and tossing), but two problems emerge as we try to do this. One is that drawing a boundary round the system can prove difficult or even meaningless (e.g. for football in Figure 17.1, because football is not just a team playing with a ball but is a whole culture). The other is that an interaction can be of two different systems, for example the cooking of the egg might be for food (biological system) or to show how it is done (educational system) or both. A third problem is with Bunge's ontology: it is too limited for our use, dissolving the difference between, for example phoning, playing football, raising one's hat, which are all social in nature.

These problems occur because Bunge's ontology, and others like it such as Hartmann's (1952) 'strata', Habermas' (1986) types of social action and Maslow's (1943) needs, are ontologies of types of thing or action – the 'what' of the cosmos – rather than ways or modes in which things can be or occur – the 'how' of the cosmos. What is needed is a comprehensive delineation of different 'how's, or ways in which things can be and occur. (Strictly, this should not be called an ontology.)

## 17.2.2 Dooyeweerd's aspects

Dooyeweerd (1955/1984) is one philosopher who might offer the view we need. He drew up a set of 15 aspects or law-spheres, which delineate the 'how', the various meaningful ways in which things can be, occur and interact. His set of aspects has a wider range than the ontologies above (see Basden (2008, p. 66) for a comparison). Moreover, Dooyeweerd held that things or activities exhibit many aspects, in a way that will be useful here. Each aspect has a distinct kernel meaning that defines ways in which things may be meaningful to us. Table 17.2 shows these, and also how the interactions listed in the Introduction might link with some aspects.

Dooyeweerd's 15 may be seen as a superset of the other categorisations mentioned earlier in the text and is able to differentiate most of the activities in the right-hand diagram, as shown in the third column of that figure in ways that allow activities to be meaningful in several ways.

Although Dooyeweerd emphasised that all things and activities do in fact exhibit all aspects (so cooking an egg might be primarily biotic in function, but it might also have a social function with guests, an aesthetic function in the enjoyment of food, a lingual function in education, a formative function as a

*Table 17.2   Aspects and their kernel meanings*

| Aspect | Meaning kernel | How it relates to Figure 17.1, etc. |
|---|---|---|
| Quantitative | To do with quantity, amount | |
| Spatial | To do with continuous extension, space | |
| Kinematic | To do with movement; flowing movement | Yoyo, football and egg tossing involve flowing movement |
| Physical | To do with fields, forces, energy + mass | The yoyo works because of gravity and momentum |
| Biotic/organic | To do with life functions | The egg is meaningful as food Dumbbells are meaningful to build up muscles |
| Psychic/sensitive | To do with sense, feeling, emotion | |
| Analytical | To do with distinguishing | |
| Formative | To do with history, culture, technology: shaping, achievement, goals, ends, means, techniques, structures | The keyboard is meaningful for achieving things |
| Lingual | To do with symbolic signification, language | Phoning is meaningful as communication. The book is meaningful as information. The keyboard is meaningful for writing. Cooking an egg may be a demonstration |
| Social | To do with social interaction and institutions | Raising one's hat is meaningful as social etiquette |
| Economic | To do with frugality, resources, management | The keyboard is meaningful for work |
| Aesthetic | To do with harmony, surprise, fun and beauty | Painting is meaningful as the aesthetic activity of art. The football and yoyo are meaningful as play |
| Juridical | To do with what is due; rights and responsibilities, justice, appropriateness | Reprimanding or rewarding children |
| Ethical | To do with self-giving love, e.g. generosity | Caring for others |
| Faith/pistic | To do with belief, vision, aspiration, commitment, loyalty, creed and religion | Religious worship |

skill and so on), he proposed that there is usually one aspectual function that 'leads' the activity. The leading aspect is the one that most defines the way it is meaningful to us, and the other aspects support it. Dooyeweerd saw aspects not only as categories but also as the underlying cosmic mechanism of meaningful diversity, in that they are the transcendental conditions that enable everything to be, occur and interact.

Dooyeweerd did not claim that his suite of aspects is in any way a final, complete list; indeed he claimed the opposite (Dooyeweerd, 1955/1984, vol. II, p. 556). But it is useful to us here because it is not only more comprehensive than others, but also more grounded in everyday experience, and more soundly based in philosophy, including critical philosophy (Basden and Wood-Harper, 2006). We will adopt it here to gain insight into the diversity of contributions in this work.

## 17.3   Understanding diversity in the chapters

Dooyeweerd's aspects will be used here as an analytical tool. The main aspects of the application areas mentioned in each chapter will be identified, with the links between them. Secondary aspects are ignored.

### 17.3.1   Aspectual profiles of chapters

Figure 17.2 depicts these main aspects of each chapter, identified by the number given in Table 17.2. For most chapters at least two aspects are shown, linked by an arrow, the meaning of which is explained in the text below.

A brief explanation of why it is reasonable to identify certain aspects with each chapter is given in the following text; the reader might find it useful to refer to the aspect kernels in Table 17.2.

Chapter 1-M (Montebelli *et al.*) discusses how the 'non-neural' bio-regulative states of an agent affect sensori-motor activity. This is naturally expressed in Figure 17.2 by an 'affects' arrow from the biotic to the analytic aspect.

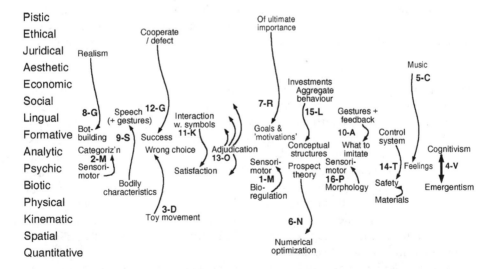

*Figure 17.2   Aspectual profiles of application areas*

Chapter 2-M (Montesano *et al.*), using the notion of affordance, investigates how a robot can categorise object properties of a sensori-motor kind (visual and tactile), identify relationships between them so as to imitate a human performing simple actions. Categorisation is an analytic functioning and sensori-motor functioning is psychic, and so this chapter's interest may be expressed by an arrow between these two aspects meaning 'allows us to infer'.

Chapter 3-D (Durán) applies non-linear dynamical systems theory to simulating the behaviour of an infant who looks in the wrong place for a toy, even though they have seen the toy put in a different place (the A-not-B error noted by Piaget). From the researcher's perspective, the stimulus given to the infant is movement and the error is one of wrong discrimination, so we may plot the interest of the chapter as how kinematic affects analytic functioning.

Chapter 4-V (Vernon *et al.*) is a different kind of chapter, which does not include any cognitive modelling approach as such. Rather, it contrasts two paradigms, cognitivist and emergentist. Here we do not take sides, but only try to express the contrast by an arrow that visually drives two aspects apart. Cognitivism stresses abstract conceptualising and separation from the environment (analytic aspect), while emergentism stresses the 'embodied and embedded' nature of the agent, which is a strongly biotic theme.

Chapter 5-C (Coutinho and Cangelosi) investigates several models of how music affects emotions. In aspectual terms, music is primarily of the aesthetic aspect, though it involves many others, and emotions are of the psychic aspect. So we may express the link discussed in the chapter by an 'affects' arrow from the aesthetic to the psychic aspect.

Chapter 6-N (Nefti *et al.*) argues that the very promising PSO method of numerical optimization lacks the ability to handle risk, and proposes that insights from a psychological theory can provide this ability. So this chapter may be expressed by an 'improves' arrow from the psychic to quantitative aspect.

Chapter 7-R (Ruini *et al.*) stresses the distinction between strategic and tactical levels of behaviour. But what is meant by 'strategic' and 'tactical'? The specific application concerns (artificial) organisms moving towards food or water or away from predators under 'motivations' of hunger, etc., and thus the two levels might seem to refer to the psychic and biotic aspects. The authors' interest seems wider, however, the link between goals that direct behaviour (formative aspect) and what it is that affects choice between them when they conflict. Choice among goals presupposes a belief, often tacit, by the organism about what is of ultimate importance (such as expressed by 'I must stay alive to reproduce'). Such beliefs are pistic in nature. If this does indeed reflect the authors' interest then it may be expressed in Figure 17.2 by an 'affects' arrow from the pistic to formative aspect.

Chapter 8-G (Gemrot *et al.*) presents an approach to building bots for games, with the aim that they are as realistic as possible. This kind of realism, which enhances the quality of games, is constituted in giving due respect to reality, which is the juridical aspect. Bot-building is a formative activity. So a link is shown ('affects') from the juridical to the formative aspect.

Chapter 9-S (Simko and Cummins) looks at how the bodily characteristics of organs affect how speech is actually carried out, with reference to similar work related to gestures. To speak of bodily characteristics is meaningful in the biotic aspect, with a few features such as mass being meaningful in the physical aspect, while speech and gestures, as language, are concepts that are meaningful in the lingual aspect. So the theme of this chapter may be represented aspectually by an 'affects' arrow from biotic to lingual aspect.

Chapter 10-A (Alissandrakis *et al.*) discusses how robots choose what, of humans, to imitate, by a cyclical communication process that consists of human gestures to demonstrate what to do and feedback from the robot about what they have learned. This may be expressed in Figure 17.2 by an 'affects' arrow from the lingual to the analytic aspect.

Chapter 11-K (Kutar and Basden) presents the cognitive dimensions framework as a way of evaluating the degree of satisfaction users experience when interacting with information or symbols. This may be expressed by an 'affects' arrow from the lingual aspect to that of satisfaction, the psychic.

Chapter 12-G (Grinberg *et al.*) discusses how the Prisoners' Dilemma type of games may be approached with a view to successful outcome. Success is of the formative aspect. Central to the Prisoners' Dilemma is the ethical distinction between self-giving (cooperation) and self-interest (defection). So the interest of this chapter may be expressed by an 'affects' arrow from ethical to formative.

Chapter 13-O (Oussalah and Strigini) addresses the general problem of adjudication between solutions offered by different agents. Such adjudication is largely of the analytic aspect here, despite the juridical overtones of the word used. Unlike other chapters, this one seeks to generalise across types of application (physical, sensory, social, etc.) so this is expressed in Figure 17.2 by multiple 'can apply to' arrows radiating from the analytic aspect.

Chapter 14-T (Tsagarakis *et al.*) is concerned that the hard, rigid actuators of conventional robots threaten the safety of the human body, and propose using soft, compliant ones instead. It discusses both material properties and control mechanisms for this. This may be expressed in Figure 17.2 by two 'affects' arrows converging on the biotic aspect from the physical and formative.

Chapter 15-L (Lovric *et al.*) aims to identify the types of conceptual structures that arise from aggregate financial decisions. We may express this as an 'allows us to infer' arrow from the social and economic aspects (aggregate and financial) to the analytic and formative aspects (concepts and structures).

Chapter 16-P (Pfeifer and Gomez) shows by numerous examples drawn from the natural world how appropriate morphology can greatly reduce the complexity of computation needed for good sensory or motor functioning. The theme is that the biotic aspect directly improves the psychic functioning, as well as indirectly by providing a substratum for it.

## 17.3.2   Patterns emerging

This aspectual picture provides various ways of discerning characteristics of the book as a whole, some of which merely support what we might already suspect, while some disclose new and surprising things.

The first thing that is obvious is the rich diversity of types of application of cognitive modelling. That they are diverse is obvious from a cursory reading of the chapters, but an aspectual analysis reveals how rich the diversity is. Thirteen of Dooyeweerd's 15 aspects are represented. The wide range of aspects represented in the collection means that it offers a usefully wide range of exemplars. In some multi-aspectual edited works of this kind, one aspect predominates, and the others feel like outliers, just bolted on for no real reason. Although certain aspects – the biotic, psychic, analytic and formative – occur more frequently, there is no sense that others are 'bolted on'. Rather, these central aspects are almost always linked with one of the others.

A second thing we might notice, related to the above, is that the majority of inter-aspect links in the chapters are those that involve impact ('affects') or the coming-into-being of something different ('allows us to infer'), which are two expressions of embodied interaction with the world. The whole work is thus revealed to be no merely static collection but has a dynamic character, almost alive with potential.

A third observation discloses types of interaction with, and embodiment in, the world. The central aspects, which are mentioned in most chapters, are aspects of our individuality as human beings. Aspects of individuality are those whose laws concern activity or behaviour as an individual, and include the biotic aspect (laws of enabling an organism to remain distinct from its environment), the sensitive aspect (laws of sentience and response), the analytic aspect (laws concern conceptualisation, distinction and categorisation) and the formative aspect (laws of intentional formation, design and achievement). The laws of the physical aspect, by contrast, concern fields, etc. which pervade individuals, and those of postformative aspects largely presuppose social or inter-individual functioning. This suggests that there are at least four kinds of interaction or embodiment that an individual can experience with the world: biotic, sensory, conceptual and intentional, and that they are fundamentally different from each other such that each may be researched in its own right even when in combination. It may be helpful, therefore, to see the 'cognitive body' that is build up by interaction as not just biological, but also psychological, analytical and intentional. Another interesting proposal, which deserves research attention, comes from Dooyeweerd's contention that we function in all aspects: that all of these aspects of individuality work in harmony.

The fourth observation is that in most cases the aspect of individuality is linked to an aspect of application or world. Each of the four may in principle interact with any of the aspects of the world. The environment with which the 'body' interacts exhibits many more aspects, not only the biotic to formative, but also the quantitative, spatial, kinemantic, physical, lingual, social aspects, and aspects of resources, beauty, harmony and fun, responsibilities, and it is ethical and spiritual in nature. This picture of interactivity and embodiment not only displays the diversity of types of interaction with which this chapter began, but also provides a basis for addressing (understanding, discussing and studying) the diversity of embeddedness in the world, of enaction and of embodiment.

## 17.3.3  Pointers to future work

Some application aspects, and so some types of interaction or embodiment, are not as well represented in this collection as others. Though the quantitative aspect pervades discussion of modelling techniques, only one paper discusses our interaction with the world of numbers. Although it has been argued that one chapter is interested in the pistic functioning of beliefs that lead to choice of goals, this is rather indirect; interaction with, and embodiment in, the world of beliefs and commitments needs much fuller and more explicit research. Likewise, the ethical aspect is present only in the very limited form of colla-boration versus defection, and the richness of generosity and self-giving (or their opposites) experienced in everyday life deserves much fuller treatment. A similar thing could be said in relation to the juridical, aesthetic, economic and social aspects. Thus there is considerable potential for future research in how cognition is embodied in all these aspects of life.

A second, and perhaps more fundamental, suggestion is that the nature of cognition and how it is built up might differ in fundamental ways for each aspect of the world that is meaningful when interaction occurs. To put it another way, the different types of application should not be seen as merely requiring different models to be built up, but as requiring different ways of building models. This is perhaps more fundamental than most chapters in this work realise, and will be discussed at some length.

In Figure 17.1 different rules apply to each type of interaction (football, cooking, etc.). These are not just different rules but different kinds of rule, because each kind of interaction is meaningful in different ways. This gives two levels of rules. If a cognitive agent is to enactively learn and construct a set of rules for interacting, then it already presupposes meta-rules that guide it in the construction of its set, which encapsulate kinds of rationality that lie behind the rules. Not only are the rules diverse, but so might be the meta-rules, as is already reflected in this collection in the difference between neural and semantic nets.

In a classic article, Winch (1958) argued that there are different kinds of rationality (specifically those directing the physical sciences, social science and religion). Dooyeweerd extended this idea and gave it a concrete form in his theory of aspects. Table 17.3 summarises the rationality of each aspect, and with what it is concerned; the third column is explained later in the text. The force of these aspectual rationalities may be understood if we reflect on situa-tions in everyday life and see the variety of ways in which things 'make sense' or in which we justify our actions.

*Table 17.3    Aspectual rationalities*

| Aspect | Rationality concerned with | Modelling approach |
|---|---|---|
| Quantitative | Arithmetic, statistics, calculus | |
| Spatial | Geometry, topology | |
| Kinematic | Movement, mechanics | |
| Physical | Causality over fields | |
| Biotic/organic | Health, growth, reproduction and speciation | |
| Psychic/sensitive | Responses to stimuli and mental states | Connectionism |
| Analytical | Discrimination, categorisation, logic | Connectionism object-orientation |
| Formative | Plans and actions to achieve goals and intentional change | Object-orientation |
| Lingual | Representation, cross reference, communicative rationality (Habermas, 1986) | |
| Social | Respect and role | |
| Economic | Frugal management of resources | |
| Aesthetic | Harmonising, playing and humour | |
| Juridical | Ensuring due; legal argument | |
| Ethical | Being generous | |
| Faith/pistic | Commitment, relationship with the absolute | |

The suggestion here is that when a cognitive agent constructs its set of rules for interacting with the world in a way that is meaningful in a certain aspect, the meta-rules for such construction should be appropriate to that aspect's rationality, and that meta-rules of other aspects are inappropriate. For example when constructing a model of physical interaction, expect to include causal relationships and fields, but when constructing a model of interaction with symbols, expect relationships of cross-reference, expect that some of them will be implicit rather than explicit, but do not expect to include causal relationships. Some sets of aspectual meta-rules might prove less relevant to specific types of interaction than others, but in the general case every set should be made available within the agent. This has much to do with affordance and a small example of it in operation can be found in Montesano *et al.*'s chapter in this work. They provided meta-rules, in the simplified form of vectors of properties, for a robot to follow in making up its rules for interaction. The robot was guided to consider both physico-spatial and psychic properties and, as expected, found that the psychic property (colour) was irrelevant for physical tasks and did not contribute to the set of rules it built up.

More generally, both connectionist and object-oriented approaches presuppose their own specific types of meta-rules, as indicated in the third column of Table 17.3. Connectionist (neural net) meta-rules assume rationality of the psychic aspect insofar as they operate by links between input and output, and the analytic aspect insofar is, they presuppose vectors of distinct inputs and outputs. The meta-rules of object-orientation or semantic nets assume rationality of the analytic aspect insofar as they presuppose distinct, categorised objects, and the formative aspect insofar as they operate by methods.

If we assume we can build rules using meta-rules of a different aspect then something important will be lost and the behaviour of the resultant agent will be unrealistic, unnatural or clumsy. Thus neither connectionist nor object-oriented approaches are, on their own, appropriate for constructing models or sets of rules for the lingual or spatial aspects, for example. The meta-rules for the lingual aspect are concerned not only with very complex non-linear structures in the continuous stream of text or speech, but also with what Basden and Klein (2008) call text-meaning (semantics) and especially with what they call life-meaning, which includes allusions, idioms and cultural connotations and illocutionary intention. The meta-rules for the spatial aspect are concerned with continuous extension, such as when a shape is expanded and, as Basden (2008) shows, this can result in structures that were not allowed for. Neither connectionism nor object-orientation are appropriate for these and, if applied, can make the building up of the model (whether by enaction or design) cumbersome and error-prone. Instead, cognitive agents should have encapsulated within them meta-rules that recognise such lingual things as allusion, idiom, connotation and illocutionary intention, spatial things such as continuous extension, and similar things germane to each and every aspect. That this is technically possible is evidenced by, for example the powerful text-understanding mechanisms employed on the Internet and geometric reasoners of the type originally suggested so long ago by Funt (1980).

So the second strategic suggestion for future research is that enactive cognitive agents should be furnished with the whole set of types of aspectual rationality, encapsulated as meta-rules for the building up of sets of rules (models) when they interact with the world. To furnish agents only with neural or semantic net formalisms, as most chapters here do, does them a disservice, and it will be an interesting challenge to expand to other aspects. This will be a long-term research programme. It is likely to require exploration of meta-rules for each aspect in turn, followed by research into how to integrate them into an overall system. Perhaps Oussalah and Strigini's chapter on adjudication might contribute to the integration?

The benefits of this approach accrue not only to those working under the enactive paradigm, however, but also to those working under the traditional paradigm, where the model in the artificial agent is designed and written by human knowledge representation activity. Basden (2008) has discussed in some detail how the aspects may be used in a manner similar to the above to devise knowledge representation formalisms that are appropriate for each aspect. The benefits include faster, more natural modelling of real-world multi-aspectual complexity, which is less error-prone and generates models with greater robustness and longevity. Nevertheless, in line with the prevailing attitude in this collection, the focus has been on enriching the enactive, 'embodied' paradigm.

A third suggestion for future research is that the aspectual approach outlined in this chapter might contribute to meeting the challenges posed in Vernon's chapter, which must be faced when adopting the embodied/enactive approach.

- Challenge 1 is that understanding how cognitive capabilities develop by building on sensory-motor capabilities is not straightforward. This challenge may be accounted for by the analytic aspect being irreducible to the psychic aspect. However, Dooyeweerd maintained that each aspect has inherent connections of dependency and analogy with others so his discussions of how the psychic aspect 'anticipates' the analytic might contribute to meeting this challenge.
- Challenge 2 states that the speed of developing cognitive capabilities is limited to the speed of interaction with environment. This argument, however, is at the level of rules rather than meta-rules; so it may be that providing the agent with high-quality sets of meta-rules, as proposed above, can speed this up.
- Challenge 3 is to incorporate motivations that underpin goals, and the interplay between motivations of different types. Motivation makes goals meaningful and Vernon identifies two types, social and explorative. Dooyeweerd would suggest there are many more than two ways of being meaningful (viz. aspects), thus extending the challenge, but his extensive exploration of inter-aspect relationships might help meet it.
- Challenge 4 is to extend the self-construction of world models beyond sensory-motor perception–action, to what Vernon calls the 'more abstract knowledge' so important to real cognitive systems. Dooyeweerd not only recognised the 'more abstract' aspects, but has explored their types, ranging from analytic to faith types. Providing agents with meta-rules for all these might help meet this challenge.
- Challenge 5 is that new adaptive materials must be integrated into morphological models of cognition. This requires new ways of thinking because traditional physics has presupposed distinct entities but this is inappropriate when using compliant materials as discussed in Tsagarakis *et al.*'s chapter. Dooyeweerd's elucidation of kernel meanings of aspects would suggest that distinction is not to be found in the physical aspect itself but in the analytic aspect that has traditionally been uppermost in the act of thinking about physical reality. He thus clarifies the nature of the difficulty. Attending to the kernel meanings of aspects might open up not only the possibility of using adaptive materials but also other new approaches to embodied cognitive systems.

## 17.4    Conclusion

In moving away from a rationalistic view of cognition and AI towards a more 'embodied' view, we find that the important issue is not that body is opposed to mind, but rather the existential notion of interaction with the world. That this is so has been recognised increasingly over the past decade. But, accepting that it is so thrusts upon us the next challenge: the diversity of types of meaningful interaction. In our 'embodied' or existential state we interact in many types of ways, each of which can influence the types of cognitive modelling that are undertaken. So it is time to begin addressing seriously the nature of the diversity of ways in which we interact with the world.

In the field of cognitive modelling (including AI) these types of interaction are the types of application area, and they affect the very core of modelling itself. This collection of chapters provides a useful variety of types of application area and it can thereby provide a good study of meaningful diversity and the way diversity can affect modelling.

We have examined the potential of Dooyeweerd's (1955/1984) notion of modal aspects as a basis for understanding the nature of diversity. Aspectual thinking is common whenever thinkers wish to differentiate things that cannot be reduced to each other but must be considered separately during analysis, and many suites of aspects (or levels, strata, genera, action types, needs, etc.) have been proposed. Dooyeweerd's suite has certain advantages over most others, of being more comprehensive and more soundly based in both philosophy and everyday experience. His theory of aspects also allows things to function in more than one aspect in a coherent way.

We wanted to know whether and how all the chapters relate to each other. We can see that the chapters cover almost the entire range of Dooyeweerd's aspects. In almost all cases, at least two aspects were found, giving a richer picture than mere aspectual categorisation would have done. Each chapter is seen to encapsulate a dynamic, in which one aspect affects another, one being an aspect of individuality and the other an aspect of interaction with the world.

The aspects of our individuality include the biotic aspect of organic life, the psychic aspect of sentience, the analytic aspect of conceptualisation, the formative aspect of intentional structuring or processing. So this kind of analysis might enrich the notions of body and embodiment, to involve at least four aspects, all working in harmony as instruments do in an orchestra.

We wanted to know how to continue research in the future. One avenue is to investigate how these aspects of individuality or embodiment can work in harmony. Another is to recognise that each aspect of individuality interacts with aspects of the world that make the interaction meaningful, and to investigate the nature of embodied interaction in each such aspect; to increase generality, it might be particularly useful to study interactions found in majority-world situations. After this, it would be important to investigate the nature of multi-aspectual embodied interaction.

We wanted to know how to cope with possible new types of application and interaction not well represented here. The proposal is to furnish artificial cognitive agents with meta-rules that encapsulate the basic type of rationality and functioning of every aspect. No longer need researchers be restricted to neural or semantic net approaches, but a future can be envisaged in which a variety of types of meta-rules are integrated together, each encapsulating a different aspect of the world. This might contribute to addressing the five challenges for the embodied interaction paradigm that Vernon poses in his chapter.

This collection of chapters, entitled *Advances in Cognitive Systems*, not only deals with cognitive techniques and technologies, promotes an enactive-embodied paradigm, but, in the diversity of applications that it encompasses, it tentatively points a way towards the issue that is hastening to meet us very soon, the issue of diversity itself.

# References

Basden, A. *Philosophical Frameworks for Understanding Information Systems*. 2008. Hershey, PA: IGI Global.

Basden, A. & Klein, H.K. 'New research directions for data and knowledge engineering: a philosophy of language approach'. *Data & Knowledge Engineering*. 2008; **67**(2): 260–285.

Basden, A. & Wood-Harper, A.T. 'A philosophical discussion of the root definition in *soft* systems thinking: an enrichment of CATWOE'. *Systems Research and Behavioral Science*. 2006, vol. 23, pp. 61–87.

Bunge, M. 'Treatise on basic philosophy', vol. 4. *Ontology 2: A World of Systems*. 1979. Boston, MA: Reidal.

Clouser, R. *The Myth of Religious Neutrality: An Essay on the Hidden Role of Religious Belief in Theories*. 2005. Notre Dame, IN: University of Notre Dame Press.

Dooyeweerd, H. *A New Critique of Theoretical Thought*, vol. I–IV. 1955/1984, 1975 edn. Jordan Station, Ontario: Paideia Press.

Funt, B.V. 'Problem-solving with diagrammatic representations'. *Artificial Intelligence*. 1980; **13**(3):201–230.

Habermas, J. 'The theory of communicative action', vol. 1. *Reason and the Rationalization of Society* (tr. T. McCarthy). 1986. Cambridge, England: Polity Press.

Hart, H. *Understanding Our World: An Integral Ontology*. 1984. Lanham, MD: University Press of America.

Hartmann, N. *The New Ways of Ontology*. 1952. Chicago, IL: Chicago University Press.

Heidegger, M. *Being and Time* (tr. J. Macquarrie, B. Robinson ). 1927/1962. Oxford, UK: Blackwell Publishing.

Maslow, A. 'A theory of human motivation'. *Psychological Review*. 1943, vol. 50, pp. 370–396.

Pfeifer, R. & Bongard, J. *How the Body Shapes the Way We Think: A New View of Intelligence*. 2007. Cambridge, MA: Bradford Books, MIT Press.

Winch, P. *The Idea of a Social Science and its Relation to Philosophy*. 1958. London: Routledge and Kegan Paul.

# Index